Fundamentals of Statistical Thermodynamics

SERIES IN THERMAL AND TRANSPORT SCIENCES
Gordon J. Van Wylen, Coordinator

FUNDAMENTALS OF CLASSICAL THERMODYNAMICS
Gordon J. Van Wylen and Richard E. Sonntag

FUNDAMENTALS OF STATISTICAL THERMODYNAMICS
Richard E. Sonntag and Gordon J. Van Wylen

FLUID MECHANICS
Arthur G. Hansen

THE TRANSPORT OF HEAT AND MASS
John A. Clark

Fundamentals of STATISTICAL THERMODYNAMICS

RICHARD E. SONNTAG & GORDON J. VAN WYLEN

Department of Mechanical Engineering
University of Michigan

JOHN WILEY & SONS, INC. NEW YORK · LONDON · SYDNEY

Library of Congress Catalog Card Number: 65-27654
Printed in the United States of America

To Pat and Margaret

Preface to the Series in Thermal and Transport Sciences

This Series in Thermal and Transport Sciences had its origin in a number of discussions regarding curriculum among engineering faculty members of the University of Michigan. We believed that by integrating the teaching of thermodynamics, fluid mechanics, and heat and mass transfer at the undergraduate level, we could present this material more effectively and efficiently. There was, however, a need for carefully integrated textbooks; and my colleagues Professors John A. Clark, Arthur G. Hansen, Richard E. Sonntag, and I have undertaken to write a series of basic undergraduate textbooks in these fields.

We decided at the outset that each book should be complete in itself so that it could be used quite independently of the series. At the same time, however, it was our goal to provide such correlation between the books that their use in a series of courses would provide an integrated coverage of thermodynamics, fluid mechanics, and heat transfer. Thus, it would be possible to achieve the educational advantages inherent in such an approach.

It was also evident that a series such as this would provide an excellent foundation for the preparation of a number of applied or specialized books which might involve all three of these areas. With a uniform approach and definitions in these basic books, the transitions from the basic courses to these applied and specialized topics would be accomplished smoothly and efficiently. We also anticipate that a number of advanced textbooks in related areas will be included in the series. Thus this series will contain the basic books in thermodynamics, fluid mechanics, and heat transfer, as well as applied and advanced books in these areas.

The first decision we faced in writing the basic books concerned classical and statistical thermodynamics. It was our conviction that the series should cover both these topics, and that the classical thermodynamics should precede the statistical thermodynamics. Professor Sonntag and I are co-authors of *Fundamentals of Classical Thermodynamics* and

Fundamentals of Statistical Thermodynamics, with Professor Sonntag as principal author of the latter book and myself of the former.

The second decision we faced concerned the sequence of the series as a whole. We decided the order of topical presentation would be classical thermodynamics, fluid mechanics, and heat and mass transfer, and that each book would be written on the assumption that the student has covered the material in the prerequisite courses. Statistical thermodynamics would follow classical thermodynamics in this order of presentation and parallel the subject matter in fluid mechanics and heat transfer. Schematically this is as follows.

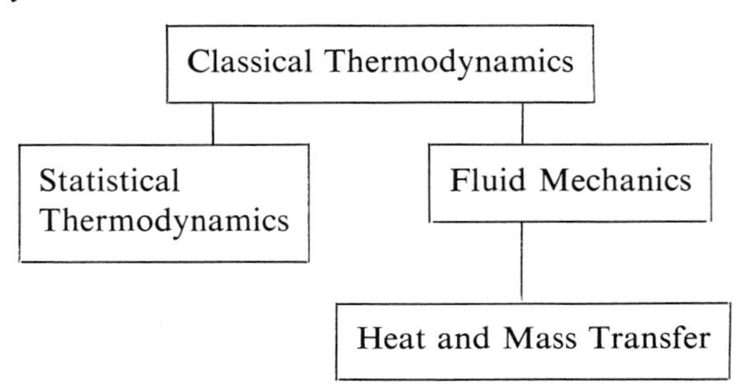

Professor A. G. Hansen is the author of *Fluid Mechanics* and Professor J. A. Clark is the author of *The Transport of Heat and Mass.*

Our original goal was to use the same symbols throughout the four books. However, we have realized that each of these fields does have a certain body of widely accepted symbols, and therefore in a limited number of cases we deemed it best to follow the conventions of each of the three fields and use different symbols for the same quantities in the books. In any case, uniformity of concepts and definitions is of much more importance and of greater benefit to the student than uniformity of symbols. We have attempted, therefore, to establish this uniformity in fundamental principles throughout the series. The concepts of system and control volume have been used throughout. Every effort has been made to maintain a consistent set of definitions and use of terms in each of the four basic texts.

As authors we extend our appreciation to our colleagues and to Mr. A. R. Beckett and his associates at John Wiley and Sons; and as editor, I thank each of the authors for the stimulating experience of working with them.

G. J. Van Wylen

March, 1965

Authors' Preface

The subject of statistical thermodynamics and the related fields of statistical mechanics and quantum mechanics have become increasingly important to engineers in recent years. Research and technology have expanded into a number of diverse areas that have made the study of such topics and devices as plasmas, rarefied gas dynamics, magneto-hydrodynamics, thermoelectric generators and refrigerators, ion engines, thermionic and solar converters, nuclear systems, lasers, semiconductors, superconductivity, ortho- and para-hydrogen, and liquid helium of great interest for present and potential application. A thorough study of many of these topics requires, to varying degrees, a description and analysis from the statistical microscopic as well as from the macroscopic point of view. In addition, a study of statistical thermodynamics is beneficial in promoting a greater understanding of the foundations, laws, properties, and applications of thermodynamics.

This book on statistical thermodynamics is the second volume to be published in the "Series in Thermal and Transport Sciences" and follows our book, *Fundamentals of Classical Thermodynamics.* As indicated in the Preface to the Series, this book is intended for students having a background in classical thermodynamics, although it can also be used in a carefully sequenced but simultaneous presentation with classical thermodynamics.

In writing this book, we have attempted so to organize and present the material as to provide considerable flexibility with respect to course content, level of presentation, and student background and maturity in mathematics and science. We have also endeavored to retain a distinct engineering perspective throughout the text and problems. A number of the topics and devices mentioned above are discussed in this book, and the groundwork is laid for a subsequent, perhaps independent, study of many others. Much of this material has been used in note form for several years in both undergraduate and graduate courses for engineering students. We have assumed that the reader does not have a background in quantum mechanics or advanced mathematics.

The basic purposes of our presentation are to formulate the physical problem, to develop the mathematical models relevant to this physical problem, to determine the required information through the application of quantum (or classical) mechanics, and to study the applications of the results, all for equilibrium systems of independent particles. We then discuss the more general systems of dependent particles and introduce the theory and treatment of non-equilibrium systems.

Chapter 1 serves as an introduction to the subject. The required mathematical topics are treated in Chapter 2, which includes a coverage of mathematical probability and statistics, and also certain relevant mathematical procedures and formulae. It is intended that this material be covered briefly in class, the objective of this chapter being to form a foundation for the development of the course material to follow. Chapter 3 deals with the development of the mathematical models for systems of independent particles and includes applications to certain classical systems. Chapter 4 then forms the important bridge between the microscopic and macroscopic viewpoints, through a discussion of the equilibrium state and the laws and properties of thermodynamics. The subject of quantum mechanics is treated in Chapter 5, with particular emphasis directed toward evaluation of quantum state energies for various modes of energy. Fairly detailed solutions of the Schrödinger wave equation are included for certain molecular modes. A detailed discussion of these solutions is probably more appropriate for graduate students. For undergraduates, a discussion of the results of these solutions and their significance and use is, in general, sufficient. At this point we also consider the atomic and molecular electronic states, the Pauli exclusion principle, and give a brief description of band theory of solids.

Chapters 6 to 8 are concerned with the properties and behavior of systems of independent particles, and they deal in turn with monatomic gases, solids, and diatomic and polyatomic gases. Included in these chapters are certain special topics such as the photon gas, electron gas, and the ortho-para hydrogen problem. Chapter 9 treats the subject of chemical equilibrium from both the macroscopic and the microscopic viewpoints. The macroscopic development has been included in order to demonstrate the parallelism of the two approaches, and to serve as a brief review for those students whose background in this area is insufficient. This chapter concludes with a discussion of ionization of gases.

Chapter 10 introduces the theory of ensembles, the statistical mechanics for systems of dependent particles. It has been our experience that the student gains a better understanding of this material after having studied thoroughly the independent particle problem, because this enables him to follow the analogies between the two developments. Fluctuations in energy and density are also considered at this point. Development of the equation of state for pure substances and mixtures is the subject of Chapter 11; this is a rather specialized topic intended for more advanced students. Virial coefficients, intermolecular potential functions, and an introduction to the theories of liquids are also studied in this chapter. The final chapter constitutes an introduction to the thermodynamics of irreversible processes and includes such topics as entropy production, the phenomenological relations, and the Onsager reciprocal relations. Also discussed here are a simplified theory of transport processes and properties and the thermoelectric effects.

It is, of course, very difficult to acknowledge the valuable assistance of the many people who have contributed in one way or another to the preparation of this book. Certainly, we are deeply indebted to our co-authors in this series, Professors A. G. Hansen and J. A. Clark, to our colleagues Professors W. Mirsky, H. Merte, and G. E. Smith, and to our many students for their helpful criticisms, suggestions, and advice; to Mrs. Bernice Ogilvy and her associates in the department office for their assistance in the preparation of the notes on which this manuscript is based; and especially to our families for their continual encouragement and understanding during these past years.

We hope that this introduction to statistical thermodynamics will stimulate and encourage further study in this and related fields in the various engineering disciplines. Comments and suggestions from our readers will be appreciated.

Richard E. Sonntag
Gordon J. Van Wylen

Ann Arbor, Michigan
October, 1965

Contents

Symbols xvii

Chapter 1
Introduction 1

1.1 Fields of Application 1
1.2 Units 3
1.3 Classical Mechanics 4
1.4 Quantum Mechanics 5
1.5 Energy Storage and Degrees of Freedom 7
1.6 Statistical Mechanics and Thermodynamics 9

Chapter 2
Mathematics 11

2.1 Definition of Probability 11
2.2 Conditional and Compound Probabilities and Independent Events 15
2.3 Total Probability and Mutually Exclusive Events 21
2.4 Permutations, Combinations, and Repeated Trials 25
2.5 Distribution Functions, Mean Values, and Deviations 34
2.6 Special Combinatorial Problems 41
2.7 Lagrange Multipliers 45
2.8 Stirling's Formula 47
2.9 Euler-Maclaurin Summation Formula 48

Chapter 3
Statistical Mechanics for Systems of Independent Particles 53

3.1 Introduction 53
3.2 Boltzmann, Bose-Einstein, and Fermi-Dirac Statistics 56
3.3 The Equilibrium Distribution 60
3.4 Identification of the Multipliers 66

3.5 The Partition Function 70
3.6 The Maxwell-Boltzmann Velocity Distribution 72
3.7 The Equilibrium Distribution for the Components of a Mixture 78

Chapter 4
Thermodynamics

4.1 Internal Energy and Specific Heat 85
4.2 The First Law of Thermodynamics 87
4.3 Entropy 89
4.4 The Second Law of Thermodynamics 92
4.5 The Third Law of Thermodynamics 97
4.6 Thermodynamic Properties 99
4.7 Bose-Einstein and Fermi-Dirac Statistics 103
4.8 Entropy from Information Theory 109
4.9 The Properties of an Ideal Gas Mixture 114

Chapter 5
Quantum Mechanics

5.1 The Bohr Theory of the Atom 119
5.2 Wave Characteristics of Electrons and the Heisenberg Uncertainty Principle 121
5.3 The Schrödinger Wave Equation 123
5.4 Translation 126
5.5 Application of the Wave Equation to Molecules 130
5.6 Rigid Rotator 131
5.7 Harmonic Oscillator 137
5.8 General Rotation-Vibration 142
5.9 Electronic States of Atoms and Molecules 146
5.10 The Pauli Exclusion Principle 158
5.11 Band Theory of Solids 160

Chapter 6
Monatomic Gases

6.1 Contributions to the Partition Function and to Properties 166
6.2 Translation 169
6.3 An Alternative Evaluation of the Translational Partition Function 173
6.4 Electronic Levels 174
6.5 The Photon Gas 177

Chapter 7
Monatomic Solids

7.1 The Harmonic Oscillator 186
7.2 The Einstein Solid 188
7.3 The Debye Solid 190
7.4 The Electron Gas in a Metal 196

Chapter 8
Diatomic and Polyatomic Gases

8.1 The Diatomic Gas—Simple Internal Model 202
8.2 Rotation 204
8.3 Vibration 209
8.4 Electronic Ground Level and Chemical Energy 212
8.5 Diatomic Gases—General Internal Model 213
8.6 Homonuclear Diatomic Molecules at Low Temperature 219
8.7 The Polyatomic Gas 227

Chapter 9
Chemical Equilibrium

9.1 Chemical Equilibrium Constants 239
9.2 Equilibrium Composition 246
9.3 Simultaneous Reactions 249
9.4 Developments from Statistical Mechanics 254
9.5 Ionization 259

Chapter 10
Statistical Mechanics for Systems of Dependent Particles

10.1 The Canonical Ensemble 268
10.2 The System of Independent Particles as a Special Case 274
10.3 Fluctuations in Internal Energy 276
10.4 The Grand Canonical Ensemble 278
10.5 Fluctuations in Density 283

Chapter 11
The Behavior of Real Gases and Liquids

11.1 The Virial Equation of State 289
11.2 The Virial Coefficients 294
11.3 Intermolecular Potential Functions and Evaluation of the Virial Coefficients 297

11.4 The Virial Equation of State for a Mixture 304
11.5 The Liquid Phase 309

Chapter 12
Irreversible Processes

12.1 Forces, Flows, and Entropy Production 317
12.2 Transport Processes and Properties 322
12.3 Coupled Flows and the Phenomenological Relations 327
12.4 The Onsager Reciprocal Relations 330
12.5 The Thermomechanical Effect and Thermomolecular Pressure Difference 334
12.6 Thermoelectric Effects 337

Appendix

A.1 Constants and Conversion Factors 349
A.2 Atomic Weights 350
A.3 Series and Integrals 351
A.4 Harmonic Oscillator Functions 352
A.5 Constants for Diatomic Molecules 358
A.6 Constants for Polyatomic Molecules 359
A.7 The Reduced Second and Third Virial Coefficients for the Lennard-Jones (6–12) Potential 360
A.8 Force Constants for the Lennard-Jones (6–12) Potential from Experimental Virial Coefficient Data 361

Some Selected References 363

Answers to Selected Problems 365

Index 367

Symbols

A	area
	Helmholtz function
a	Helmholtz function per mole
	number of atoms in a molecule
a_i	generalized non-equilibrium variable
B	second virial coefficient
B_ε	rigid rotator function
B^*	rotation-vibration function
b	Morse function constant
$b_1, \ldots$	cluster integrals
b_0	intermolecular potential constant
C	third virial coefficient
C_p	constant-pressure specific heat per mole
C_v	constant-volume specific heat per mole
c	velocity of light
D	Morse function constant—total dissociation energy
	atomic term symbol
D_ε	rotational stretching constant
D_0	observed dissociation energy
d	electron state symbol
F	atomic term symbol
f	electron state symbol
G	Gibbs function
g	Gibbs function per mole
$\mathbf{g}_j$	jth energy level degeneracy
H	enthalpy
H_v	Hermite polynomial
h	Planck's constant
	enthalpy per mole
h°	enthalpy of formation
I	moment of inertia
	information
	electric current
$\mathbf{J}$	atomic total angular momentum number
J_i	generalized flow of type i

$\mathbf{j}$	molecular rotational quantum number
K	equilibrium constant
k	Boltzmann constant
$\mathbf{k}_x, \mathbf{k}_y, \mathbf{k}_z$	x-, y-, z-directional translational quantum numbers
$\mathbf{L}$	atomic orbital angular momentum number
L_{ik}	phenomenological coefficient
$\mathbf{l}$	electron azimuthal quantum number
M	molecular weight
$\mathbf{M}_s$	molecular spin number
m	mass of a particle
$\mathbf{m}_l$	electron magnetic quantum number
$\mathbf{m}_n$	nuclear spin quantum number
m_r	molecular reduced mass
$\mathbf{m}_s$	electron spin quantum number
N	number of particles
N_0	Avogadro's number
n	number of moles
$\mathbf{n}$	electron principal quantum number
P	pressure
	mathematical probability
	atomic term symbol
P_j^{λ}	associated Legendre function
p	electron state symbol
Q	heat
$Q_1, \ldots$	configurational integrals
Q^*	heat transport parameter
R	gas constant
r	radius
r_ε	molecular equilibrium separation
S	entropy
	atomic term symbol
S^*	entropy transport parameter
s	entropy per mole
	electron state symbol
T	temperature
t	time
	rotation-vibration variable
U	internal energy
u	internal energy per mole
u_{v_0}	zero-point energy per mole
V	volume
V	velocity
v	specific volume (per mole)
$\mathbf{v}$	molecular vibration quantum number
v_F	free volume
W	work

w thermodynamic probability—number of states
X_k generalized force
x_ε anharmonicity constant
x^* rotation-vibration variable
Z single-particle partition function
Z_G grand partition function
Z_N system partition function

Script Letters

${}_N\mathscr{C}_M$ combinations of N, M at a time
$\mathscr{D}$ coefficient of diffusion
$\mathscr{E}$ electrical EMF
$\mathscr{K}$ thermal conductivity
${}_N\mathscr{P}_M$ permutations of N, M at a time
$\mathscr{R}$ electrical resistance
$\mathscr{U}$ uncertainty
$\mathscr{Z}$ nuclear charge
$\mathscr{z}$ activity function

Greek Letters

α undetermined multiplier
vibrational stretching constant
α_{AB} Seebeck coefficient
β undetermined multiplier
γ rotation-vibration constant
Δ molecular term symbol
δ rotation-vibration constant
ϵ energy of a state or level
intermolecular potential constant
η number of systems in an ensemble
viscosity
θ angle
θ_r characteristic rotational temperature
θ_v characteristic vibrational temperature
Λ mean free path
λ wavelength
absolute activity
second molecular rotational number
μ chemical potential
ν frequency
stoichiometric coefficient
π molecular term symbol
π_{AB} Peltier coefficient
ρ_N number density
Σ molecular term symbol

σ	intermolecular potential constant
σ_A	Thomson coefficient
Φ	potential energy
	molecular term symbol
ϕ	angle
φ	pair potential energy
	work function
Ψ	wave function
ω	wave number
ω_ε	equilibrium vibrational wave number
ω^*	rotation-vibration constant

Subscripts

e	electronic
i	energy state
j	energy level
mp	most probable (equilibrium)
n	nuclear
r	rotation
t	translation
v	vibration

Marks over Symbols

—	mean value
·	rate

1 Introduction

In this chapter, we introduce the subject of statistical thermodynamics and discuss some of its present applications in the field of engineering. The basic hypotheses and principles of classical mechanics, quantum mechanics, and statistical mechanics and thermodynamics are also briefly presented.

1.1. Fields of Application

In recent years, the engineering applications of statistical thermodynamics and the related subjects of quantum mechanics and statistical mechanics have grown at an increasingly rapid rate. Statistical thermodynamics and these related subjects have contributed to our basic understanding of the nature and behavior of substances. In addition, with research and technology moving toward both extremes of temperature, it is more and more frequently necessary to adopt the microscopic point of view in the analysis of various observed phenomena as well as in the optimization of the design of systems and devices.

A number of applications at high temperatures can be cited. The development of the plasma and ion engines for rockets, which will probably be used extensively as the propulsion systems for deep space probes, is already under way. The direct production of electric power from high temperature plasmas by magnetohydrodynamic converters is another promising field, and many other widely diverse potential applications of magnetohydrodynamics can be cited.

At the opposite end of the temperature scale, the field of cryogenics has expanded very rapidly in recent years, and has involved, in several areas, the application of statistical thermodynamics and quantum mechanics. Certain of the phenomena that are observed in liquid and solid hydrogen and liquid helium, the two lightest elements and those having the lowest boiling points, can be explained and analyzed only from a microscopic or molecular point of view. Superconductivity, the phenomenon involving the complete absence of electrical resistivity in certain elements and alloys at very low temperatures, and various electronic devices such as infrared detectors, which are operated at very low temperatures, are other examples of the applications of the microscopic viewpoint at cryogenic temperatures.

A knowledge of quantum mechanics is necessary to the understanding of semiconductor materials, vital in electronics and in the growing application to thermoelectric power generators and refrigerators. The gas and crystal masers and lasers now under development, which show promise of almost limitless application in communications and other fields, are also quantum mechanical devices. The subject of quantum mechanics had its origins in Planck's theory of radiation, and studies in the field of radiation today are closely linked to quantum mechanics.

Most applications of thermodynamics involve a knowledge of thermodynamic properties, which can be determined from the specific heat and P-V-T behavior of the substance. One method of determining specific heat employs data obtained from spectroscopic measurements; by procedures that are based on statistical thermodynamics, procedures to be developed in later chapters of this book, the specific heat at zero pressure (the ideal gas specific heat) can be calculated. Absolute entropies, common-base enthalpy values, and equilibrium and ionization constants, all of which are required in the analysis of processes involving chemical reactions, can also be determined from spectroscopic data by techniques that will be developed in our study of statistical thermodynamics. Considerable effort has been put forth in the development of the equation of state based on the molecular viewpoint. This involves the use of intermolecular potential functions to determine the P-V-T behavior of real gases and mixtures. The behavior of the solid and liquid phases of matter is also being intensively studied from the microscopic viewpoint. The field of non-equilibrium thermodynamics, which is of great importance in the analysis of finite-rate processes and in the calculation of thermal conductivity and viscosity and diffusion coefficients, is based on the equations of statistical mechanics and utilizes intermolecular potential functions for the evaluation of the transport properties.

Many of these applications will be discussed in some detail during the course of our development of statistical thermodynamics. In addition, a study of statistical thermodynamics provides an insight into the behavior of substances and the laws of classical thermodynamics, thus giving us a more thorough and comprehensive understanding of the subject of thermodynamics.

1.2. Units

The metric system of units is used in essentially all investigations of the behavior of substances at the atomic level. Because the English system of units is commonly used in making many engineering calculations, we review briefly the metric system and the more important constants. A more complete set of constants and conversions is included in the Appendix. In the cgs system, the basic units of length, mass, and time are the centimeter, gram, and second, respectively. Other units of importance to us are: the Angstrom unit, Å (a unit of length),

$$1\text{Å} = 10^{-8} \text{ cm}$$

and an alternative unit of mass, the gm mole, a mass in grams numerically equal to the molecular weight of the substance.

The unit of force in this system is the dyne, the force required to accelerate a mass of 1 gm at the rate of 1 cm/sec^2, so that, from Newton's second law of motion, $F = ma$, we find that

$$1 \text{ dyne} = 1 \text{ gm} \times 1 \text{ cm/sec}^2$$

The basic unit of energy is the erg, corresponding to the work done by a force of 1 dyne acting through a distance of 1 cm. Therefore,

$$1 \text{ erg} = 1 \text{ dyne} \times 1 \text{ cm} = 1 \text{ gm} \times 1 \text{ cm}^2/\text{sec}^2$$

Other quantities of energy are given in terms of the joule and calorie, where

$$1 \text{ joule} = 10^7 \text{ ergs}$$

and

$$1 \text{ cal} = 4.184 \text{ joules}$$

Pressure is measured in absolute atmospheres,

$$1 \text{ atm} = 1{,}013{,}250 \text{ dynes/cm}^2 = 760 \text{ mm Hg}$$

and the temperature scale is the Kelvin or absolute centigrade scale,

$$T\,(^\circ\text{K}) = T\,(^\circ\text{C}) + 273.15$$

Volumes are given in cubic centimeters, cc or cm^3, or in liters, where

$$1 \text{ lit} = 1000.028 \text{ cm}^3$$

Consequently, in this system, the gas constant is

$$R = 1.98726 \frac{\text{cal}}{\text{gm mole-K}} = 0.082055 \frac{\text{lit-atm}}{\text{gm mole-K}}$$

A number of other constants are of particular interest. Avogadro's number, N_0, the number of molecules per gram mole, is

$$N_0 = 6.023 \times 10^{23} \frac{1}{\text{gm mole}}$$

Boltzmann's constant k, the gas constant per molecule, is

$$k = \frac{R}{N_0} = 1.38044 \times 10^{-16} \text{ erg/K}$$

Planck's constant h is

$$h = 6.62517 \times 10^{-27} \text{ erg-sec}$$

and the velocity of light, c, is

$$c = 2.998 \times 10^{10} \text{ cm/sec}$$

The ratio hc/k is frequently of significance. From the constants given above, this ratio is found to be

$$\frac{hc}{k} = 1.4388 \text{ cm-K}$$

One other constant, the atomic mass unit, is of particular importance to us. The mass of an atom per unit atomic weight is

$$1.6598 \times 10^{-24} \text{ gm}$$

These and other units, conversion factors, and constants are summarized in Table A.1 of the Appendix. The atomic weights of the elements are given in Table A.2.

1.3. Classical Mechanics

The laws and equations of classical or Newtonian mechanics are satisfactory for most calculations regarding macroscopic systems, since we are commonly concerned with systems of enormous size compared with the dimensions of atoms or molecules. To appreciate this statement fully, let us consider a box 1 inch on a side and filled with a monatomic gas at room temperature and pressure. Such a system is not particularly large for a classical thermodynamic system.

Using Avogadro's number and the ideal gas equation of state, we find that there are approximately 10^{20} atoms in this volume. We therefore expect no difficulties if we assume that the gas is continuous for the purpose of macroscopic calculations. This is the approximation normally made in classical thermodynamics.

If instead we adopt a microscopic point of view, consider the individual atoms in the box, and assume that the motion of each can be described by the equations of classical mechanics, we can, at least in theory, predict the future behavior of the system, given the exact state of each particle at some instant of time.† We assume that the gas is ideal, i.e., the atoms have no influence on each other except during collisions. We assume also that the atoms are distinguishable from one another, so that we can determine the exact positional coordinates x, y, z, and velocity components V_x, V_y, V_z (or corresponding components of momentum) for each individual atom at any instant of time. It should be pointed out that classical mechanics places no restriction upon the values of velocity, so that an atom can conceivably possess any value of kinetic energy $\frac{1}{2} mV^2$ at a particular time.

Since the state of each individual atom is specified by the determination of six coordinates (three of position and three of velocity or momentum), a single point in a six-dimensional space having such coordinates would completely describe the location and motion of the atom. Such a space is called a phase space, this particular type being termed μ-space. It follows that to specify exactly the state of the system of 10^{20} particles requires the determination of 6×10^{20} coordinates. A single point in a space of 6×10^{20} dimensions would then exactly fix the location and motion of every particle in the system, and consequently the exact state of the system itself. This type of phase space is called a γ-space. Given the required 6×10^{20} coordinates, we could predict the future behavior of the system, collisions of atoms with the walls and with other atoms, with corresponding changes in direction and momentum (and energy). Admittedly, this becomes a completely hopeless computational problem, the only solution to which is to make calculations of a statistical nature—average values and deviations from the averages for quantities of interest to us. This is the subject of statistical mechanics.

Our discussion has been limited here to a monatomic gas with energy consisting solely of translation of the atoms of which the gas is comprised. As will be discussed in Section 1.5, a molecule in general may possess energy in various ways and it has some number of degrees of freedom, f ($f = 3$ for the three directions of translation in our example). Consequently, a corresponding μ-space is a phase space of $2f$ dimensions, and a γ-space is one of $2Nf$ dimensions.

1.4. Quantum Mechanics

The quantum theory of radiation was proposed by Planck at the beginning of this century in an attempt to explain the mechanism of

† See, for example, C. Kittel, *Elementary Statistical Physics*, John Wiley and Sons, New York, 1958, Appendix C, p. 219.

thermal radiation theoretically. Acceptance of this radical departure from the electromagnetic theory began to grow only after Einstein was able to explain the photoelectric effect in terms of the new theory. The science grew rapidly after Bohr's application of the quantum theory to Rutherford's model of the atom in 1913. The resulting theory enjoyed great successes for a period of years, and is a very useful model even today. The concept of electrons revolving about a central nucleus in only certain allowed orbits and consequently possessing only discrete values of energy is simple to visualize, and the model is relatively easy to use and sufficiently accurate for many calculations. However, it gradually became apparent that certain phenomena could not be explained on the basis of the Bohr model, and the "new" quantum mechanics evolved in the mid 1920's, in the form of wave mechanics by Schrödinger and matrix mechanics by Heisenberg. The equations and results of quantum mechanics that are necessary for our purposes will be developed in Chapter 5. For the present, it is necessary for us only to be aware of a few of the hypotheses and results of quantum mechanics.

There are several basic differences between the assumptions of classical mechanics and those of quantum mechanics. In classical mechanics it is assumed that like particles are distinguishable from one another; in quantum mechanics like particles are assumed to be indistinguishable. Furthermore, in classical mechanics it is assumed that the position and momentum of a particle at any instant of time can be simultaneously determined (known) exactly. Quantum mechanics, on the other hand, is based upon the fact that there is always some minimum uncertainty in their product according to the relation

$$\Delta x \times \Delta(m\mathsf{V}_x) \cong h$$

in which Δx is the uncertainty in the position x, $\Delta(m\mathsf{V}_x)$ is the uncertainty in x-direction momentum, and h is Planck's constant. This fundamental principle of quantum mechanics is called the Heisenberg uncertainty principle. We note that, if the uncertainty in the position of the particle is made very small, then the uncertainty in the particle's momentum becomes relatively large, and vice versa. Our inability to determine these quantities simultaneously with precision is not due to the use of poor instrumentation, but is instead of a fundamental nature. Since Planck's constant h is so extremely small, we do not ordinarily observe this fundamental uncertainty principle in everyday macroscopic measurements.

According to quantum mechanics, a particle has associated with it a wave function, which is interpreted in terms of the probability of locating this particle in a certain place at a given time. The corresponding differential equation then has valid solutions only for certain discrete energy states (in certain cases—translation and usually rotation—these discrete energy values lie so close together that the assumption of continuous energy, and classical mechanics, is reasonable). These discrete energy states, called the quantum states of the particle, are specified by a set of quantum numbers. It is often true that different combinations of these numbers specifying the energy state result in exactly or almost exactly the same magnitude or level of energy. As a result, we often find it convenient to speak of energy levels, each of which has a certain number of quantum states. The number of different quantum states that have the same energy level is termed the degeneracy of the energy level, a term that we shall use frequently in our subsequent developments.

We shall also have occasion to refer to the Pauli exclusion principle, which in basic form states that no two electrons in a given atom can have identical values for all their quantum numbers. The Pauli exclusion principle in more general form concerns the symmetry of the wave function with respect to an interchange of coordinates of two identical particles, and it is found to apply to certain particles and not to others.

1.5. Energy Storage and Degrees of Freedom

In our study of statistical thermodynamics, it will be necessary to keep in mind the particular manner in which energy is possessed by molecules or atoms.

Let us consider a gas contained in a tank at a certain pressure and temperature. We assume that it is permissible to divide the energy of this system into three types: the potential energy resulting from the intermolecular forces for the molecular configuration at any instant of time; the kinetic energy of translation of the molecules inside the tank; and the energy internal to the individual molecules.

The first of these is impossible to determine exactly, even theoretically, because we do not know the exact configuration and orientation of the molecules at any time, nor do we know the exact intermolecular potential function. As we shall discover later, at low to moderate densities the molecules are relatively widely spaced, on the average, so that only two-molecule or two- and three-molecule interactions contribute significantly to the configurational energy. In such a region, the problem can be handled quite satisfactorily for reasonably simple molecules. Furthermore, at very low densities the average distances between molecules become so large that even these contributions become negligible, and we may assume the configurational energy to be zero. Consequently, we have in this case a system of independent particles (an ideal gas) and therefore we are able to concentrate our efforts on the problems of evaluating the contributions of translational and internal modes.

For the translational energy, which depends only upon the velocities of the molecules of which the system is comprised, it is necessary to use the equations of mechanics, either quantum or classical. The internal contribution is more difficult to evaluate because, in general, it may result from a number of contributions. We consider first the simple case of a monatomic gas, helium for example, in which the particles to be considered are simply the helium atoms. These atoms possess electronic energy as a result of both orbital angular momentum of the electrons about the nucleus and angular momentum of the electrons spinning on their axes. In the absence of an external field, the energy contribution of the latter is negligible, but it does affect the degeneracy of an energy level. Excited nuclear states are so high in energy (temperature) as to be of no concern to us here, but nuclear spin must be included in a few special cases.

For molecules comprised of two or more atoms each, the situation becomes increasingly more complex. In addition to having electronic states, a molecule may rotate about its center of gravity and the atoms may vibrate with respect to one another, perhaps with several degrees of freedom. Interactions between rotation and vibration are generally small and are normally accounted for in the form of a correction term.

In discussing the number of degrees of freedom, f, for these energy modes, we shall consider the very common case in which all the molecules are in the ground or lowest accessible electronic and nuclear states. Since each atom has 3 degrees of freedom, x, y, z, the number of degrees of freedom for a molecule comprised of a atoms is then

$$f = 3a$$

For a monatomic gas $a = 1$; hence $f = 3$, as discussed previously. In a diatomic molecule $a = 2$, and there are 6 degrees of freedom. Three of these are the translation of the molecule as a whole in the x, y, and z-directions, and two are for rotation. Let us take the origin at the center of gravity of the molecule, and the y-axis along the molecule's internuclear axis, as shown in Figure 1.1. The molecule will then have an appreciable

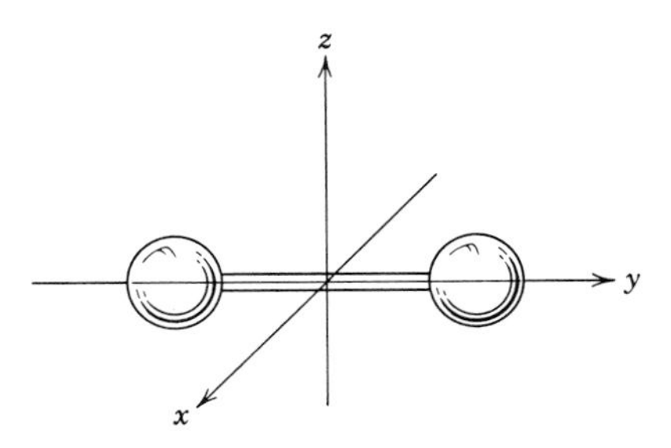

Figure 1.1 The coordinate system for a diatomic molecule.

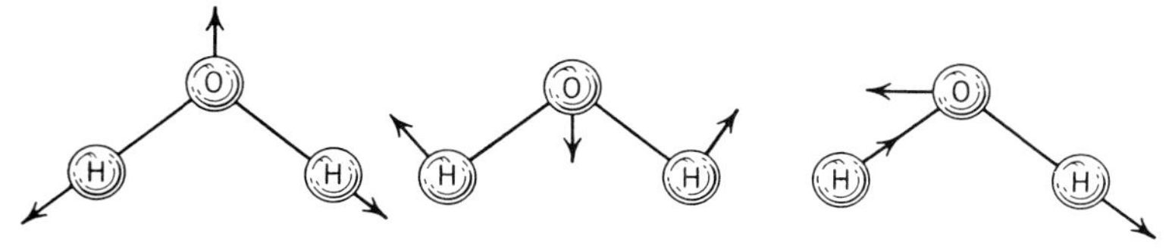

Figure 1.2 The three principal vibrational modes for the water molecule.

moment of inertia about the x-axis and the z-axis, but not about the y-axis. The sixth degree of freedom of the molecule is that of vibration, stretching of the bond joining the atoms.

A polyatomic molecule, in addition to the three translational degrees of freedom of the center of gravity of the molecule, has either two or three for rotation, depending upon whether the molecule is linear or non-linear. The number of vibrational degrees of freedom is accordingly

$$3a - 5 \quad \text{linear}$$

or

$$3a - 6 \quad \text{non-linear}$$

For example, let us consider the H_2O molecule, which is non-linear and has a bond angle of 104.45°. There are consequently three vibrational modes, as represented in Figure 1.2. As shown from left to right in Figure 1.2, these can be termed bond stretching, bending, and twisting modes, respectively. The frequencies of these vibrational modes can be determined with precision by spectroscopic measurement.

In some complex non-linear molecules, in addition to rotation of the molecule as a whole, there may also be a rotation of parts of the molecule with respect to other parts, called internal or hindered rotation. Then the number of rotational degrees of freedom is increased from 3 to 4. Consequently, the number of observed vibrational modes is decreased from $(3a - 6)$ to $(3a - 7)$.

1.6. Statistical Mechanics and Thermodynamics

If we are to determine the behavior of a macroscopic system by examining that of its constituent particles, either through the probability calculations of quantum mechanics or the approximate equations of classical mechanics, we realize that our calculations must be of a statistical nature. Since such systems contain billions of particles, it is unthinkable to attempt to solve the set of equations for the individual particles; however, statistical averages should prove to be extremely precise. That is, deviations from average values of quantities calculated should be negligibly small, provided that the system exists at a state of equilibrium. Such procedures, then, form the foundation of the subject of statistical mechanics. Application of these methods to thermodynamic systems and problems constitutes the subject of statistical thermodynamics. Indeed, the entire subject of classical or macroscopic thermodynamics is founded on the assumption of a macroscopic state of equilibrium, which, from a microscopic viewpoint, is simply the most probable state consistent with the restraints on the system. As a result, although the methods of statistical thermodynamics would be invaluable even if used only for the calculation of thermodynamic properties, statistical thermodynamics also provides insight into the nature of many of the phenomena considered in classical thermodynamics.

The relatively new field of non-equilibrium mechanics and thermodynamics, which is often referred to as irreversible thermodynamics, and which is still in the process of development, deals with the explanation and evaluation of systems and processes departing from equilibrium. This subject is based upon the results of equilibrium statistical mechanics and thermodynamics and the theory of fluctuations of quantities from their equilibrium values. The many fields of application include finite-rate chemical reaction, the transport processes, and irreversible processes in general.

2 *Mathematics*

In the preceding chapter, we have seen that the calculations we are able to make from a microscopic viewpoint regarding the state of a macroscopic system comprised of billions and billions of particles are necessarily of a statistical nature. Before we proceed with the development of the principles of statistical mechanics, some of the mathematical tools and methods that are basic to the subject, particularly statistics and probability theory, are introduced. In order to give the reader a firm base and general understanding of the fundamental principles and background of statistics and probability theory, these concepts are discussed in considerably more detail than is necessary for the development of the expressions of statistical mechanics. Those already familiar with these concepts may wish to omit Sections 2.1 through 2.5.

2.1. Definition of Probability

Interest in mathematical probability dates back to ancient history, but formal study of the subject is generally considered to have begun in Europe during the seventeenth century with the correspondence between Pascal and Fermat. Along with others, they were concerned with the problem of predicting outcomes of certain games of chance. A general interest in the subject developed among mathematicians and scientists, and the complex modern theories of probability and statistics gradually evolved. Although present-day applications have been extended to widely diverse fields, it is beneficial for purposes of instruction to develop the subject in an elementary manner, using simple examples of predicting outcomes of games of chance.

Suppose that we toss a fairly weighted coin and inquire about the chances of its landing heads or tails. Since we have no information leading us to believe that one side or the other is more likely to land face up, we can only conclude that the two are equally likely to occur on any toss. We might then say that, for any toss selected at random, the probability of heads occurring is 1/2 and the probability of tails is 1/2.

Suppose that we toss 100 coins simultaneously. Intuitively, we expect that the number of heads occurring will be between, say, 40 and 60, but probably not exactly 50. It would be an unusual occurrence if 95 of the coins landed heads, but certainly not an impossible one. We naturally expect that the result more often will be in the general vicinity of 50 per cent heads. If the 100 coin toss is repeated 100 or 1000 times, it is reasonable to expect the average frequency of occurrence of heads to be close to 50, or the relative frequency of heads close to 1/2. For any such random (chance) event, the relative frequency of the event approaches some stable value (in our example, 1/2) as the number of trials becomes very large.

This, then, forms the basis for our definition of mathematical probability. Considering a random event A, the probability $P(A)$ of that event is defined as the limiting value of the relative frequency of occurrence of $A, f(A)$, as the number of trials becomes very large, or

$$f(A)/N \rightarrow P(A) \quad \text{as } N \text{ becomes large} \tag{2.1}$$

Frequently it is possible to calculate probabilities *a priori*, so that it is unnecessary to perform series of experimental tests. For such cases, it is convenient to keep in mind the early classical definition of probability, which is:

If there are N mutually exclusive, equally likely outcomes of an experiment, M of which result in a certain event A, then the probability of A is given by the ratio M/N. That is,

$$P(A) = M/N \tag{2.2}$$

The term "mutually exclusive," meaning that no two of the N outcomes can occur simultaneously, will be discussed in a later section.

The classical definition of probability alone is sufficient for the analysis of simple problems. In complex situations, however, it may prove extremely difficult or even impossible to divide the experiment into equally likely outcomes. Consequently, it may be necessary to assign probabilities based on the analysis of experimental data, according to our definition, Eq. 2.1. Inasmuch as the two definitions must be analogous for a truly random experiment, it is advantageous to think of mathematical probability both ways, in terms of our definitions, Eq. 2.1 and Eq. 2.2.

Consistent with our definitions are the following:

$$P(A) = 0 \quad \text{constitutes an impossibility} \tag{2.3}$$

$$P(A) = 1 \quad \text{constitutes a certainty} \tag{2.4}$$

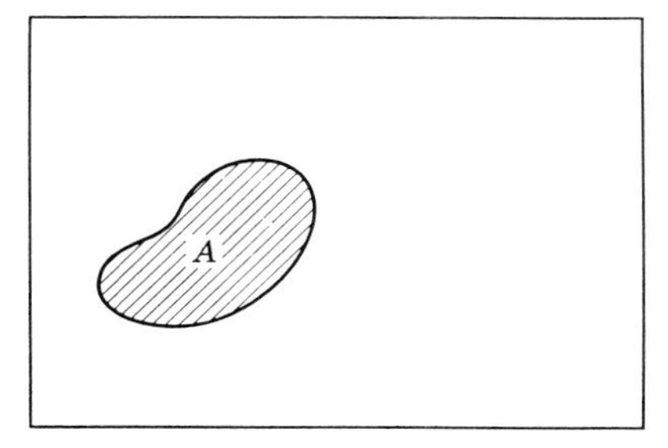

Figure 2.1 Outcomes of an experiment favorable to event A.

so that, in general,

$$0 \leq P(A) \leq 1 \tag{2.5}$$

for any random event A.

It is also helpful to the understanding of probabilities to use some of the concepts of set theory. Suppose that, in a given experiment, each of the possible outcomes can be represented by a point somewhere inside the box (called the sample space) shown in Figure 2.1. Those resulting in event A fall inside the area A, and those resulting in other events fall elsewhere in the box. Now, if both the box and A have been set up so that points are evenly distributed everywhere, that is, there is a uniform point density, then the ratio of the area A to the total area should give the probability of event A.

Now let us define the event "not A" by the symbol A^*, the occurrence of any possible event except A. From our previous definitions,

$$P(A) + P(A^*) = 1 \tag{2.6}$$

In terms of Figure 2.1, the area of the box not included inside A constitutes or represents the event A^*. The events A and A^* are commonly referred to as complementary events.

We now have three ways of looking at or thinking of probability. To attain a firm grasp of the subject, it is important to develop the habit of using all three in the analysis of a problem. It is unnecessary and often inappropriate to use all three explicitly, but in order to develop a consistent and general approach we should at least think in terms of each method.

Example 2.1

Suppose that we toss a fairly weighted die (for our purposes, a cube with faces numbered 1–6). What is the probability that (*a*) the 3 will land face up? (*b*) the face landing up will be a number greater than 3? (*c*) anything but 3 will land face up?

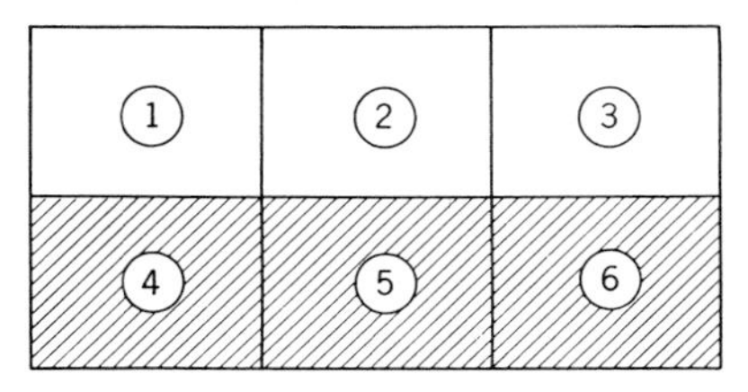

Figure 2.2 Illustration for Example 2.1*b*.

(*a*) Having no information to the contrary, we conclude that the six faces are equally likely to land face up. The six possible outcomes N must also be mutually exclusive, since no two can land up at the same time. If the event A is the face numbered 3 landing up, then $M = 1$, as there is only one way for A to occur. Therefore, from Eq. 2.2, $P(A) = M/N$,

$$P(3) = 1/6$$

Note also that the probability of each of the other numbers occurring is 1/6.

(*b*) As in part (*a*), there are 6 equally likely possibilities, $N = 6$. In this case, the desired event B is any number greater than 3, which could be satisfied by the appearance of a 4, 5, or 6. Consequently, $M = 3$, and from Eq. 2.2, $P(B) = M/N$,

$$P(>3) = 3/6 = 1/2$$

(*c*) Again, $N = 6$. Now the desired event C is anything but a 3, which is the same as A^*. This problem can be analyzed in two ways. First, event C can be satisfied by the appearance of a 1, 2, 4, 5, or 6 so that $M = 5$, and $P(C) = M/N$,

$$P(3^*) = 5/6$$

Alternatively, from Eq. 2.6, $P(A) + P(A^*) = 1$,

$$P(3^*) = 1 - P(3) = 1 - 1/6 = 5/6$$

In the majority of cases, it is somewhat simpler to use the latter approach.

In this example, we have used the second method of looking at probability, the intuitive or classical approach, and perhaps have even formed mental pictures of the problem. For example, part (*b*) can be represented as in Figure 2.2, in which each area includes only one point, since there is but one way for each result to occur. The points, or possible results, are uniformly distributed within the sample space, with dividing lines drawn accordingly. The shaded area of the bottom half of the box represents results favorable to event B, and the probability of B equals 1/2, as determined in the example.

How have we made use of our first definition of probability, that in terms of relative frequency? We did not actually conduct a series of experiments on the die, or even toss it once. We have merely speculated about the outcome of a possible toss. However, this definition of probability must have been considered when it was assumed that each of the six faces is equally likely to land face up, in accordance with the statement that the die is fairly weighted. Perhaps this conclusion was drawn on the basis of past experience. Indeed, it might be gratifying to perform a lengthy test on the die to decide whether it is in fact fairly weighted.

2.2. Conditional and Compound Probabilities and Independent Events

While in the preceding section we concerned ourselves only with simple probabilities, we now extend the analyses to more complex situations. Consider the situation in which we have some information about the experiment before conducting the test. This information, which will have some influence on our probability predictions, may possibly be in the form of a restriction placed upon the outcome, or it may concern results of trials already completed.

Let us assume that event B is dependent in some manner upon another event A. Of importance in such a case is the conditional probability of B, denoted by the symbol $P(B/A)$, which is to be read as the probability of B given the fact that A has occurred. A simple example will help to clarify the concept of conditional probability.

Example 2.2

A standard deck of playing cards consists of 13 cards in each of 4 suits, 2 of which are red (hearts and diamonds), and 2 of which are black (clubs and spades). Each suit contains the cards 2, 3, 4, . . . , 10, J, Q, K, A. Two cards are drawn from a well-shuffled deck and placed face down on the table.

(*a*) Without looking at either card, determine the probability that the second card drawn is an ace.

(*b*) Suppose the first card drawn is turned over and found to be the ace of diamonds. How does this knowledge influence the prediction of part (*a*)?

The understanding of the difference in these two situations is an important concept in probability theory.

(*a*) We are concerned here only with the second card drawn from the deck. The fact that one other card has already been placed apart from the remainder of the deck has no influence on our prediction regarding the second, since we do not know what the first card is. That is, there was no information conveyed by the first draw. Thus, the second card could equally well be any of the cards in the deck, so $N = 52$. Of these, four would result in the desired event B, the draw of an ace, giving

$$P(B) = P(\text{ace}) = M/N = 4/52 = 1/13$$

(*b*) In this case, we have received the information that the card of interest to us cannot possibly be the ace of diamonds, but could be any of the remaining 51. The number resulting in the desired event B has also been reduced by 1, as one of the four aces is gone. Let event A be the fact that the second card can be any card except the ace of diamonds. Event B is still the draw of an ace on the second draw. Event A conveys to us the information that there are only 51 possibilities, of which 3 are favorable to event B. Then the probability of event B given event A, $P(B/A)$, is

$$P(B/A) = P(\text{ace/not ace of diamonds}) = 3/51 = 1/17$$

A comparison of parts (*a*) and (*b*) of this example demonstrates quite clearly that the conveyance of information regarding a situation may have a strong influence on the prediction of probabilities. At this point we must be very careful about drawing too general a conclusion, however, as frequently we may receive information that is not pertinent to the analysis of a particular situation. Such cases will be demonstrated later in this section.

We have just considered the probability of an event B given the fact that a prior event A has occurred. Going back one step further poses a logical question: What is the probability that both A and B will occur? In the language of mathematical logic, the occurrence of both would be called the intersection of A and B, the reason for which is made evident by our pictorial representation in Figure 2.3. In this figure, all possible outcomes of the experiment are represented by uniformly spaced points in the sample space. Those resulting in event A are included in the horizontally shaded area, and those resulting in B by the vertically shaded area. It

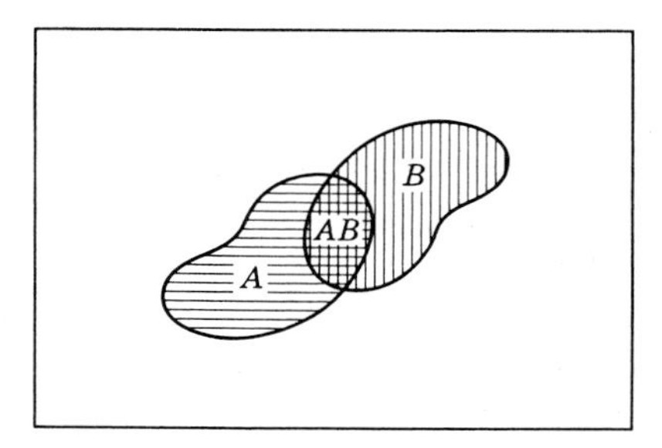

Figure 2.3 Representation of the intersection of two events.

follows that the occurrence of both A and B is given by the cross-hatched area. To help in determining the probability of this event, denoted by $P(AB)$ and often referred to as a compound probability, consider the manner in which the probability $P(B/A)$ must be represented on the diagram. The given fact that A has occurred restricts us to area A in Figure 2.3. Now the occurrence of B in addition is limited to that portion of area A which is also favorable to B, that is, the vertically shaded portion. A uniform point density having been specified, the quantity $P(B/A)$ must be given by the ratio of the cross-hatched area AB to the horizontally shaded area A. Since

$$\left(\frac{\text{area } AB}{\text{area box}}\right) = \left(\frac{\text{area } A}{\text{area box}}\right) \times \left(\frac{\text{area } AB}{\text{area } A}\right)$$

it follows that the compound probability is given by

$$P(AB) = P(A) \times P(B/A) \tag{2.7}$$

Remember that such area ratios are merely other ways of saying relative frequencies. Hence the definition above might just as well have been made in terms of frequencies of occurrence of the events. It is probably helpful, however, to have available a mental image of the concepts involved.

Example 2.3

Two cards are drawn from a well-shuffled deck. What is the probability that (*a*) both are aces? (*b*) neither is an ace?

(*a*) Let event A be the draw of any ace on the first card, and event B the draw of any ace on the second card. As in Example 2.2,

$$P(A) = 4/52 = 1/13$$
$$P(B) = 4/52 = 1/13$$

But

$$P(B/A) = 3/51 = 1/17$$

Therefore, from Eq. 2.7,

$$P(AB) = P(A) \times P(B/A) = (1/13)(1/17) = 1/221$$

(*b*) In this case, we are concerned with the complementary events A^* and B^*, the draw of anything but an ace on the first and second draws, respectively. Thus

$$P(A^*) = 48/52 = 12/13$$
$$P(B^*) = 48/52 = 12/13$$

But

$$P(B^*/A^*) = 47/51$$

Therefore

$$P(A^*B^*) = P(A^*) \times P(B^*/A^*) = (12/13)(47/51) = 188/221$$

Since the sum of the probabilities of all possible results must equal unity, we conclude from Eq. 2.6 that the probability that an ace will be drawn at either the first draw or the second, but not both, is

$$1 - 1/221 - 188/221 = 32/221$$

This will be confirmed in Section 2.3.

It is interesting to note here that, using exactly the same approach as in deriving Eq. 2.7, we can show that the probability of AB is also given by

$$P(AB) = P(B) \times P(A/B) \tag{2.8}$$

If Eqs. 2.7 and 2.8 are combined, another relation of interest is obtained:

$$\frac{P(B/A)}{P(B)} = \frac{P(A/B)}{P(A)} \tag{2.9}$$

Equation 2.9 is a simplified version of Bayes' theorem; it demonstrates how the prediction of the probability of event B is influenced by information regarding event A.

It was pointed out earlier that there are situations in which information regarding an event A is received, but it has no bearing on the estimation of the probability of another event B. In such a case, B is said to be independent of A. That is,

$$P(B/A) = P(B) \tag{2.10}$$

If event B is independent of A, then, from Eqs. 2.9 and 2.10,

$$P(A/B) = P(A) \tag{2.11}$$

so that A must also be independent of B. If events A and B are independent of each other, then their compound probability is given by

$$P(AB) = P(A) \times P(B) \tag{2.12}$$

Although we can all think of many cases in which two given events are obviously independent, very often it may be a difficult task to decide whether two events are or are not independent. In such instances, Eq. 2.12 may be used as a test for independency. If the events are independent, then Eq. 2.12 must be satisfied. The following example demonstrates this point.

Example 2.4

A card is drawn at random from a well-shuffled deck. Let event A be the draw of an ace, and event B the draw of a spade. Are events A and B independent?

The answer to this question is perhaps not obvious. Let us calculate the required probabilities and test for the independence of A and B by the use of Eq. 2.12.

$$P(A) = 4/52 = 1/13$$
$$P(B) = 13/52 = 1/4$$

Also, since the event AB is the draw of the ace of spades, of which there is but one in the deck,

$$P(AB) = 1/52$$

Alternatively, if we solve for $P(AB)$ from Eq. 2.12,

$$P(AB) = (1/13)(1/4) = 1/52$$

which is the same, and we conclude that A and B are independent events. As a point of interest, if we are told that the card is an ace (event A), then, since there are four aces in the deck, one of which is the ace of spades, the probability of event B, given A, is

$$P(B/A) = 1/4$$

We find that the information conveyed by event A is not relevant to the evaluation of the probability of event B. Also, if we are told that the card is a spade (event B), then, since there are 13 spades in the deck, one of which is the ace, the probability of event A, given B, is

$$P(A/B) = 1/13$$

which is equal to $P(A)$ in accordance with Eq. 2.11, and the information conveyed by the occurrence of B does not cause a re-evaluation of the probability of A.

Let us extend our consideration of independent events to a more general situation, that of three events, mutually independent. In order for three events, A, B, C, to be independent,

$$P(ABC) = P(A) \times P(B) \times P(C) \tag{2.13}$$

and, in addition, the events must be pairwise independent:

$$\begin{aligned} P(AB) &= P(A) \times P(B) \\ P(AC) &= P(A) \times P(C) \\ P(BC) &= P(B) \times P(C) \end{aligned} \tag{2.14}$$

Equation 2.13 can be deduced by writing

$$P(ABC) = P(AB) \times P(C) \tag{2.15}$$

and applying Eq. 2.12 in the form of the first of the set 2.14. It is helpful to construct a representative pictorial diagram to see the difference between AB and ABC. Since ABC can be initially divided into any of the three pairs AB, AC, or BC, it follows that the last two equations of 2.14 are also necessary conditions for the independence of A, B, and C.

Example 2.5

A fairly weighted coin is tossed three times in succession. What is the probability that all three tosses will land heads?

Let the events A, B, C be the appearance of heads at the first, second, and third tosses, respectively. In this experiment, we realize intuitively that the three events are independent, since the coin has no memory. For each toss, there are only two possible results, heads or tails; hence

$$P(A) = P(B) = P(C) = 1/2$$

Consequently, from Eq. 2.13,

$$P(ABC) = P(3 \text{ heads}) = (1/2)(1/2)(1/2) = 1/8$$

To remove any doubts about the independency of the three events, we construct a table, as shown below, of all the possible outcomes of the three-toss experiment. We find that there are eight possible results of the experiment representing all possible combinations of either heads or tails for each of the three tosses. Only one of the eight possibilities results in all three heads, giving a probability of 1/8. Since this is in agreement with the value calculated from Eq. 2.13, we conclude that the events are independent. Problems similar to this one will be studied in greater detail in Section 2.4.

Outcome No.	Toss No.		
	1	2	3
1	H	H	H
2	H	H	T
3	H	T	H
4	H	T	T
5	T	H	H
6	T	H	T
7	T	T	H
8	T	T	T

Whether or not two events are independent of each other depends in some cases upon the method of conducting the experiment. Reference to

Example 2.3 makes it apparent that event B is not independent of A because

$$P(B/A) \neq P(B)$$

That is, having been given information about the first card drawn, we were able to eliminate one possibility when speculating about the second. Such an experiment corresponds to sampling without replacement. Notice from the following example how the experiment is changed if the sampling is made with replacement.

Example 2.6

A card is drawn at random from a well-shuffled deck, its denomination and suit are noted, and then it is replaced. A card is then drawn at random again (possibly the same one). What is the probability that both are aces?

Let event A be the draw of an ace on the first draw, and B the draw of an ace on the second. As in Example 2.3,

$$P(A) = P(B) = 4/52 = 1/13$$

However, this time we have replaced the card in the deck after the first draw, so that the information received by that outcome is of no value to us when making the second draw. It has been stated that the second draw was made at random. We might say that we did not note where the first card was inserted into the deck. If this is true, then the second draw must be independent of the first, or

$$P(B/A) = P(B)$$

so that

$$P(AB) = P(A) \times P(B) = (1/13)(1/13) = 1/169$$

and the probability of drawing an ace both times is somewhat greater than before. However, if we have any idea about where the card was replaced (whether it was near the top of the deck, for example), then we have received at least a certain amount of information from the replacement, and B is not strictly a random event. The probability in such a case would be greater than 1/169 and would depend upon how much information was received during the replacement of the first card.

2.3. Total Probability and Mutually Exclusive Events

In Example 2.3, we calculated as 1/221 the probability of drawing an ace on each of two draws without replacement, and as 188/221 that for not drawing an ace either time. We also concluded that the probability of drawing an ace on either the first draw or the second but not both must be 32/221, since that is the only remaining possibility. That outcome might

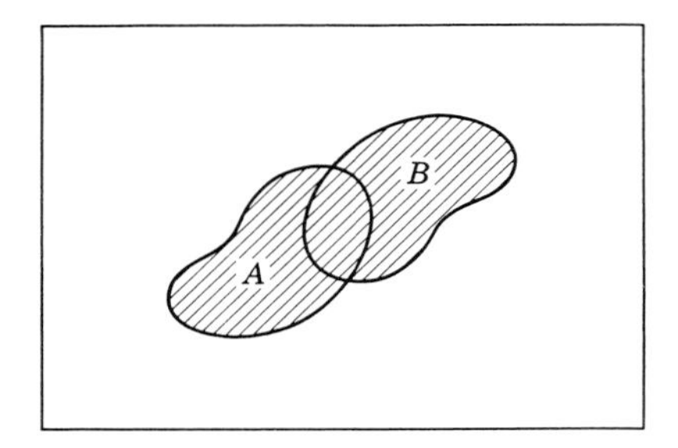

Figure 2.4 Representation of the union of two events.

be referred to as A or B but not both, in which case the "or" is termed the "exclusive or." It is more commonly of interest to consider the "inclusive or" which means event A or event B or both. Unless otherwise specified, we shall assume "or" to mean the "inclusive or."

In Example 2.3, if we desire to calculate the probability of A or B, denoted by the symbol $P(A + B)$, its value is found to be

$$P(A + B) = 1/221 + 32/221 = 33/221$$

since $(A + B)$ denotes the realization of event A or event B or both. In the language of mathematical logic, this is termed the union of A and B. To clarify the concept, let us construct a diagram of uniform point density, the points denoting the possible outcomes of the experiment. We refer again to Figure 2.3 and recall the areas representing A, B, and AB. Now, the event $(A + B)$ is represented by the area,

$$\text{area}_{A+B} = \text{area}_A + \text{area}_B - \text{area}_{AB}$$

It has been necessary to subtract area_{AB} from the sum to avoid counting it twice, as is evident in the figure. Dividing by the area of the box, the respective area ratios can be written as probabilities:

$$P(A + B) = P(A) + P(B) - P(AB) \tag{2.16}$$

Of course, the derivation of Eq. 2.16 is identical with that above if relative frequencies of occurrence are used. If the diagram is redrawn with the area $(A + B)$ shaded, as in Figure 2.4, the name union becomes self-explanatory.

Example 2.7

A card is drawn at random from a well-shuffled deck. What is the probability that it is a spade or an ace?

Let event A be the draw of a spade, event B the draw of an ace. Then

$$P(A) = 13/52$$
$$P(B) = 4/52$$
$$P(AB) = 1/52$$

so that

$$P(A + B) = 13/52 + 4/52 - 1/52 = 16/52$$

Checking this result by counting, we find that any of the 13 spades, including the ace and also the other 3 aces, will satisfy the outcome $(A + B)$, giving a total of 16 favorable results. Note that we do not make the mistake of counting the ace of spades twice, the draw of that card being the event AB.

It is not difficult to construct a representative diagram for three mutually intersecting events A, B, and C and conclude that for the union

$$P(A + B + C) = P(A) + P(B) + P(C) - P(AB) - P(AC) - P(BC) + P(ABC) \quad (2.17)$$

We return now to the term "mutually exclusive," which was used in connection with the classical definition of probability. If two events are mutually exclusive, they cannot occur together; that is, the two do not intersect. This can be shown quite simply by the representation of two such events as in Figure 2.5. If two events A and B are mutually exclusive, then it follows that

$$P(AB) = 0 \quad (2.18)$$

and, as a result of Eq. 2.18,

$$P(A + B) = P(A) + P(B) \quad (2.19)$$

in which the plus sign now denotes the "exclusive or." It is important to keep in mind that Eq. 2.19 is a special case of the general equation 2.16 and applies only to mutually exclusive events. We also note that, for

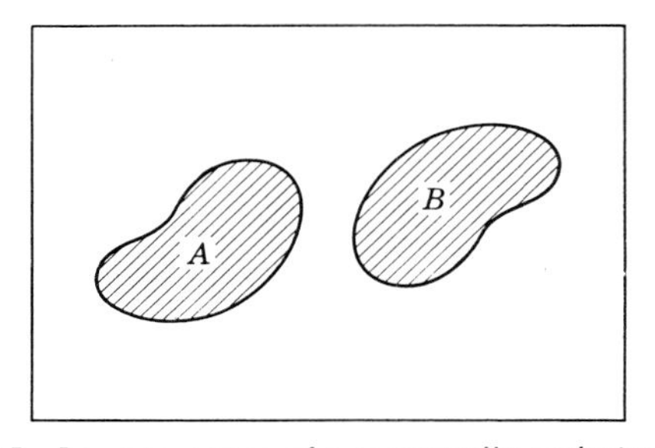

Figure 2.5 Representation of two mutually exclusive events.

three mutually exclusive events, the right side of Eq. 2.17 reduces to the sum of the individual simple probabilities.

Example 2.8

A fairly weighted die is tossed once. What is the probability that it will land (*a*) either a 3 or even? (*b*) either a 4 or even?

(*a*) If event A is the appearance of a 3 and B an even number, then A and B are mutually exclusive, since they cannot both be satisfied.

$$\begin{aligned} P(A) &= 1/6 \\ P(B) &= 3/6 \\ P(AB) &= 0 \end{aligned}$$

Therefore

$$P(A + B) = 1/6 + 3/6 - 0 = 4/6 = 2/3$$

Checking by counting the possible outcomes satisfying $(A + B)$, we find that there are four, the appearance of a 2, 3, 4, or 6.

(*b*) If event A is the appearance of a 4 and B an even number, then A and B in this case are not mutually exclusive.

$$\begin{aligned} P(A) &= 1/6 \\ P(B) &= 3/6 \\ P(AB) &= 1/6 \end{aligned}$$

so that

$$P(A + B) = 1/6 + 3/6 - 1/6 = 3/6 = 1/2$$

Checking this result, the outcomes satisfying $(A + B)$ are the numbers 2, 4, or 6, a total of three.

Let us now re-examine the situation described in part (*a*) of Example 2.2, which involved drawing two cards from a deck. There it was reasoned intuitively that, since we had received no information from the first draw, the probability that the second card drawn was an ace must be 1/13. Let event A be the draw of an ace at the first draw, and B the draw of an ace at the second. Obviously, A^* must be the draw of anything but an ace on the first draw. Since A and A^* are complementary and mutually exclusive, it follows that

$$\begin{aligned} P(B) &= P(AB + A^*B) \\ &= P(AB) + P(A^*B) \\ &= P(A) \times P(B/A) + P(A^*) \times P(B/A^*) \end{aligned}$$

Since there is a 4/52 chance that the unnoted first card drawn is an ace,

$$P(A) = 4/52$$
$$P(A^*) = 48/52$$
$$P(B/A) = 3/51$$
$$P(B/A^*) = 4/51$$

and

$$P(B) = (4/52)(3/51) + (48/52)(4/51) = 4/52 = 1/13$$

as we reasoned earlier.

There is a natural tendency to confuse the terms independent and mutually exclusive by thinking in terms of the inappropriate phrase "have nothing to do with each other" for both. This is, of course, a very poor choice of words. Examining the equation used to test for independence, Eq. 2.12,

$$P(AB) = P(A) \times P(B)$$

we realize that, if two events are said to be independent then they cannot also be mutually exclusive unless at least one of them is impossible, with a probability of 0. Conversely, if two possible ($P \neq 0$) events are said to be mutually exclusive, then they cannot also be independent. For example, if A and B are mutually exclusive, then $P(B/A) = 0$, so that, if $P(B) \neq 0$, we conclude that A and B are not independent.

2.4 Permutations, Combinations, and Repeated Trials

In the preceding sections, we have calculated probabilities in cases for which it was not difficult to determine the number of possible outcomes of an experiment either through intuitive reasoning or by direct counting. It is important at this point to develop analytical means for evaluating more complex situations.

First, let us define a permutation of a group of objects as any specific arrangement or ordering of the members of the group. For example, consider the three letters ABC. There are six arrangements or permutations of the three letters:

$$\begin{array}{ccc} ABC & BAC & CAB \\ ACB & BCA & CBA \end{array}$$

In such a problem, it is a simple matter to list all the possibilities, but suppose we asked for the number of permutations of all the letters in the alphabet? It would be an extremely tedious, if not a hopeless, task to attempt to list all the possibilities.

In an effort to develop formulas for calculating permutations, let us analyze what has been done in listing the arrangements of the three letters ABC. In writing down any arrangement, there are three choices for the first letter, either A, B, or C. After this decision has been made, however, there are only two letters remaining, so that there are two possible choices for the second member of the arrangement. Once that has been selected, there is but one letter remaining and therefore only a single choice for the third letter in the arrangement. Thus we find that there are

$$3 \times 2 \times 1 = 6$$

possible arrangements or permutations of the group of three letters. Looking at this problem in terms of sampling, we might alternatively say that it is possible to pick six distinguishable samples of three letters from this group.

Let us now find the number of permutations of the group of four letters $ABCD$. As before, in writing down any permutation, there are four choices for the first, three for the second, two for the third, and only one for the fourth. Consequently, there are

$$4 \times 3 \times 2 \times 1 = 24$$

permutations of the four letters. We find that this answer is correct if we list all the possibilities:

$ABCD$	$BACD$	$CABD$	$DABC$
$ABDC$	$BADC$	$CADB$	$DACB$
$ACBD$	$BCAD$	$CBAD$	$DBAC$
$ACDB$	$BCDA$	$CBDA$	$DBCA$
$ADBC$	$BDAC$	$CDAB$	$DCAB$
$ADCB$	$BDCA$	$CDBA$	$DCBA$

By extending these analyses to increasingly more complex examples, we conclude that, for N objects, there are N choices for the first member of any arrangement, $(N - 1)$ choices for the second, . . . , and one for the Nth, so that the number of permutations of N objects considered as a group, ${}_N\mathscr{P}_N$, is

$${}_N\mathscr{P}_N = N(N-1)(N-2)\cdots 3 \times 2 \times 1 = N! \qquad (2.20)$$

In the symbol used for the number of permutations, the subscript N before the symbol $\mathscr{P}$ denotes the number of objects in the group. The subscript N after the symbol specifies that all N objects were considered together, or we might say that the sample consisted of N objects. We have also introduced the factorial symbol $N!$ defined as the product of all integers from 1 to N. It is of interest to note that

$$N \times (N - 1)! = N! \tag{2.21}$$

Consequently, if Eq. 2.21 is to be valid for $N = 1$, we must also define

$$0! = 1 \tag{2.22}$$

which introduces no difficulties.

Up to this point, our concern has been only with the possible arrangements of all the objects of which the group is comprised. Suppose we consider now the possibility of picking only a portion of the group. For example, the number of permutations of two letters from the group $ABCD$ is given by

$$\begin{array}{cccc} AB & BA & CA & DA \\ AC & BC & CB & DB \\ AD & BD & CD & DC \end{array}$$

or a total of 12. Analyzing the process, we find that there are four choices for the first letter and three for the second, giving the total number of possible arrangements

$$4 \times 3 = 12$$

as we found by direct listing.

Extending this reasoning to the general case, we may think of this process as the number of ways of picking a sample of M objects out of a group of N without replacement ($M \leq N$), or, in other words, the number of permutations of N objects taken M at a time without repetition. There are N choices for the first object, $(N - 1)$ choices for the second, . . . , and $(N - M + 1)$ choices for the Mth object, so that

$${}_N\mathscr{P}_M = N(N-1)(N-2)\cdots(N-M+2)(N-M+1) \tag{2.23}$$

where the symbol ${}_N\mathscr{P}_M$ is read as the number of permutations of N objects taken M at a time.

Equation 2.23 can be converted to a more useful form by using the factorial expression

$$\begin{aligned} N! &= N(N-1)\cdots(N-M+1)(N-M)(N-M-1)\cdots 3 \times 2 \times 1 \\ &= N(N-1)\cdots(N-M+1) \times (N-M)! \end{aligned} \tag{2.24}$$

Comparing Eq. 2.23 and Eq. 2.24,

$${}_N\mathscr{P}_M = \frac{N!}{(N-M)!} \tag{2.25}$$

We see that, for $M = N$, Eq. 2.25 reduces to Eq. 2.20 (since $0! = 1$), and the two are found to be consistent.

Equation 2.25 may be viewed in the following manner: Of the $N!$ permutations of N objects, there are $(N - M)!$ permutations of the $(N - M)$ objects remaining after M have been selected. Therefore the number of permutations when only M are selected from the group of N, ${}_N\mathscr{P}_M$, is given by

$$(N - M)! \times {}_N\mathscr{P}_M = {}_N\mathscr{P}_N$$

In these considerations, we have not allowed repetition of symbols, or, in terms of sampling, the sampling has been made without replacement. Suppose that repetition (or replacement) had been permitted. Now, for the problem of selecting two letters from the group $ABCD$, there would be four choices for the first, and also four for the second, or

$$4 \times 4 = 4^2 = 16$$

possible ways of choosing two of the four letters. In addition to the twelve given previously, there would now also be the four doubles AA, BB, CC, DD. In general, the number of ways of picking a sample of M objects from a group of N with replacement, or the number of permutations of N objects taken M at a time with repetition, is N^M instead of that given by Eq. 2.25.

Example 2.9

Two distinguishable dice are rolled simultaneously. How many possible outcomes are there? How many of these result in a total of 7? of 10? What are their probabilities?

There are two dice, with six possible outcomes for each. Thus the total number of possibilities is

$$6^2 = 36$$

Of the 36, 6 result in a total of 7, these being 6–1, 5–2, 4–3, 3–4, 2–5, 1–6. Therefore, the probability of getting a total of 7 must be

$$P(7) = 6/36 = 1/6$$

On the other hand, there are only three ways of getting a 10, these being 6–4, 5–5, 4–6, so the probability of a 10 is

$$P(10) = 3/36 = 1/12$$

There are other aspects to be considered. Suppose that, in the consideration of selecting a sample of M objects from a group of N, we are not interested in the order of the objects within the sample drawn. For example, in the problem of selecting two letters from the group $ABCD$, if order is not considered there are only the six results

$$\begin{array}{lll} AB & BC & CD \\ AC & BD & \\ AD & & \end{array}$$

instead of the 12 results given previously. That is, all such results as AB and BA are no longer considered as being different. Since we are now interested only in selections of objects rather than arrangements, we use the terminology "number of combinations" instead of "number of permutations." For the preceding problem the number of combinations of four objects taken two at a time is 6. In a manner similar to our explanation following Eq. 2.25, we conclude that the number of combinations of four letters taken two at a time should be the number of permutations of four objects taken two at a time divided by the 2! permutations of the two objects selected, so as to discount the order within the sample. Then the number of combinations is given by

$$\frac{(4!/2!)}{2!} = \frac{4!}{2!2!} = 6$$

as found above by counting.

The number of combinations of N objects taken M at a time (i.e., the number of ways of choosing a sample of M objects from a group of N objects where the order within the sample is not to be considered) is given by

$$_N\mathscr{C}_M = \frac{_N\mathscr{P}_M}{M!} = \frac{N!}{M!(N-M)!} \tag{2.26}$$

Since the order within the sample of M is not to be considered, the number of combinations is given by the number of permutations of N objects taken M at a time divided by $M!$, the number of permutations of the M objects within the sample.

Example 2.10

In how many ways can two cards be drawn from a standard deck of 52, if (*a*) order is not considered? (*b*) order is considered?

(*a*) If the order is not considered, then, from Eq. 2.26, there are

$$_{52}\mathscr{C}_2 = \frac{52!}{2!(52-2)!} = \frac{52 \times 51}{2} = 1326$$

different pairs that can be drawn. Comparing this problem with Example 2.3, we note that, of the 1326 possible pairs, only six are pairs of aces, this being the number of combinations of the four aces taken two at a time, given by

$$_{4}\mathscr{C}_{2} = \frac{4!}{2!(4-2)!} = 6$$

Since there are 1326 possible outcomes, six of which result in a pair of aces, the probability of that result is

$$P(2 \text{ aces}) = 6/1326 = 1/221$$

which is in agreement with Example 2.3.

(*b*) If order is to be considered, then, from Eq. 2.25, there are

$$_{52}\mathscr{P}_{2} = \frac{52!}{(52-2)!} = 52 \times 51 = 2652$$

possible outcomes, exactly twice that of part (*a*). Each of the results of part (*a*) must in this case be considered as two separate results to account for ordering of the pair. That is, 10H — 6C is different from 6C — 10H, where H and C refer to hearts and clubs.

Again comparing with Example 2.3, of the 2652 possibilities including order, the number resulting in a pair of aces is

$$_{4}\mathscr{P}_{2} = \frac{4!}{2!} = 12$$

and the corresponding probability

$$P(2 \text{ aces}) = 12/2652 = 1/221$$

is the same as found before.

The preceding example must have raised questions about which quantity is of more practical interest, permutations or combinations. This is a problem that must be faced in every analysis and one that should not be taken lightly. The only answer to be given is that each case is individual, and each time we must ask ourselves whether or not ordering is of importance. For example, in one form of the card game of poker, each player is dealt five cards before the game begins. In such a case, the order in which the five cards in a hand were dealt to a player is of no significance, the only thing of importance being the resulting group. On the other hand, in another form of the same game, each player is dealt only two cards at the beginning, and receives the third, fourth, and fifth individually at later times during the course of the game itself. In this case, the order in which the cards are received is certainly of concern to a player, for, as he receives additional information with each card, he attempts to re-evaluate his chances for eventually winning the game. At the end of the game, the ordering has no effect on determining who is the winner, but order does influence the players during the game. A player who did not look at any of his cards until he had received all five would obviously be at a tremendous disadvantage in the long run. We can all, no doubt, think of many additional illustrations in which ordering within a sample or group is either of primary concern or of no importance whatsoever. There are also, of course, many complex situations in which such a decision is not at all obvious.

One other combinatorial problem is of interest to us at this time. Suppose that not all the objects making up a group are different. For example, how many permutations are there for the group of letters $AABB$? The possible arrangements are

$$\begin{array}{ll} AABB & BBAA \\ ABBA & BAAB \\ ABAB & BABA \end{array}$$

or a total of six. The analysis is somewhat more difficult than for the situations examined previously. Let us for the moment assume that the two A's are not identical, but are instead distinguishable from each other as A_1 and A_2, and similarly that the B's are distinguishable as B_1 and B_2. For such a group of four letters taken all together, the number of permutations is

$$_4\mathscr{P}_4 = 4! = 24$$

Now let us drop the subscripts on the two A's. Having done so, we realize that the number of permutations given above is no longer correct, but must be divided by 2!, the number of permutations of the two A's that are no longer distinguishable. If now we drop the subscripts on the B's as well, we must again divide by 2! to discount the number of permutations of the pair of B's. The number of permutations of the objects $AABB$ is therefore

$$_4\mathscr{P}_{2,2} = \frac{4!}{2!2!} = 6$$

which is in agreement with the value found above by direct counting. By considering other examples, the analysis above may be generalized to the following: The number of permutations of N objects, M_1 of one kind, M_2 of a second kind, . . . , M_k of a kth kind, $_N\mathscr{P}_{M_1,M_2,\ldots,M_k}$, where

$$\sum_k M_k = N$$

is

$${}_N\mathscr{P}_{M_1, M_2, \ldots, M_k} = \frac{N!}{M_1!M_2! \cdots M_k!} = \frac{N!}{\prod\limits_k M_k!} \tag{2.27}$$

The symbol Π in Eq. 2.27 will be used extensively to denote the continued product over all values of k.

From our discussion of combinations, it follows that Eq. 2.27 also represents the number of combinations of N objects taken $M_1, M_2, \ldots, M_k$ at a time, ${}_N\mathscr{C}_{M_1, M_2, \ldots, M_k}$. That is,

$${}_N\mathscr{C}_{M_1, M_2 \ldots, M_k} = \frac{N!}{\prod\limits_k M_k!} \tag{2.28}$$

is the number of ways of dividing a group of N distinguishable objects into groups of $M_1, M_2, \ldots, M_k$, where the order within each of the groups is not to be considered. This result follows from a repeated application of Eq. 2.26 to each of the groups $1, 2, \ldots, k - 1$.

Let us now utilize the results of this section in evaluating probabilities in repeated trials. This matter has been considered briefly in previous sections, but only in an informal manner. In Example 2.5, it was found that the probability of obtaining three heads in three tosses of a coin is 1/8. From the table given there, we note that the probabilities of 3, 2, 1, or 0 heads in three tosses are 1/8, 3/8, 3/8, 1/8, respectively. In evaluating these probabilities, we have not been concerned with the order in which the heads occur, only with the number occurring in three tosses. We might say that the probability of a result (say 2 heads out of three tosses) equals the probability of getting that result in one way times the number of ways that the result can occur.

In general, consider an experiment with two outcomes, success or failure. A success may be the appearance of heads in the toss of a coin, for example; in this case the appearance of tails constitutes a failure. Or, in tossing a die, the appearance of a 3 may be the success, any of the five other results being considered a failure. For any trial, let the probability of success be p. Then the probability of failure would be $(1 - p)$. For N trials, the probability of achieving N successes is p^N, since the trials are independent. Similarly, the probability that all N trials will result in failures is $(1 - p)^N$. It follows that the probability of achieving M successes immediately followed by $(N - M)$ failures is

$$p^M \times (1 - p)^{N-M}$$

The probability for M successes and $(N - M)$ failures in any specific order is also given by this expression, since the outcome of each of the N trials is specified as being a success or a failure. However, it may be possible to achieve M successes in N trials in more than one way, as has been discussed earlier in this section, and commonly we are not interested in the order in which the success appeared during the experiment, but only in the final number of successes. Thus we should multiply the probability of achieving M successes one way by the number of ways M successes out of N can occur without regard for order (the number of combinations of N objects taken M at a time). Consequently, it follows from Eq. 2.26 that the probability $P(M)$ of achieving M successes in any unspecified order out of N trials is

$$P(M) = \frac{N!}{M!(N-M)!}(p)^M(1-p)^{N-M} \tag{2.29}$$

Example 2.11

What is the probability that in six tosses of a fairly weighted die the number 3 will appear twice or less?

The probability of getting a 3 on any toss is 1/6. From Eq. 2.29 the probability that 3 will not appear in six tosses is

$$P(0) = \frac{6!}{0!(6-0)!}(\tfrac{1}{6})^0(\tfrac{5}{6})^6 \cong 0.3349$$

Similarly, the probability that the number 3 will appear once is

$$P(1) = \frac{6!}{1!(6-1)!}(\tfrac{1}{6})^5(\tfrac{5}{6}) \cong 0.40188$$

and that for 3 appearing twice is

$$P(2) = \frac{6!}{2!(6-2)!}(\tfrac{1}{6})^2(\tfrac{5}{6})^4 \cong 0.20094$$

Thus the probability that 3 will occur twice or less is

$$P(\leq 2) = P(0) + P(1) + P(2) \cong 0.93772$$

Example 2.12

A fairly weighted coin is tossed six times in succession. What is the most probable number of heads? What is the probability of the most probable number?

The probability for obtaining heads on any single toss is 1/2. Then, from Eq. 2.29, for $N = 6$,

$$P(M) = \frac{6!}{M!(6-M)!}(\tfrac{1}{2})^M(\tfrac{1}{2})^{6-M}$$

where M is the number of heads. If all the possible values of M are evaluated, the probabilities are

M	$P(M)$
0	0.015625
1	0.09375
2	0.234375
3	0.3125
4	0.234375
5	0.09375
6	0.015625

The most probable number of heads is seen to be 3, which has a probability of 0.3125.

2.5. Distribution Functions, Mean Values, and Deviations

Suppose that we analyze the results of the preceding section, and in particular discuss Example 2.12 in detail. The number of heads, M, resulting in the six-coin-toss experiment may be viewed as a variable, one that may take on only the discrete values 0, 1, 2, . . . , 6. If we performed that experiment a great many times, we would expect the results to be distributed with a relative frequency close to the calculated values of $P(M)$. If these values are plotted on a histogram, the probabilities are

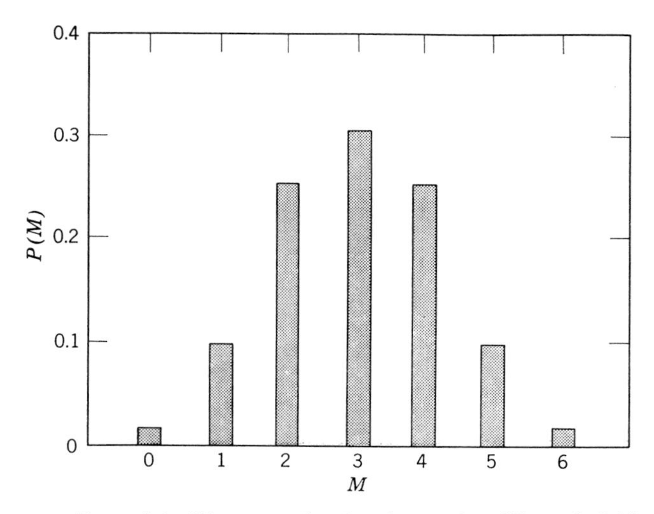

Figure 2.6 Histogram showing the results of Example 2.12.

seen to be symmetric about the most probable result, 3 heads, as shown in Figure 2.6.

The function $P(M)$ may also be called a distribution function, and the distribution specified by Eq. 2.29 is called the binomial distribution. When the probability function is used for this purpose, the distribution function is said to be normalized, since

$$\sum_M P(M) = 1 \tag{2.30}$$

Suppose that the six-coin-toss experiment is performed R times, where R is a very large number. The number of times that M heads will appear is $R \times P(M)$. The average number of heads occurring per trial, $\bar{M}$, which is also called the arithmetic mean value, is the sum of $M \times R \times P(M)$ over all values of M, with the total divided by the number of experiments R. That is,

$$\bar{M} = \frac{1}{R} \sum_M M \times R \times P(M) = \sum_M M \times P(M) \tag{2.31}$$

so that the mean value of M is independent of the number of trials, assuming that the number is sufficiently large. It can be shown by using Eq. 2.31 that the mean value of M for Example 2.12 is 3.0, the same as the most probable result.

In a similar manner, we could determine the mean value of the quantity M^2, which is designated $\overline{M^2}$, and is found to be

$$\overline{M^2} = \sum_M M^2 \times P(M) \tag{2.32}$$

a quantity that will be found to be of interest to us. Note that the bar indicates the mean value of the entire quantity beneath it. For Example 2.12, the mean value of M^2 is found by Eq. 2.32 to be 10.50, which we note is not the same as the quantity $(\bar{M})^2$.

Mean values such as those given by Eq. 2.31 or Eq. 2.32 supply some information but do not indicate the shape of the distribution. For this reason, it is necessary to have some knowledge about deviations from the mean value. We consider two such deviations here, first the average deviation from the mean value, which is given by

$$\overline{M - \bar{M})} = \sum_M (M - \bar{M}) \times P(M) = \sum_M M \times P(M) - \bar{M} \times \sum_M P(M)$$

Using Eq. 2.30 and Eq. 2.31, and the fact that $\bar{M}$ is a constant, this reduces to

$$\overline{(M - \bar{M})} = \bar{M} - \bar{M} = 0 \tag{2.33}$$

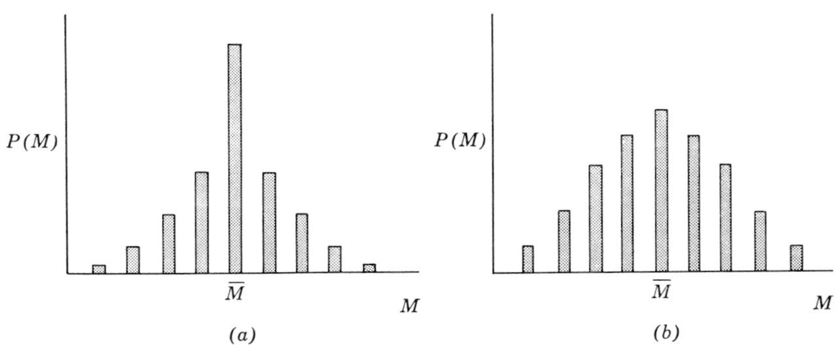

Figure 2.7 Histograms for distributions with different standard deviations. (*a*) Small value of σ; (*b*) large value of σ.

The second deviation of interest to us is the mean-square deviation, also called the variance, σ^2, which can be expressed as

$$\begin{aligned}\sigma^2 &= \overline{(M - \bar{M})^2} = \sum_M (M - \bar{M})^2 \times P(M) \\ &= \sum_M (M^2 - 2M\bar{M} + \bar{M}^2) \times P(M) \\ &= \sum_M M^2 \times P(M) - 2\bar{M} \sum_M M \times P(M) + \bar{M}^2 \sum_M P(M)\end{aligned}$$

Substituting Eqs. 2.30, 2.31, and 2.32 into this equation yields

$$\sigma^2 = \overline{(M - \bar{M})^2} = \overline{M^2} - 2\bar{M} \times \bar{M} + \bar{M}^2 = \overline{M^2} - \bar{M}^2 \qquad (2.34)$$

A term commonly used in statistics is the standard deviation, σ, the square root of the variance given in Eq. 2.34. For Example 2.12, the variance is found from Eq. 2.34 to be

$$\sigma^2 = 10.50 - (3.0)^2 = 1.50$$

and the standard deviation is the square root of this number.

Figure 2.7 demonstrates the significance and utility of the standard deviation, σ. Both distributions have the same mean value $\bar{M}$. However, the distribution in Figure 2.7*a* has a small value of σ, while that shown in Figure 2.7*b* has a large σ. Thus the standard deviation (or variance), together with the mean value, gives a meaningful description of the distribution in a variable.

Example 2.13

Assume that the experiment of Example 2.11, which involved the tossing of a die six times, has been performed a great many times with the distribution of results found to be that given by Eq. 2.29. Plot a histogram of the results, and calculate the mean and the variance.

The experiment involves tossing a die six times and observing the number of times that the face 3 appears. The probabilities for this occurring 0, 1, and 2 times have already been found to be approximately

$$P(0) = 0.3349$$
$$P(1) = 0.40188$$
$$P(2) = 0.20094$$

Similarly, using Eq. 2.29,

$$P(3) = 0.05358$$
$$P(4) = 0.00804$$
$$P(5) = 0.00064$$
$$P(6) = 0.00002$$

The histogram of these results is shown in Figure 2.8. This experiment is seen to result in a binomial distribution that is not symmetric (is asymmetric).

Using the values of $P(M)$ just determined and Eq. 2.31, the mean value is found by summation to be

$$\bar{M} = 1.00$$

which is seen to be the same as the most probable number of 3's. Similarly, using Eq. 2.32,

$$\overline{M^2} = 1.8333$$

Substituting these values into Eq. 2.34 yields

$$\sigma^2 = 1.8333 - 1.00 = 0.8333$$

In both the examples studied, we found that the most probable result and the mean value were the same. It should be pointed out that this is

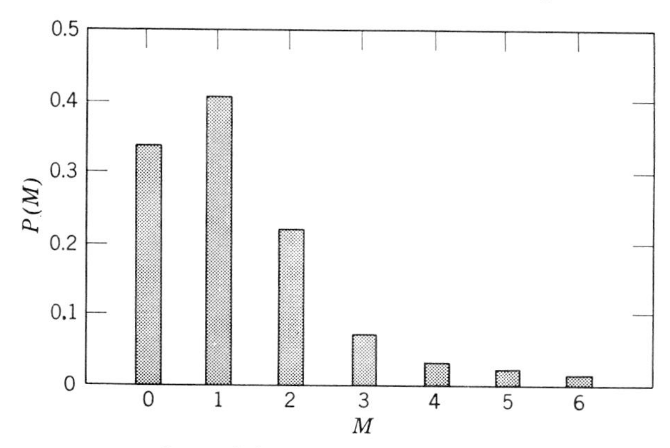

Figure 2.8 Histogram for Example 2.13.

not necessarily always true. In fact, we could find a general expression for the mean value for all distributions given by Eq. 2.29, that is, the binomial distribution. If the expression for $P(M)$ given by Eq. 2.29 is substituted into Eq. 2.31, the result is

$$\bar{M} = N \times p \tag{2.35}$$

where N is the number of trials in the experiment and p is the probability for success on a given trial. For Example 2.12, the experiment consisted of six tosses or trials, and the probability of heads in any trial was 1/2. Thus, if the experiment is repeated many times, the mean number of heads should be $6 \times 1/2$ or 3, as found before by summation. For Example 2.13, the mean number of appearances of a 3 in six tosses of a die is $6 \times 1/6$ or 1. However, suppose the experiment consisted in tossing the die seven times. In the long run, the mean number of appearances of a 3 would be $7 \times 1/6$ or $1\frac{1}{6}$ times, whereas, if all the probabilities are calculated from Eq. 2.29, the most probable result is found to be 1.

Similarly, if Eq. 2.29 is substituted into Eqs. 2.31 and 2.32, and if the resulting expressions are used to find the variance from Eq. 2.34, the result can be shown to be, again for the binomial distribution only,

$$\sigma^2 = Np(1 - p) \tag{2.36}$$

The values calculated for the two examples are seen to be in agreement with Eq. 2.36.

We must keep in mind that, in the majority of practical applications, the approach is necessarily the opposite of that considered so far in this section. Commonly, the *a priori* probabilities and distribution functions are not known, and the role of statistics thus lies in the analysis of a set of experimental data, in terms of calculated mean values and standard deviations. In such analyses, the probabilities in Eqs. 2.31, 2.32, and 2.34 are simply the observed relative frequencies of occurrence, in accordance with our original definition of probability.

Let us now consider two such cases that are of particular interest. There are many physical problems where N is very large and p very small (and the exact value is not necessarily known). Consequently, if we recall the results of Example 2.13 and imagine extending N to successively larger values and p to successively smaller ones, we realize that only relatively small values of M would be of any practical significance. Let us develop an approximation to the binomial formula for such a case. When N is much larger than M, each term in Eq. 2.24 is essentially equal to N, so that

$$\frac{N!}{(N - M)!} = N(N - 1) \cdots (N - M + 1) \cong N^M \tag{2.37}$$

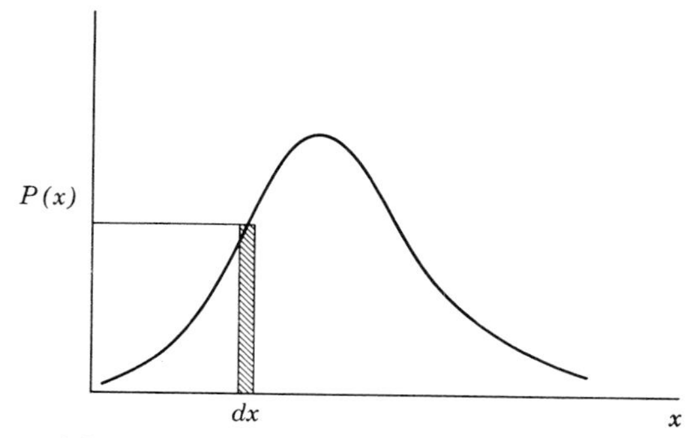

Figure 2.9 Representation of a continuous distribution function.

and

$$(1 - p)^{N-M} \cong (1 - p)^N$$

Therefore Eq. 2.29 can be closely approximated by

$$P(M) = \frac{(Np)^M}{M!}(1 - p)^N = \frac{(\bar{M})^M}{M!}(1 - p)^{(1/p)\bar{M}} \tag{2.38}$$

As p becomes very small, it can be shown that

$$(1 - p)^{1/p} \cong e^{-1}$$

Therefore, for large N and small p,

$$P(M) = \frac{(\bar{M})^M}{M!} e^{-\bar{M}} \tag{2.39}$$

Equation 2.39 expresses the Poisson distribution function. From Eq. 2.36, the variance for this distribution is found to be

$$\sigma^2 = Np = \bar{M} \tag{2.40}$$

This means simply that for the Poisson distribution the distribution is sharply peaked when the mean value is small and is broad when the mean value is large.

In the other case of interest to us, the possible values of the discrete variable are so close together that M may be approximated very closely by a continuous variable x. The corresponding distribution function is now continuous, as, for example, that shown in Figure 2.9, which may be thought of as the limiting case of a histogram like that in Figure 2.6 or 2.8. Now, for a continuous variable, the probability that the variable will have a value in the small interval between x and $x + dx$ is, as shown in Figure 2.9,

$$P(x)\, dx$$

and, since the probability is used as the distribution function, the distribution is said to be normalized. That is, when it is integrated over all possible values of x,

$$\int P(x)\,dx = 1 \tag{2.41}$$

Equation 2.41 may be viewed as the limiting case of Eq. 2.30 as the intervals between discrete values of the variable become progressively smaller. In a similar manner, the limiting case of Eq. 2.31 for the mean value of the variable is given by

$$\bar{x} = \int xP(x)\,dx \tag{2.42}$$

and, similarly, the subsequent three expressions are

$$\overline{x^2} = \int x^2 P(x)\,dx \tag{2.43}$$

$$\overline{(x - \bar{x})} = 0 \tag{2.44}$$

$$\sigma^2 = \overline{(x - \bar{x})^2} = \overline{x^2} - \bar{x}^2 \tag{2.45}$$

These expressions are all analogous to those previously derived, of course, but we have in this case the obvious mathematical advantage of being able to work with integrals instead of summations. The use of Eqs. 2.41 through 2.45 still requires that a distribution function be known, however. One such function frequently used in the analysis of experimental data is the Gauss distribution function, or normal error function, expressed as

$$P(x) = Ae^{-a(x-\bar{x})^2} \tag{2.46}$$

which gives a symmetric distribution about the mean value $\bar{x}$, in terms of the constants A and a. Let

$$y = x - \bar{x} \tag{2.47}$$

Then

$$dy = dx$$

so that Eq. 2.41 can be evaluated. The result is

$$A\int_{-\infty}^{+\infty} e^{-ay^2}\,dy = A\sqrt{\frac{\pi}{a}} = 1 \tag{2.48}$$

or

$$A = \sqrt{\frac{a}{\pi}} \tag{2.49}$$

and it is found that the probability function, Eq. 2.46, is expressed in terms of the mean $\bar{x}$ and the constant a, which are presumed to be given or found by analysis of the data.

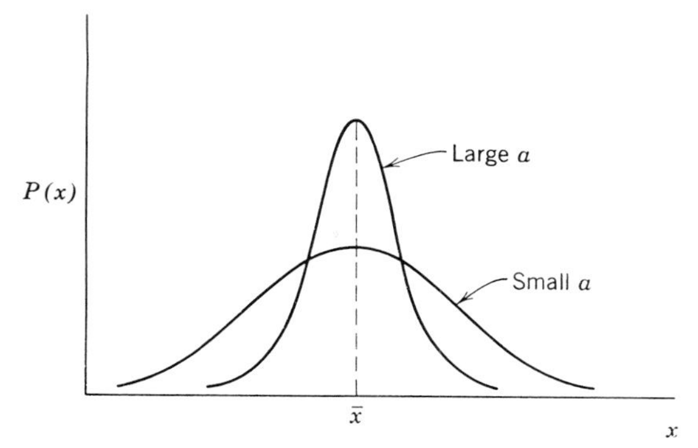

Figure 2.10 Representation of two Gaussian distributions having the same mean value.

From Eqs. 2.45, 2.48, 2.43, and 2.46, the variance is given as

$$\sigma^2 = \overline{(x - \bar{x})^2} = \overline{y^2} = A\int_{-\infty}^{+\infty} y^2 e^{-ay^2}\, dy = \sqrt{\frac{a}{\pi}}\left(\frac{1}{2a}\sqrt{\frac{\pi}{a}}\right) = \frac{1}{2a} \quad (2.50)$$

We find, from Eq. 2.50, that the variance is inversely proportional to the constant a, so that a large value of a should correspond to a sharply peaked distribution, while a small value corresponds to a broad distribution, as indicated by Figure 2.10 for two distributions having the same mean value. Figure 2.10 is similar to Figure 2.7, which showed two distributions having the same mean value for a discrete variable.

We shall find in Chapter 4 that from a microscopic point of view the classical concepts of thermodynamic equilibrium and the second law of thermodynamics are directly related to the existence of extremely sharply peaked molecular distributions.

2.6. Special Combinatorial Problems

In the preceding sections of this chapter, we have discussed briefly the fundamentals of probability and statistics necessary to provide a background for the understanding of the principles of statistical mechanics. In the following chapters, we shall be concerned with the possible distributions of elementary particles, atoms, or molecules among various quantum states. Two different assumptions must be made in the description of such distributions. The first is concerned with whether the particles are distinguishable or indistinguishable from one another, and the second with whether there is or is not a limit of one particle per state (the Pauli exclusion principle).

Let us for the present view these particles as round balls, or in general any such objects, and the various quantum states as boxes, and then analyze the problem of finding the number of ways of placing the objects, say N in number, in g (number of) boxes. The problem is complicated by the fact that the N objects may either be assumed to be distinguishable (numbered, for example) or indistinguishable, and we may or may not place a limit of no more than one object per box. The g boxes are always considered to be distinguishable, however. Thus we must consider four distinct cases.

Case A. Distinguishable objects with a limit of no more than one object per box. There are N objects and g boxes, and since there is an upper limit of one object per box, it follows immediately that

$$N \leq g$$

In placing the first object, there are g choices. Once this object has been dropped into a box, however, there are only $(g - 1)$ remaining choices for the second object. Similarly, there are $(g - 2)$ choices for the third object, . . . , and finally $(g - N + 1)$ choices for the Nth. The total number of ways of placing the N objects in the g boxes is then the same as that given by Eq. 2.23 or 2.25 for the permutation of N objects taken M at a time. Therefore

the number of ways of placing N distinguishable objects in g *distinguishable boxes with a limit of no more than one object per box* ($N \leq g$) *is*

$$\frac{g!}{(g - N)!}$$

Case B. Distinguishable objects with no restriction on the number per box. Since there are g boxes, any of which may be used any number of times, there are g choices for placing the first object, g choices for the second object, . . . , and finally g choices for the Nth, so that

the number of ways of placing N distinguishable objects in g *distinguishable boxes with no restriction on the number per box is* g^N

Case C. Indistinguishable objects with a limit of no more than one object per box. The result for this case would be identical with that for case A except that here the N objects are indistinguishable. Consequently, the result of case A must be divided by the $N!$ permutations of the N objects, to discount the ordering (or distinguishability) of the objects. The result is then the same as that given by Eq. 2.26 for the number of combinations of N objects taken M at a time, or

the number of ways of placing N indistinguishable objects in g *distinguishable boxes with a limit of no more than one object per box* ($N \leq g$) is

$$\frac{g!}{(g - N)!N!}$$

Case D. Indistinguishable objects with no restriction on the number per box. This case is considerably more difficult to analyze, and it does not correspond exactly to analyses made in previous sections. Let us represent the various g (number of) boxes by the symbols $B_1, B_2, B_3, \ldots,$ B_g, and for the moment consider the N objects to be distinguishable as represented by the symbols $0_1, 0_2, 0_3, \ldots, 0_N$. Now consider any arrangement of the list of these $(g + N)$ symbols that is completely arbitrary except that the list begins with the symbol B_1, for example,

$$B_1\ 0_2\ 0_7\ 0_5\ B_5\ B_3\ 0_4\ B_6\ 0_1\ 0_9 \ldots$$

Let us read this arrangement of symbols as specifying the following particular distribution of objects in boxes: Box 1 contains objects 2, 7, 5 (in that order); box 5 is empty; box 3 contains object 4; box 6 contains objects 1, 9; etc. After the symbol B_1, there are $(g + N - 1)!$ permutations or arrangements of the $(g + N - 1)$ symbols, and consequently that number of possible lists beginning with B_1. However, $(g - 1)!$ of these lists correspond to the same numbered objects in the same order placed in the same numbered boxes, if the relative positions of the $(g - 1)$ B's (with their following 0's) are shifted. One such shift would be

$$B_1\ 0_2\ 0_7\ 0_5\ B_3\ 0_4\ B_5\ B_6\ 0_1\ 0_9 \ldots$$

which gives exactly the same distribution as the preceding list. Therefore, the $(g + N - 1)!$ lists are reduced by the factor $(g - 1)!$ Now, recall that the objects are in reality identical and indistinguishable, so that all we are concerned with is that there are any 3 objects in no specific order in box 1, no objects in box 5, any single object in 3, etc., so that the number of lists of interest to us must further be reduced by the factor $N!$ permutations of the N objects. Consequently,

the number of ways of placing N indistinguishable objects in g *distinguishable boxes with no restriction on the number per box is*

$$\frac{(g + N - 1)!}{(g - 1)!N!}$$

Of these four possible cases, case A is the only one that will be of no special interest to us later, and it has been included here for the sake of completeness. Case B for distinguishable objects is that corresponding to the assumptions of classical mechanics, and it will be referred to as Boltzmann statistics. Cases C and D, which involve indistinguishable objects or particles, correspond to the assumptions of quantum mechanics with and without the exclusion principle; they are called Fermi-Dirac and Bose-Einstein statistics, respectively.

Example 2.14

Calculate the number of ways of placing two balls in three numbered boxes for each of the four cases considered above.

Case A. The balls are distinguishable (say one black and the other white), with no more than one ball per box. The number of possibilities is

$$\frac{g!}{(g - N)!} = \frac{3!}{(3 - 2)!} = 6$$

These six possible distributions are

Distribution Number	Box Number		
	1	2	3
1	●	○	
2	○	●	
3	●		○
4	○		●
5		●	○
6		○	●

Case B. The balls are distinguishable, but there is no limit on the number per box. The number of possibilities is in this case

$$g^N = 3^2 = 9$$

These nine distributions are

Distribution Number	Box Number		
	1	2	3
1	● ○		
2	●	○	
3	○	●	
4	●		○
5	○		●
6		● ○	
7		●	○
8		○	●
9			● ○

Case C. The balls are now indistinguishable (both black), with no more than one per box. Now distributions 1 and 2, 3 and 4, 5 and 6 of case A are no longer distinguishable, the total number of distributions that we are able to distinguish being given by

$$\frac{g!}{(g - N)!N!} = \frac{3!}{(3 - 2)!2!} = 3$$

These three distributions are

Distribution Number	Box Number		
	1	2	3
1	●	●	
2	●		●
3		●	●

Case D. The balls are again indistinguishable, but now there is no limit on the number per box. Thus distributions 2 and 3, 4 and 5, 7 and 8 of Case B are now identical, the number of possibilities thus being

$$\frac{(g + N - 1)!}{(g - 1)!N!} = \frac{4!}{2!2!} = 6$$

The six distributions are

Distribution Number	Box Number		
	1	2	3
1	● ●		
2	●	●	
3	●		●
4		● ●	
5		●	●
6			● ●

2.7. Lagrange Multipliers

In the final sections of this chapter, we shall discuss several specialized techniques and approximations that are necessary to the development and evaluation of the expressions of statistical mechanics. The first to be considered is Lagrange's method of undetermined multipliers, a formal procedure for determining a maximum point in a continuous function subject to one or more constratints.

We consider first some function $f(x, y, z)$, a continuous function of the three independent variables x, y, z. Since the function is continuous, the total differential may be expressed as

$$df = \left(\frac{\partial f}{\partial x}\right) dx + \left(\frac{\partial f}{\partial y}\right) dy + \left(\frac{\partial f}{\partial z}\right) dz \tag{2.51}$$

At the maximum (or minimum) of the function, it is necessary that

$$df = \left(\frac{\partial f}{\partial x}\right) dx + \left(\frac{\partial f}{\partial y}\right) dy + \left(\frac{\partial f}{\partial z}\right) dz = 0 \tag{2.52}$$

But, since the variables x, y, z are independent, it follows that changes dx, dy, dz are also independent. Thus the only way that Eq. 2.52 can be satisfied for such arbitrary changes is for all of the coefficients to be identically zero at the maximum. That is,

$$\frac{\partial f}{\partial x} = \frac{\partial f}{\partial y} = \frac{\partial f}{\partial z} = 0 \tag{2.53}$$

Now suppose that, instead, the variables x, y, z are not all independent but are restricted by some interrelation

$$g(x, y, z) = 0 \tag{2.54}$$

so that any two of the three variables can be chosen as the independent variables, the third being dependent according to Eq. 2.54. In this case, the argument leading to Eq. 2.53 is no longer valid, although a maximum is still given by Eq. 2.52. However, we now have a second relation

$$\left(\frac{\partial g}{\partial x}\right) dx + \left(\frac{\partial g}{\partial y}\right) dy + \left(\frac{\partial g}{\partial z}\right) dz = 0 \tag{2.55}$$

at the maximum point. Suppose we multiply the last equation by some arbitrary constant λ and add the result to Eq. 2.52. This gives the expression

$$\left(\frac{\partial f}{\partial x} + \lambda \frac{\partial g}{\partial x}\right) dx + \left(\frac{\partial f}{\partial y} + \lambda \frac{\partial g}{\partial y}\right) dy + \left(\frac{\partial f}{\partial z} + \lambda \frac{\partial g}{\partial z}\right) dz = 0 \tag{2.56}$$

Since any two of the variables can be selected as independent, let us vary dy and dz arbitrarily. This, of course, requires that the change dx will be dependent. Therefore we choose the value of λ such that

$$\left(\frac{\partial f}{\partial x} + \lambda \frac{\partial g}{\partial x}\right) = 0 \tag{2.57}$$

at the maximum. Having selected such a value for λ, the only way that Eq. 2.56 can be identically zero for any arbitrary dy and dz is for their coefficients to be identically zero, or

$$\left(\frac{\partial f}{\partial y} + \lambda \frac{\partial g}{\partial y}\right) = 0 \tag{2.58}$$

$$\left(\frac{\partial f}{\partial z} + \lambda \frac{\partial g}{\partial z}\right) = 0 \tag{2.59}$$

Similarly, any two variables can be taken as independent; then the undetermined multiplier λ is so chosen as to make the coefficient of the dependent variable equal to zero. The subsequent argument is then analogous, and the same expressions, Eqs. 2.57, 2.58, and 2.59 result.

We have considered here the simple case of a function subject to but one constraint and have consequently found it necessary to introduce one undetermined multiplier. This method can very easily be extended to determine the maximum of a function of any number of variables under any lesser number of constraints. By exactly the same procedure and entirely analogous arguments, the resulting number of undetermined multipliers is found to be equal to the number of restraining equations.

2.8. Stirling's Formula

We have seen in Section 2.6 that the combinatorial expressions of interest to us contain factorials of various quantities, for example the number of objects, N. If we attempt to apply such expressions to the problem of particles in quantum states, we find that for a system of any reasonable size (say, a cube 1 mm on a side), the numbers involved become extremely large. Consequently, a direct evaluation of factorials is completely out of the question, and we must develop an approximation to $N!$ that is valid for large N.

It follows from an integration by parts that

$$\int_0^\infty x^N e^{-x}\, dx = N \int_0^\infty x^{N-1} e^{-x}\, dx \tag{2.60}$$

From the similarity of the two integrals, we note that the integral on the right side of Eq. 2.60 could in turn be expressed using this same rule. When we repeat the integration by the same rule over and over, it soon becomes apparent that, if N is an integer, the integral on the left side of Eq. 2.60 is equal to $N!$ Let us approximate this integral for large N. For the variable x, substitute the expression

$$x = N + \sqrt{N}\, y \tag{2.61}$$

Now, since

$$x^N = e^{N \ln x} \tag{2.62}$$

the integrand can be written

$$x^N e^{-x}\, dx = \sqrt{N}\, e^{-N} e^{N \ln (N+\sqrt{N}y) - \sqrt{N}y}\, dy \tag{2.63}$$

Expanding the logarithm, for large N,

$$\ln (N + \sqrt{N}\, y) = \ln N + \ln \left(1 + \frac{y}{\sqrt{N}}\right) = \ln N + \frac{y}{\sqrt{N}} - \frac{y^2}{2N} + \cdots \tag{2.64}$$

If we substitute Eq. 2.64 into Eq. 2.63 and integrate, the result should be approximately $N!$ for large values of N. Therefore

$$N! \cong \sqrt{N}\, e^{-N} N^N \int_{-\sqrt{N}}^{+\infty} e^{-y^2/2}\, dy \tag{2.65}$$

But, if N is large, an examination of the integrand of Eq. 2.65 shows that the contribution from $-\infty$ to $-\sqrt{N}$ is negligible, so that

$$N! \cong \sqrt{N}\, e^{-N} N^N \int_{-\infty}^{+\infty} e^{-y^2/2}\, dy \cong N^N e^{-N} \sqrt{2\pi N} \tag{2.66}$$

For our purposes, the logarithm of $N!$ is of interest. Equation 2.66 can be written in the form

$$\ln N! \cong N \ln N - N + \tfrac{1}{2} \ln 2\pi N \tag{2.67}$$

But, for large values of N,

$$\tfrac{1}{2} \ln 2\pi N \ll N$$

and

$$\ln N! \cong N \ln N - N \tag{2.68}$$

This equation is known as Stirling's formula, it will be used extensively in our development of statistical thermodynamics, because we are concerned with systems in which N is large.

2.9. Euler-Maclaurin Summation Formula

In Section 2.5, we discussed a special case of distribution functions in which discrete values of the variable are very close together, so that the variable may be reasonably approximated by a continuous variable, the probability function then becoming a continuous function. For such a case, summations over all values of the variable were then replaced by integrations.

In succeeding chapters, we shall be concerned with the task of summing certain functions over all possible discrete energy levels. It will be found that in certain cases these levels lie very close together, so that the functions might be considered continuous with the summations replaced by integrals. As an example, consider some function $f(j)$, where j is an integer, and the function is such that $f(j)$ changes only very slightly from j to $j + 1$. If in addition the function is such that $f(j) \to 0$ for large j, then

$$\sum_{j=a}^{\infty} f(j) = \sum_{j=a}^{\infty} f(j)\,\Delta j \approx \int_{x=a}^{\infty} f(x)\,dx \tag{2.69}$$

If we assume also that the derivatives of $f(x) \to 0$ for large x, then it may be more accurate to represent the summation of Eq. 2.69 by a series including values of $f(x)$ and its derivatives at the lower limit a, such as

$$\sum_{j=a}^{\infty} f(j) = \sum_{j=a}^{\infty} f(j)\,\Delta j = \int_{x=a}^{\infty} f(x)\,dx + C_0 f(a) + C_1 f^{1}(a) + C_2 f^{11}(a) + C_3 f^{111}(a) + C_4 f^{1V}(a) + C_5 f^{V}(a) + \cdots \tag{2.70}$$

To evaluate the constants in Eq. 2.70, let us consider a simple function that meets our requirements, such as the function

$$f(j) = e^{-mj} \tag{2.71}$$

summed from 0 to ∞. For $a = 0$,

$$\begin{aligned} f(0) &= 1 \\ f^{1}(0) &= -m \\ f^{11}(0) &= +m^2 \\ f^{111}(0) &= -m^3 \\ &\cdots\cdots\cdots \end{aligned} \tag{2.72}$$

Furthermore, we know that

$$\int_0^{\infty} e^{-mx}\,dx = \frac{1}{m} \tag{2.73}$$

and

$$\sum_{j=0}^{\infty} e^{-mj} = \frac{1}{1 - e^{-m}} \tag{2.74}$$

Now, if we expand e^{-m} in a series,

$$e^{-m} = 1 - m + \frac{m^2}{2} - \frac{m^3}{6} + \frac{m^4}{24} + \cdots \tag{2.75}$$

and substitute the series into Eq. 2.74 and carry out the division,

$$\sum_{j=0}^{\infty} e^{-mj} = \frac{1}{m} + \frac{1}{2} + \frac{m}{12} - \frac{m^3}{720} + \frac{m^5}{30{,}240} + \cdots \tag{2.76}$$

Substituting Eqs. 2.76, 2.73, and the set 2.72 into Eq. 2.70, and comparing coefficients, we find that the constants are

$$\begin{aligned} C_0 &= +\frac{1}{2} \\ C_1 &= -\frac{1}{12} \\ C_2 &= 0 \\ C_3 &= \frac{1}{720} \\ C_4 &= 0 \\ C_5 &= -\frac{1}{30{,}240} \\ &\cdot\ \cdot\ \cdot\ \cdot\ \cdot\ \cdot \end{aligned} \tag{2.77}$$

so that Eq. 2.70 can be written

$$\begin{aligned} \sum_{j=a}^{\infty} f(j) = \int_a^{\infty} f(x)\,dx + \frac{1}{2} f(a) \\ - \frac{1}{12} f^{1}(a) + \frac{1}{720} f^{111}(a) - \frac{1}{30{,}240} f^{V}(a) + \cdots \end{aligned} \tag{2.78}$$

Of course, having considered only one particular function, we should not conclude that Eq. 2.78 is valid in general. If this formula is tested against other functions, however, it is in fact found to be a general expression for functions satisfying the specified requirements. Equation 2.78 is a special case of the Euler-Maclaurin summation formula where the upper limit is infinite.

PROBLEMS

2.1 A card is drawn at random from a well-shuffled deck of 52 playing cards. Determine the probability of drawing:

(*a*) An ace.
(*b*) A heart.
(*c*) A black king.

2.2 A box contains 4 white balls and 3 black balls. What is the probability that the second ball drawn is white, if:

(*a*) The color of the first is not known?

(*b*) The first is white?

2.3 There are 10 coins in a box, 4 pennies, 3 nickels, 2 dimes, and 1 quarter. One coin is then drawn out, and its denomination is not disclosed.

(*a*) What is the probability that the second coin drawn will be a nickel?

(*b*) If it is now learned that the first coin drawn is not a penny, what is the probability that the second will be a nickel?

2.4 In a certain manufacturing process it is known from previous experience that 70% of the finished parts are of acceptable quality.

(*a*) If 2 parts are selected at random, what is the probability that at least 1 will be acceptable?

(*b*) How many parts must one select in order to be 99% sure that at least 1 of them will be acceptable?

2.5 A system is comprised of electrical components, each of which is 99.9% reliable. If any component fails, the system fails. Find the probability that the system will operate properly, if the system is comprised of:

(*a*) 10 components.

(*b*) 500 components. (Use the binomial series.)

2.6 A card is drawn at random from a well-shuffled deck. What is the probability that it is:

(*a*) A spade or a red card?

(*b*) A spade or a black ten?

2.7 Three types of missiles are fired at a certain target. Type A has a probability of 0.1 of scoring a hit; type B, a probability of 0.3; and type C, a probability of 0.5. Find the probabilities of 0, 1, 2, or 3 hits.

2.8 (*a*) How many 7-digit telephone numbers are there, if the only restriction is that the first digit cannot be a 0 or a 1?

(*b*) How many license plates can be made using 2 letters followed by 4 digits? By using 6 digits?

2.9 How many 3-letter words can be constructed from the letters in the alphabet if no repetition of letters is allowed? How many combinations of letters taken 3 at a time are there? Explain why these results are different.

2.10 Find the number of ways that 4 pennies, 3 nickels, 2 dimes, and 1 quarter can be arranged in a line.

2.11 A shipment of 50 manufactured parts has been received from the factory. It has been decided that, if 10% are defective, the shipment should be returned. Instead of testing all the parts, however, a sample will be taken, and, if every part in the sample is reliable, the entire shipment will be accepted without further tests. Evaluate this testing procedure if the sample consists of:

(*a*) 5 parts.

(*b*) 10 parts.

2.12 A fairly weighted coin is tossed 10 times. Calculate the probabilities for each of the possible numbers of heads. How are these probabilities affected if the coin is weighted to land heads 60% of the time?

2.13 A tennis match is won by the first player to win 3 sets. Assume that players A and B are evenly matched.

(*a*) Calculate the probabilities that player A will win the match in 3, 4, or 5 sets, respectively.

(*b*) If player A wins the first set, how are the probabilities changed?

2.14 In a certain transistor production process, it has been found that, on the average, 5% of the transistors do not meet the design specifications. In a sample of 60 selected at random, how many defective ones would be expected? With what standard deviation?

2.15 Consider the binomial distribution for a very large number of trials, N, in which the simple probability p of a success does not become small. Such a distribution can reasonably be represented by a Gaussian distribution having the same mean and standard deviation as the binomial distribution.

(*a*) Find an expression for the probability of achieving M successes in N trials for such a distribution.

(*b*) Develop an expression in terms of the error function [erf (x) is tabulated in Table A.3] for the probability of finding the variable within $\pm k\sigma$ of the mean value. Evaluate this probability for $k = 1$, 2, and 3.

2.16 (*a*) Using the same approximation as in Problem 2.15, show that the probability of getting at least M successes in N trials is given in terms of the error function.

(*b*) A coin is observed to land heads 5200 times in 10,000 tosses. Is it reasonable to believe that the coin is fairly weighted?

2.17 Repeat Example 2.14 for the case of 3 balls and 3 boxes.

2.18 Five distinguishable balls are dropped at random into 3 boxes. What is the probability that 3 balls will be in the first box, 2 in the second, and 0 in the third, if there is no limit on the number per box?

2.19 The experiment of Problem 2.18 is repeated with indistinguishable balls. Is the probability the same?

2.20 As a simple preliminary to the problem of mixing two substances, consider the following experiment. Box I contains 2 white balls, and box II contains 2 black balls. A ball is taken at random from box I and dropped into II, resulting in an "intermediate result" of 1 ball in I and 3 in II. The trial is completed by taking a ball at random from II and dropping it in I. This two-step trial is to be repeated many times.

(*a*) Calculate the probabilities for each of the possible intermediate and final configurations for the first trial.

(*b*) Repeat part (*a*) for the second and third trials.

(*c*) What limits are approached as the number of trials becomes large?

2.21 As an example of the method of undetermined multipliers, consider the following simple problem. A rectangle has a base x and height y. Use the method of undetermined multipliers to find the values of x and y that give the maximum area, if $x + y = 8$.

2.22 Evaluate 10! by direct multiplication and compare with the value given by Eq. 2.66. Then calculate the logarithm of 10! and compare with the approximations of Eqs. 2.67 and 2.68.

3 *Statistical Mechanics for Systems of Independent Particles*

Classical thermodynamics is based on the assumption that the system under consideration exists in a state of thermodynamic equilibrium. If we wish to examine this system from the microscopic point of view and evaluate the distribution of particles among various energy levels resulting in the equilibrium state, it is necessary to use the mathematical models and techniques that were developed in the previous chapter. It is assumed in the present development that the particles of which the system is comprised are independent (for example, an ideal gas); that is, intermolecular forces are considered to be negligibly small. Therefore, as discussed in Section 1.5, we are concerned only with the energy contributions of the translational and internal modes for the individual particles. The system energy then is merely the sum of the energies of the individual particles. The general case, the statistical mechanics for systems of dependent particles, will be developed in Chapter 10.

3.1. Introduction

The problem of expressing the distribution of particles or atoms among energy levels is relatively complex. To avoid encountering all the complicating factors at once, let us begin by analyzing an extremely simplified system.

Consider a system comprised of only four particles which may be distributed among only four possible energy levels. These levels have the relative values of energy ϵ_j of 0, 1, 2, 3, and the total energy of the system is equal to 3. This means that the only distributions of particles among levels permitted are those that result in the given total energy. We need not at this point concern ourselves with the way in which the particles possess this energy, and we assume that each of the levels is equally likely. The only possible distributions of particles among energy levels consistent with the statement of the problem are shown in the accompanying table, where the distributions are labeled I, II, and III. These are the only distributions of the four particles resulting in a total energy of the system equal to 3.

Energy Level ϵ_j	Number of Particles in Each Energy Level		
	Distribution I	Distribution II	Distribution III
3	1	0	0
2	0	1	0
1	0	1	3
0	3	2	1

In distribution I, one of the particles is in energy level 3, and all the others are in level 0. If the four particles are distinguishable, as W, X, Y, Z for example, then this distribution can arise in four different ways, since any one of the four particles can be the one in level 3. We shall use the symbol w for the number of ways a given distribution can arise, so that

$$w_{\mathrm{I}} = 4$$

This result can also be found from Eq. 2.28, as

$$w_{\mathrm{I}} = {}_4\mathscr{C}_{3,0,0,1} = \frac{4!}{3!0!0!1!} = 4$$

Similarly, for distribution II, any of the particles W, X, Y, Z can be the one in energy level 2. However, for each of these four possibilities, there are three possibilities for the particle in level 1, with the remaining pair in level 0. Therefore distribution II can arise in 12 ways or, from Eq. 2.28,

$$w_{\mathrm{II}} = {}_4\mathscr{C}_{2,1,1,0} = \frac{4!}{2!1!1!0!} = 12$$

The reasoning for distribution III is the same as that for distribution I or, using Eq. 2.28,

$$w_{\mathrm{III}} = {}_4\mathscr{C}_{1,3,0,0} = \frac{4!}{1!3!0!0!} = 4$$

Thus we conclude that there is a total of 20 possible arrangements of distinguishable particles among energy levels resulting in a system energy equal to 3, and that 12 of these correspond to distribution II, which is therefore the most probable distribution.

In this analysis, we assumed that the particles were distinguishable and carried the labels W, X, Y, Z. Suppose that, instead, the particles are indistinguishable from one another. Then we can no longer say that distribution I can arise in 4 different ways, because it is not possible to distinguish between the particles. All that is known is that some particle is in level 3 and the other three are in level 0. Also, since we cannot distinguish between the various arrangements resulting in distribution II or between those resulting in distribution III, we must say that, for indistinguishable particles,

$$w_{\mathrm{I}} = w_{\mathrm{II}} = w_{\mathrm{III}} = 1$$

In our simple system of four particles and four energy levels we have discussed the possible distinguishability or indistinguishability of particles, but we have made two restrictions. We have not considered the energy mode, for example translation, rotation, etc., discussed in Section 1.5. This matter will not be discussed at length until after we have more fully developed the subject of quantum mechanics and studied the various energy modes. The other restriction made in our example was the assumption that the energy levels were equally probable, or equally accessible to the particles. This assumption is not in general a reasonable one and should now be corrected.

A fundamental hypothesis of statistical mechanics is that individual quantum states are equally likely. Since at each level of energy there may be a number of equally likely quantum states, the number of states varying in some manner from one level to another, it follows that the energy levels are not necessarily equally likely. Therefore each level of energy ϵ_j must be assigned some statistical weight g_j, which is the number of quantum states having that energy. This statistical weight g_j is commonly referred to as the degeneracy of the energy level ϵ_j. It should also be pointed out here that the quantum states at a particular energy level may in some cases have very slightly different values of energy, so that there is actually an energy range $\Delta\epsilon$ rather than a single unique value of energy at each level. This introduces no difficulty, however, as long as $\Delta\epsilon$ is negligibly small compared to the differences in energy between one level and another.

Thus, energy levels are not in general equally likely, and it is apparent that our previous example would have been somewhat more reasonable if the levels had had different statistical weights, say 3, 3, 4, 4, respectively. Let us construct a model to represent such a case. Visualize the individual

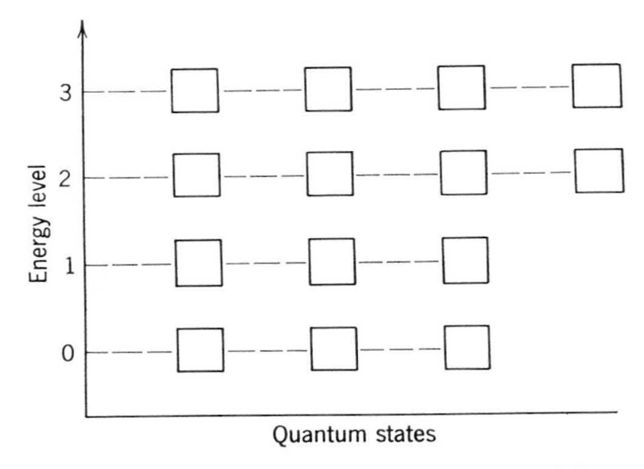

Figure 3.1 Schematic representation of quantum states at different levels of energy.

quantum states as cells or boxes, with the energy of each corresponding to its height, as shown in Figure 3.1.

Thus there are three cells at energy 0, three at 1, and four each at energies of 2 and 3. In accordance with our discussion of a small range $\Delta\epsilon$, note that each of the cells at a given level is not at exactly the same height, or energy, but very nearly so. Now the problem of distributing particles among energy levels, each having a certain number of equally likely states, may be thought of as a problem of dropping balls into boxes of equal size which are arranged as shown in Figure 3.1. This complicates our analysis considerably.

3.2. Boltzmann, Bose-Einstein, and Fermi-Dirac Statistics

Rather than proceeding directly to the problem described above, let us develop general expressions for the quantity w discussed previously, considering the three cases of interest discussed briefly in Section 2.6. During the development of these expressions, it may be helpful to the reader to keep in mind a model such as proposed in Figure 3.1.

Boltzmann Statistics. For the Boltzmann statistical model, it is assumed that the particles are distinguishable and that there is no limit on the number of particles per quantum state. The particles are to be distributed in some manner among various energy levels ϵ_j, each of which has a degeneracy or number of quantum states g_j. First, the number of ways in which N distinguishable particles can be divided into groups (levels), with N_1 particles in the first group, N_2 in the second, . . . , is, according to Eq. 2.28,

$$\frac{N!}{\prod_j N_j!}$$

However, for each of these groups or levels, j, the number of ways that the N_j particles in that group can be placed in the g_j cells (quantum states) of that group with no restriction on the number of particles per cell, is, from Section 2.6,

$$g_j^{N_j}$$

Therefore the number of distinguishable states of the system consistent with a given distribution of particles among energy levels or, in other words, the number of ways that a given distribution can arise is

$$w = \left(\frac{N!}{\prod_j N_j!}\right)\left(\prod_j g_j^{N_j}\right) = N! \prod_j \left(\frac{g_j^{N_j}}{N_j!}\right) \tag{3.1}$$

The quantity of interest, w, is frequently called the thermodynamic probability, which is not the same as mathematical probability. The latter quantity, for a given distribution, would be given by the thermodynamic probability w for that distribution divided by the sum of the values of w for all possible distributions.

Bose-Einstein Statistics. This model assumes that the particles constituting the system under consideration are indistinguishable, and that there is no limit on the number of particles per quantum state. Of the N particles in the system, N_1 are in the first energy level, N_2 in the second level, etc., for a given distribution. Since we cannot distinguish which particles are in which levels but only the numbers in each, we are not concerned with interchanges of particles between levels such as given by Eq. 2.28. However, for the N_j particles in the jth energy level, the number of ways of placing the N_j particles in the g_j cells or quantum states of that level is, from case D of Section 2.6,

$$\frac{(g_j + N_j - 1)!}{(g_j - 1)!N_j!}$$

Therefore the thermodynamic probability for a given distribution of the number of ways of achieving a given distribution of particles among energy levels is given by the product of this quantity over all the levels, or

$$w = \prod_j \frac{(g_j + N_j - 1)!}{(g_j - 1)!N_j!} \tag{3.2}$$

Note the contrast to Eq. 3.1, which applies to distinguishable particles.

Fermi-Dirac Statistics. The third statistical model of interest to us assumes, as does Bose-Einstein statistics, that the particles are indistinguishable, but in this case there is a limit of no more than one particle per quantum state. As in the previous case, for a given distribution of N_1 particles in the first energy level, N_2 in the second, etc., we are not concerned with interchanges of particles between levels, since the particles are assumed to be indistinguishable. However, for the N_j particles in the jth energy level, the number of ways of placing the N_j particles in the g_j quantum states of that level where there can be no more than one particle per quantum state is, from case C of Section 2.6,

$$\frac{g_j!}{(g_j - N_j)!N_j!}$$

Thus the thermodynamic probability of a given distribution of particles among energy levels is the product over all levels, or

$$w = \prod_j \frac{g_j!}{(g_j - N_j)!N_j!} \tag{3.3}$$

To clarify the differences in these three statistical models, we now analyze the problem suggested at the end of the preceding section.

Example 3.1

A system comprised of four particles has four energy levels with relative energy values of 0, 1, 2, 3. The total energy of the system is 3, and the statistical weights of the four levels are 3, 3, 4, 4, respectively. Determine the thermodynamic probability for each of the possible distributions, assuming (*a*) Boltzmann statistics, (*b*) Bose-Einstein statistics, (*c*) Fermi-Dirac statistics.

The possible distributions of particles among energy levels resulting in a total energy of 3 are the same as in the example considered in Section 3.1. These three distributions are repeated in the accompanying table, and the statistical weights of the energy levels are given.

Energy Level, ϵ_j	Weight, g_j	Number of Particles in Each Energy Level		
		Distribution I	Distribution II	Distribution III
3	4	1	0	0
2	4	0	1	0
1	3	0	1	3
0	3	3	2	1

(*a*) For Boltzmann statistics, using Eq. 3.1 for distinguishable particles,

$$w = N! \prod_j \left(\frac{g_j^{N_j}}{N_j!} \right)$$

For the three possible distributions,

$$w_{\mathrm{I}} = 4!\left(\frac{3^3}{3!}\right)\left(\frac{3^0}{0!}\right)\left(\frac{4^0}{0!}\right)\left(\frac{4^1}{1!}\right) = 432$$

$$w_{\mathrm{II}} = 4!\left(\frac{3^2}{2!}\right)\left(\frac{3^1}{1!}\right)\left(\frac{4^1}{1!}\right)\left(\frac{4^0}{0!}\right) = 1296$$

$$w_{\mathrm{III}} = 4!\left(\frac{3^1}{1!}\right)\left(\frac{3^3}{3!}\right)\left(\frac{4^0}{0!}\right)\left(\frac{4^0}{0!}\right) = 324$$

(*b*) For Bose-Einstein statistics, from Eq. 3.2 for indistinguishable particles,

$$w = \prod_j \frac{(g_j + N_j - 1)!}{(g_j - 1)!N_j!}$$

$$w_{\mathrm{I}} = \left(\frac{5!}{2!3!}\right)\left(\frac{2!}{2!0!}\right)\left(\frac{3!}{3!0!}\right)\left(\frac{4!}{3!1!}\right) = 40$$

$$w_{\mathrm{II}} = \left(\frac{4!}{2!2!}\right)\left(\frac{3!}{2!1!}\right)\left(\frac{4!}{3!1!}\right)\left(\frac{3!}{3!0!}\right) = 72$$

$$w_{\mathrm{III}} = \left(\frac{3!}{2!1!}\right)\left(\frac{5!}{2!3!}\right)\left(\frac{3!}{3!0!}\right)\left(\frac{3!}{3!0!}\right) = 30$$

(*c*) For Fermi-Dirac statistics, from Eq. 3.3 for indistinguishable particles with a maximum of one per quantum state,

$$w = \prod_j \frac{g_j!}{(g_j - N_j)!N_j!}$$

$$w_{\mathrm{I}} = \left(\frac{3!}{0!3!}\right)\left(\frac{3!}{3!0!}\right)\left(\frac{4!}{4!0!}\right)\left(\frac{4!}{3!1!}\right) = 4$$

$$w_{\mathrm{II}} = \left(\frac{3!}{1!2!}\right)\left(\frac{3!}{2!1!}\right)\left(\frac{4!}{3!1!}\right)\left(\frac{4!}{4!0!}\right) = 36$$

$$w_{\mathrm{III}} = \left(\frac{3!}{2!1!}\right)\left(\frac{3!}{0!3!}\right)\left(\frac{4!}{4!0!}\right)\left(\frac{4!}{4!0!}\right) = 3$$

For each of the statistical models, distribution number II is seen to be the most probable distribution, since in each case that distribution can occur in the greatest number of ways.

3.3. The Equilibrium Distribution

In Example 3.1, we found that determination of the most probable distribution by direct evaluation of the thermodynamic probability for each of the possible distributions consistent with the system energy is a relatively complex task, even for the simple system discussed there. Now, since an ordinary macroscopic system of interest to us is so much more complex than that of our example, a similar approach to the problem is most certainly out of the question. We must therefore develop general expressions or methods for determining thermodynamic probability of distributions and also for finding the most probable distribution when the system has large numbers of energy levels and particles.

Of the tremendous number of possible distributions in a macroscopic equilibrium system, only a relatively small number have appreciable values of thermodynamic probability. These distributions are all very nearly identical and result in the same macroscopic behavior (within experimental limits) of the system. Thus we speak of a single most probable distribution, as representative of this small group, and as representative of the thermodynamic equilibrium state. In doing so we realize that, from a microscopic viewpoint, equilibrium is of a dynamic character, and that there are minor and continual fluctuations in each direction about this most probable distribution. We note also that large scale fluctuations (distributions considerably or macroscopically different from the most probable) are extremely unlikely to be observed. Fluctuations about the most probable distribution will be discussed in some detail in Chapter 10 and again in Chapter 12.

Consider a system comprised of a certain number of particles, N, which are distributed in some way among the various energy levels ϵ_j. Then

$$N = \sum_j N_j \tag{3.4}$$

The internal energy U of the system is given as

$$U = \sum_j N_j \epsilon_j \tag{3.5}$$

To find the most probable distribution, we must find the particular distribution having the maximum value of w for a system of N particles and energy U.

Boltzmann Statistics. First, for Boltzmann statistics, let us, for convenience, rewrite the distribution equation 3.1 in logarithmic form:

$$\ln w = \ln N! + \ln \left(\prod_j \frac{g_j^{N_j}}{N_j!} \right)$$

Since the logarithm of a product of quantities is equal to the sum of the logarithms, it follows that

$$\ln w = \ln N! + \sum_j (N_j \ln g_j - \ln N_j!) \tag{3.6}$$

Now, assuming that N is very large and also that the N_j's (for those levels that contribute significantly to w) are very large, the factorials in Eq. 3.6 may be accurately represented by Stirling's formula, Eq. 2.68. Thus, Eq. 3.6 becomes

$$\ln w = N \ln N - N + \sum_j (N_j \ln g_j - N_j \ln N_j + N_j) \tag{3.7}$$

and, using Eq. 3.4,

$$\ln w = N \ln N + \sum_j (N_j \ln g_j - N_j \ln N_j) \tag{3.8}$$

To find the most probable distribution, the maximum of w, we differentiate Eq. 3.8 with respect to N_j and set $d(\ln w)$ equal to zero. Since N and all the g_j's are constant, it follows that

$$d(\ln w) = \sum_j (\ln g_j \, dN_j - \ln N_j \, dN_j - dN_j) = 0 \tag{3.9}$$

But

$$\sum_j dN_j = dN = 0 \tag{3.10}$$

Therefore, for the maximum of w,

$$d(\ln w) = \sum_j \ln \left(\frac{g_j}{N_j}\right) dN_j = 0 \tag{3.11}$$

Furthermore, for a given internal energy of the system,

$$dU = \sum_j \epsilon_j \, dN_j = 0 \tag{3.12}$$

The most probable distribution of particles among energy levels for Boltzmann statistics is given by Eq. 3.11 subject to the two constraints, constant N and U, that is, Eqs. 3.10 and 3.12. Consequently, the expression giving the maximum of w can be found by the method of Lagrange multipliers, as discussed in Section 2.7. Let us multiply Eq. 3.10 by some arbitrary value α, and Eq. 3.12 by some other constant β. Subtracting (for convenience) Eq. 3.11 from the sum of these results in the expression

$$\sum_j \left(\ln \frac{N_j}{g_j} + \alpha + \beta\epsilon_j\right) dN_j = 0 \tag{3.13}$$

The sum in Eq. 3.13 is taken over all energy levels, for example

$$j = 1, 2, 3, \ldots, (J-1), J$$

or a total of J levels. However, only $(J-2)$ of the N_j's are independent, because of the two restrictions. If the first two terms of the series in

Eq. 3.13 are taken as the dependent terms, the multipliers α, β can then be selected such that the coefficients of dN_1 and dN_2 are identically zero. That is,

$$\ln \frac{N_1}{g_1} + \alpha + \beta\epsilon_1 = 0 \tag{3.14}$$

$$\ln \frac{N_2}{g_2} + \alpha + \beta\epsilon_2 = 0 \tag{3.15}$$

Since the remaining $(J - 2)$ variables are independent, $dN_3, dN_4, \ldots, dN_J$ can all be varied arbitrarily. Thus the only way that Eq. 3.13 can be satisfied is for each of the coefficients to be identically zero, or

$$\ln \frac{N_j}{g_j} + \alpha + \beta\epsilon_j = 0 \tag{3.16}$$

for all values of j. This is more conveniently written in the form

$$N_j = g_j e^{-\alpha} e^{-\beta\epsilon_j} \tag{3.17}$$

This equation is called the Boltzmann distribution law, which specifies the most probable distribution of particles among energy levels. The distribution according to Eq. 3.17 is seen to be of an exponential form. Therefore, given values of the multipliers α, β, which remain to be determined, the general form of the distribution is that shown in Figure 3.2. The shape of the curve depends considerably upon the magnitudes of α and β, of course. Numerical values or expressions for determining these multipliers must be found from observed physical behavior of the system, but, before proceeding to this problem, let us derive the corresponding distribution equations for our other statistical models.

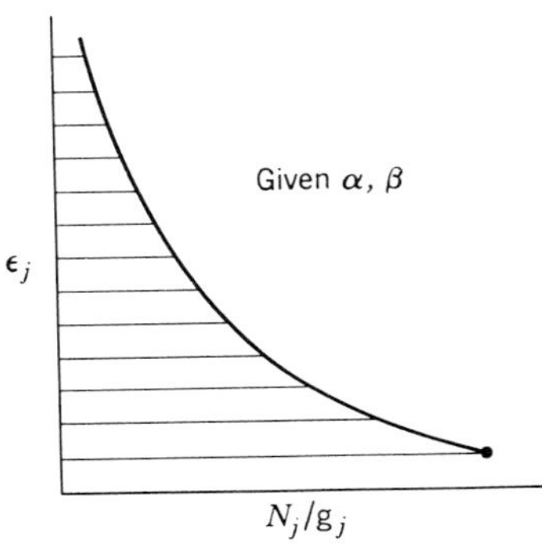

Figure 3.2 Exponential nature of the Boltzmann distribution function.

Bose-Einstein Statistics. For indistinguishable particles, the Bose-Einstein expression for thermodynamic probability w is given by Eq. 3.2. However, it will be found that, for systems of interest,

$$g_j \gg 1 \tag{3.18}$$

so that the 1's appearing in Eq. 3.2 can be neglected. The expression is then written in the logarithmic form

$$\ln w = \ln \left[\prod_j \frac{(g_j + N_j)!}{g_j! N_j!} \right]$$

$$= \sum_j \left[\ln (g_j + N_j)! - \ln g_j! - \ln N_j! \right] \tag{3.19}$$

If it is also assumed that both g_j and N_j are sufficiently large (for those levels that contribute appreciably to w), then Stirling's formula can be used to eliminate the factorials, giving

$$\begin{aligned}\ln w &= \sum_j [(g_j + N_j) \ln (g_j + N_j) - (g_j + N_j) \\ &\quad - g_j \ln g_j + g_j - N_j \ln N_j + N_j] \\ &= \sum_j [(g_j + N_j) \ln (g_j + N_j) - g_j \ln g_j - N_j \ln N_j] \end{aligned} \tag{3.20}$$

To determine the most probable distribution, we must find the maximum of Eq. 3.20 subject to the two constraints, fixed N and U, or Eqs. 3.10 and 3.12. Therefore, differentiating Eq. 3.20 and setting $d(\ln w)$ equal to zero, we obtain

$$\begin{aligned}d(\ln w) &= \sum_j [dN_j + \ln (g_j + N_j)\, dN_j - dN_j - \ln N_j\, dN_j] \\ &= \sum_j \ln \left(\frac{g_j + N_j}{N_j}\right) dN_j = 0 \end{aligned} \tag{3.21}$$

Now, if we multiply Eq. 3.10 by some value α and Eq. 3.12 by β, and subtract Eq. 3.21 from their sum, we find

$$\sum_j \left[-\ln \left(\frac{g_j + N_j}{N_j}\right) + \alpha + \beta\epsilon_j\right] dN_j = 0 \tag{3.22}$$

As with Boltzmann statistics, two of the N_j's are not independent, but their coefficients can be made identically zero by the proper choice of the multipliers α and β. Since all the remaining dN_j's are arbitrary, the coefficient of each must be zero in order that Eq. 3.22 be satisfied. Therefore

$$-\ln \left(\frac{g_j}{N_j} + 1\right) + \alpha + \beta\epsilon_j = 0 \tag{3.23}$$

for all values of j, or

$$N_j = \frac{g_j}{e^{\alpha} e^{\beta\epsilon_j} - 1} \tag{3.24}$$

for the most probable distribution. Equation 3.24 is called the Bose-Einstein distribution law and is seen to differ from Eq. 3.17 for the Boltzmann distribution only by the 1 in the denominator.

Fermi-Dirac Statistics. Our third statistical model is that for indistinguishable particles with a maximum of one particle per quantum state, the Fermi-Dirac model. Equation 3.3 for Fermi-Dirac statistics is, in logarithmic form,

$$\begin{aligned}\ln w &= \ln \left[\prod_j \frac{g_j!}{(g_j - N_j)!N_j!}\right] \\ &= \sum_j [\ln g_j! - \ln (g_j - N_j)! - \ln N_j!] \end{aligned} \tag{3.25}$$

But, using Stirling's formula, we obtain

$$\begin{aligned}\ln w &= \sum_j [g_j \ln g_j - g_j - (g_j - N_j)\ln (g_j - N_j) \\ &\qquad + (g_j - N_j) - N_j \ln N_j + N_j] \\ &= \sum_j [g_j \ln g_j - (g_j - N_j)\ln (g_j - N_j) - N_j \ln N_j] \end{aligned} \tag{3.26}$$

To find the maximum of w as given by Eq. 3.26, at fixed N, U,

$$\begin{aligned} d(\ln w) &= \sum_j [dN_j + \ln (g_j - N_j)\, dN_j - dN_j - \ln N_j\, dN_j] \\ &= \sum_j \ln \left(\frac{g_j - N_j}{N_j}\right) dN_j = 0 \end{aligned} \tag{3.27}$$

As in the previous cases, we multiply Eq. 3.10 by some α and Eq. 3.12 by β, and subtract Eq. 3.27 from their sum. The result is

$$\sum_j \left[-\ln \left(\frac{g_j - N_j}{N_j}\right) + \alpha + \beta\epsilon_j\right] dN_j = 0 \tag{3.28}$$

The multipliers α and β are so chosen as to make the coefficients of the two dependent N_j's zero. Then, to satisfy Eq. 3.28, the coefficients of all the independent N_j's must also be zero, and

$$-\ln \left(\frac{g_j}{N_j} - 1\right) + \alpha + \beta\epsilon_j = 0 \tag{3.29}$$

for all values of j, or

$$N_j = \frac{g_j}{e^{\alpha} e^{\beta\epsilon_j} + 1} \tag{3.30}$$

This equation specifies the most probable distribution for Fermi-Dirac statistics.

We know that in physical systems the particles being considered, whether they are electrons, atoms, etc., are indistinguishable and hence must obey either Bose-Einstein or Fermi-Dirac statistics. Consequently, the Boltzmann model for a system of independent particles is not really a reasonable picture of anything existing in nature and would seem to be of no use to us. However, we shall find shortly that for most gases at low to moderate density (relatively low pressure or high temperature), the number of quantum states available at any level is much larger than the number of particles in that level. That is,

$$\frac{N_j}{g_j} \ll 1 \tag{3.31}$$

for all j. From the distribution equations, 3.24 or 3.30, this is seen to be equivalent to the statement

$$e^{\alpha} e^{\beta\epsilon_j} \gg 1 \tag{3.32}$$

Therefore, under these special conditions, the -1 in the denominator of the distribution equation for Bose-Einstein statistics and the $+1$ for Fermi-Dirac statistics are negligible, and both models predict essentially the same distribution of particles among energy levels as does Boltzmann statistics.

Before concluding that our three models are identical when Eq. 3.31 is valid, however, let us see what each predicts for the thermodynamic probability under these conditions. If Eqs. 3.18 and 3.31 are reasonable, then

$$\frac{(g_j + N_j - 1)!}{(g_j - 1)!} = g_j(g_j + 1)(g_j + 2) \cdots (g_j + N_j - 1) \approx g_j^{N_j} \quad (3.33)$$

and the thermodynamic probability for a given distribution according to Bose-Einstein statistics given in general by Eq. 3.2 reduces to the form

$$w \approx \prod_j \frac{g_j^{N_j}}{N_j!} \quad (3.34)$$

Similarly, assuming Eqs. 3.18 and 3.31, and also making use of Eq. 2.24, we obtain

$$\frac{g_j!}{(g_j - N_j)!} = g_j(g_j - 1) \cdots (g_j - N_j + 1) \approx g_j^{N_j} \quad (3.35)$$

Substituting Eq. 3.35 into Eq. 3.3, we see that the thermodynamic probability for a given distribution according to Fermi-Dirac statistics also reduces to Eq. 3.34 under these conditions, as did that for Bose-Einstein statistics.

Comparing Eq. 3.34 with Eq. 3.1 for Boltzmann statistics, we find that, whereas the three models predict essentially the same distribution when Eq. 3.31 is valid, the thermodynamic probability for any given distribution is different by $N!$ for the Boltzmann statistics, or

$$w_{\text{B-E}} \approx w_{\text{F-D}} \approx \left(\frac{w}{N!}\right)_{\text{Boltz.}} \quad (3.36)$$

Of course, dividing the value of w for Boltzmann statistics, which assumed distinguishable particles, by $N!$ has the effect of discounting the distinguishability of the N particles, a procedure which has been discussed in Chapter 2.

We find that, when the assumption of Eq. 3.31 is valid, both Bose-Einstein and Fermi-Dirac statistics reduce to a common statistical model with a distribution given by Eq. 3.17 and a thermodynamic probability given by Eq. 3.34. This model, which will be the one of principal interest to us in the remainder of this text, is termed "corrected" Boltzmann statistics, the name resulting from the comparison in Eq. 3.36.

3.4. Identification of the Multipliers

In the preceding section, we developed the expressions for the equilibrium or most probable distribution of particles among energy levels and found it necessary to introduce two undetermined multipliers. Let us now attempt to eliminate these multipliers in terms of the physical behavior of substances. In doing so, we shall utilize classical mechanics, but only to the extent necessary to eliminate the multipliers.

For simplicity, we consider "corrected" Boltzmann statistics, the low density approximation to both Bose-Einstein and Fermi-Dirac statistics. For a system of N particles having internal energy U, expressed by Eqs. 3.4 and 3.5,

$$N = \sum_j N_j$$

$$U = \sum_j N_j \epsilon_j$$

the equilibrium distribution was found, in Eq. 3.17, to be

$$N_j = g_j e^{-\alpha} e^{-\beta \epsilon_j}$$

where, in each of these equations, the index j refers to energy levels and not to individual quantum states. We see that it is possible to eliminate the multiplier α in terms of β by substituting Eq. 3.17 into Eq. 3.4 to obtain

$$N = e^{-\alpha} \sum_j g_j e^{-\beta \epsilon_j} \tag{3.37}$$

or

$$e^{-\alpha} = \frac{N}{\sum_j g_j e^{-\beta \epsilon_j}} \tag{3.38}$$

which also requires that the various g_j, ϵ_j be known in order to evaluate the summation in the denominator of Eq. 3.38.

Having expressed α in terms of β and the system parameters g_j, ϵ_j, we now turn our attention to the rather difficult problem of identifying β with physical properties of the system. We note that the distribution equation 3.17 can now be written as

$$N_j = \frac{N g_j e^{-\beta \epsilon_j}}{\sum_j g_j e^{-\beta \epsilon_j}} \tag{3.39}$$

and the internal energy corresponding to the most probable distribution as

$$U = \frac{N \sum_j g_j \epsilon_j e^{-\beta \epsilon_j}}{\sum_j g_j e^{-\beta \epsilon_j}} \tag{3.40}$$

To evaluate this relation between the multiplier β and the macroscopic property internal energy, we consider the simplest possible case, a monatomic ideal gas, for which the energy is comprised solely of the kinetic energy of translation of the atoms. Using x, y, z system coordinates, any given atom has some energy ϵ_{V_x,V_y,V_z} as a result of its velocity components,

$$\epsilon_{V_x,V_y,V_z} = \tfrac{1}{2}m(V_x^2 + V_y^2 + V_z^2) \tag{3.41}$$

where m is the mass of the atom. It is important to realize that this energy ϵ_{V_x,V_y,V_z} is not the energy of a level ϵ_j, since there may be many combinations of V_x, V_y, V_z that would result in the same, or very nearly the same, energy as determined from Eq. 3.41. Since such a number is related to the statistical weight of the energy level, it should prove more convenient to rewrite Eq. 3.40 in terms of the component velocities and sum over all possible values (this procedure will be discussed more fully in Section 3.5). The resulting expression can be evaluated by making the assumption of classical mechanics, namely, that the various energy levels (or velocities) are so numerous and closely spaced that the V_x, V_y, V_z can be considered continuous variables, with the summations replaced by integrals. This will later be found to be reasonable by means of the Euler-Maclaurin summation formula and the quantum mechanical expression for translational energy. Therefore Eq. 3.40 can be written

$$U = \frac{\tfrac{1}{2}mN\iiint_{-\infty}^{+\infty}(V_x^2 + V_y^2 + V_z^2)\exp\left[-(\beta m/2)(V_x^2 + V_y^2 + V_z^2)\right]dV_x\,dV_y\,dV_z}{\iiint_{-\infty}^{+\infty}\exp\left[-(\beta m/2)(V_x^2 + V_y^2 + V_z^2)\right]dV_x\,dV_y\,dV_z} \tag{3.42}$$

Now, since V_x, V_y, V_z are independent, the triple integrals can be separated into a sum of three single integrals, each having the same form. Using the common index s for each, Eq. 3.42 becomes

$$U = \frac{\tfrac{3}{2}mN\int_{-\infty}^{+\infty}V_s^2\exp\left[(-\beta m/2)V_s^2\right]dV_s}{\int_{-\infty}^{+\infty}\exp\left[(-\beta m/2)V_s^2\right]dV_s} \tag{3.43}$$

Using the expressions for these integrals given in the Appendix, Table A.3,

$$U = \tfrac{3}{2}mN\left(\frac{\tfrac{1}{2}\sqrt{\pi 8/\beta^3 m^3}}{\sqrt{\pi 2/\beta m}}\right) = \frac{3N}{2\beta} \tag{3.44}$$

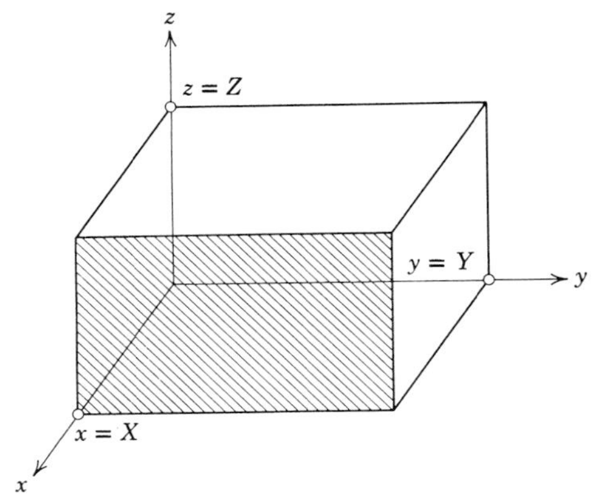

Figure 3.3 A box of dimensions X, Y, Z confining a monatomic ideal gas.

a very simple relation between the internal energy of a monatomic ideal gas and the multiplier β.

We shall now show, in a similar manner, that the internal energy of a monatomic ideal gas is a very simple function of the temperature of the system, and consequently we shall relate the multiplier β to the temperature. Consider a monatomic ideal gas contained in a box of dimensions X, Y, Z, as shown in Figure 3.3. Consider the atoms striking the wall perpendicular to the x-axis at distance $x = X$ during a unit interval of time. An atom with a given x-component of velocity V_x and any V_y, V_z must lie within a distance V_x of the wall at the beginning of the time interval (and consequently within a fraction V_x/X of the total volume) in order to strike that wall during the interval. The number of atoms in the entire box with this particular x-component velocity is denoted by N_{V_x}, and the number of these striking the X wall during the time interval is

$$\frac{V_x}{X} \times N_{V_x}$$

at which time each undergoes a change in x-direction momentum of $2mV_x$. Thus the contribution to the force exerted on the wall by atoms of a particular x-component velocity V_x is

$$\frac{2mV_x^{\,2}}{X} \times N_{V_x}$$

and the contribution to pressure is

$$\frac{2mV_x^{\,2}}{X(YZ)} \times N_{V_x} = \frac{2mV_x^{\,2}}{V} N_{V_x}$$

where V is the volume of the box. The pressure of the gas in the container is that due to the contributions of all V_x greater than zero, and assuming, V_x to be essentially continuous,

$$P = \frac{2m}{V}\int_0^\infty \mathsf{V}_x{}^2 N_{\mathsf{V}_x}\, d\mathsf{V}_x \tag{3.45}$$

If we assume the gas to be isotropic, that is, that the $+x$- and $-x$-directions are equally probable, then

$$N_{\mathsf{V}_x} = N_{-\mathsf{V}_x} \tag{3.46}$$

and

$$P = \frac{m}{V}\int_{-\infty}^{+\infty} \mathsf{V}_x{}^2 N_{\mathsf{V}_x}\, d\mathsf{V}_x \tag{3.47}$$

In accordance with our discussions of Chapter 2, the mean value of the square of the x-component velocity is given by

$$\overline{\mathsf{V}_x{}^2} = \frac{1}{N}\int_{-\infty}^{+\infty} \mathsf{V}_x{}^2 N_{\mathsf{V}_x}\, d\mathsf{V}_x \tag{3.48}$$

so that Eq. 3.47 can be conveniently written as

$$P = \frac{m}{V}(N\overline{\mathsf{V}_x{}^2}) \tag{3.49}$$

Furthermore, since the system is isotropic, the x-, y-, and z-directions are all equally probable, and

$$\overline{\mathsf{V}_x{}^2} = \overline{\mathsf{V}_y{}^2} = \overline{\mathsf{V}_z{}^2} = \tfrac{1}{3}\overline{\mathsf{V}^2} \tag{3.50}$$

so that

$$PV = \tfrac{1}{3}mN\overline{\mathsf{V}^2} = \tfrac{2}{3}N(\tfrac{1}{2}m\overline{\mathsf{V}^2}) = \tfrac{2}{3}U \tag{3.51}$$

The equation of state for an ideal gas is

$$PV = NkT \tag{3.52}$$

where k is the gas constant per atom or molecule, called the Boltzmann constant. Combining the last two equations yields

$$U = \tfrac{3}{2}NkT \tag{3.53}$$

which relates the internal energy of a monatomic ideal gas to the temperature.

Now, if we compare the two relations of Eqs. 3.44 and 3.53, we conclude that the multiplier β is inversely proportional to the temperature by the expression

$$\beta = \frac{1}{kT} \tag{3.54}$$

Although we have shown that Eq. 3.54 is valid for a monatomic ideal gas, it is also possible to demonstrate this relation under more general conditions. The completely general case will be developed in Chapter 10, but until that time we shall assume that Eq. 3.54 is valid in general.

3.5. The Partition Function

We have previously determined the equilibrium distribution for a system of particles which is at given T, V, and which obeys "corrected" Boltzmann statistics. We found that the distribution of particles among energy levels is given by Eq. 3.17,

$$N_j = g_j e^{-\alpha} e^{-\beta\epsilon_j}$$

in terms of the two undetermined multipliers α and β. In the preceding section, we eliminated the multiplier α in terms of β by Eq. 3.38,

$$e^{-\alpha} = \frac{N}{\sum_j g_j e^{-\beta\epsilon_j}}$$

and then identified β (at least for a special case) by Eq. 3.54,

$$\beta = \frac{1}{kT}$$

Thus we find that the summation appearing in the denominator of Eq. 3.38 is an extremely important quantity to the relation of the results of our mathematical model and physical reality. Since this summation is required for the elimination of the multiplier α, it will consequently appear in the relations for equilibrium-state thermodynamic properties, as for example in the expression for the internal energy, Eq. 3.40,

$$U = \frac{N \sum_j g_j \epsilon_j e^{-\beta\epsilon_j}}{\sum_j g_j e^{-\beta\epsilon_j}}$$

Using Eq. 3.54 for β, let us define a quantity that we call the partition function Z as

$$Z = \sum_j g_j e^{-\epsilon_j/kT} \tag{3.55}$$

The summation in this expression is to be made over all the j levels of energy available to the particles. In many cases, however, it proves to be more convenient to perform the summation directly over all the individual quantum states, which we refer to by the subscript i. Thus, during our later developments, it should be kept in mind that the partition function can be expressed in either of these equivalent forms:

$$Z = \sum_{j \text{ levels}} g_j e^{-\epsilon_j/kT} = \sum_{i \text{ states}} e^{-\epsilon_i/kT} \tag{3.56}$$

From the definition of the partition function, Eq. 3.55, and from Eq. 3.38 for the multiplier α, it is found that

$$e^{-\alpha} = \frac{N}{Z} \tag{3.57}$$

Furthermore, the equilibrium distribution function for "corrected" Boltzmann statistics and the corresponding internal energy of the system are given in terms of the partition function as

$$N_j = \frac{N g_j e^{-\epsilon_j/kT}}{Z} \tag{3.58}$$

and

$$U = \frac{N}{Z} \sum_j g_j \epsilon_j e^{-\epsilon_j/kT} \tag{3.59}$$

We find that the important problems of determining the equilibrium distribution and evaluating properties such as internal energy are connected intimately with the evaluation of the partition function. This in turn depends upon a knowledge of the energy levels and degeneracies for the various modes of energy. These problems will occupy a large portion of our time in the remainder of this book.

Let us at this point consider the evaluation of the partition function for one special case, that of the classical monatomic ideal gas. The energy to be considered is that of translation, as discussed in Section 3.4, and is given in terms of the velocity components V_x, V_y, V_z by Eq. 3.41,

$$\epsilon_{V_x,V_y,V_z} = \tfrac{1}{2}m(V_x^2 + V_y^2 + V_z^2)$$

It was noted previously that this energy is not that of a level; rather it is the energy for a particular set of component values V_x, V_y, V_z. Therefore, from Eq. 3.56,

$$Z = \sum_{V_x} \sum_{V_y} \sum_{V_z} g_{V_x,V_y,V_z} \exp\left[-\frac{m}{2kT}(V_x^2 + V_y^2 + V_z^2)\right] \tag{3.60}$$

The summation in Eq. 3.60 must be taken over all possible values of V_x, V_y, V_z, which according to classical mechanics are presumed to be continuous functions. Following our discussion of continuous distribution functions in Chapter 2, we conclude that the statistical weight g_{V_x,V_y,V_z} must represent the number of states in a velocity element V_x to $V_x + dV_x$, V_y to $V_y + dV_y$, V_z to $V_z + dV_z$, and is proportional to the size of the velocity element $dV_x\,dV_y\,dV_z$. This proportionality relation is given as

$$g_{V_x,V_y,V_z} = \left(\frac{m^3V}{h^3}\right) dV_x\,dV_y\,dV_z \tag{3.61}$$

This relation can be deduced by a consideration of the Heisenberg uncertainty principle (which is not a concept of classical mechanics), letting the positional indeterminacy in each direction x, y, z approach the dimensions of the system.

We now substitute Eq. 3.61 into Eq. 3.60 and integrate over all values of the velocity components. Since V_x, V_y, V_z are assumed to be independent, the result is found to be

$$\begin{aligned} Z &= \frac{m^3V}{h^3}\iiint_{-\infty}^{+\infty} \exp\left[-\frac{m}{2kT}(V_x^{\,2} + V_y^{\,2} + V_z^{\,2})\right] dV_x\,dV_y\,dV_z \\ &= \frac{m^3V}{h^3}\left[\int_{-\infty}^{+\infty} \exp\left[-\frac{m}{2kT}V_s^{\,2}\right] dV_s\right]^3 = \frac{m^3V}{h^3}\left[\sqrt{\frac{\pi}{(m/2kT)}}\right]^3 \\ &= V\left(\frac{2\pi mkT}{h^2}\right)^{3/2} \end{aligned} \tag{3.62}$$

We find that, under the assumptions of classical mechanics, the partition function for translational energy is directly proportional to the system volume and is proportional to the $\frac{3}{2}$ power of the mass and temperature.

3.6. The Maxwell-Boltzmann Velocity Distribution

Since we have just utilized the assumptions and procedures of classical mechanics to evaluate the partition function for a monatomic ideal gas, let us now extend our consideration to find the corresponding distribution of molecular velocities. These results will have numerous useful applications.

Consider a system comprised of N particles at T, V. We once again consider the most elementary case: "corrected" Boltzmann statistics, monatomic ideal gas, and classical mechanics. Of the N atoms in the system, a number, dN_{V_x,V_y,V_z}, have velocity components in the interval V_x to $V_x + dV_x$, V_y to $V_y + dV_y$, V_z to $V_z + dV_z$. The statistical weight for the velocity element $dV_x\,dV_y\,dV_z$ was given in Section 3.5, Eq. 3.61:

$$g_{V_x,V_y,V_z} = \left(\frac{m^3V}{h^3}\right) dV_x\,dV_y\,dV_z$$

for the same set of assumptions considered here. Therefore the fraction of atoms, $dN_{V_x,V_y,V_z}/N$ in this velocity element is given by the equilibrium

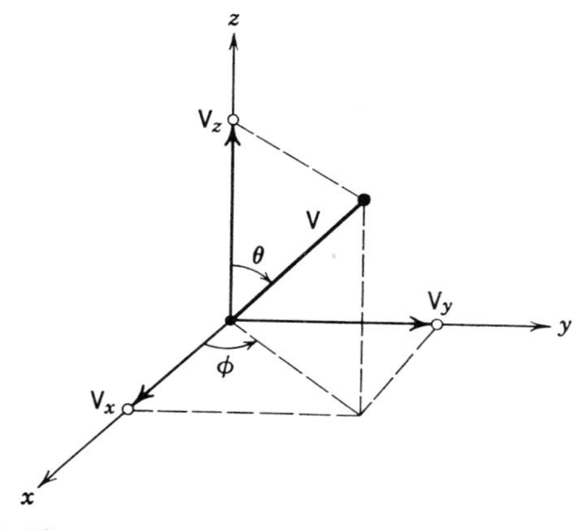

Figure 3.4 The cartesian and spherical coordinate systems.

distribution equation (3.58) as

$$\frac{dN_{V_x,V_y,V_z}}{N} = \frac{(m^3V/h^3)\exp\left[-\dfrac{m}{2kT}(V_x^2 + V_y^2 + V_z^2)\right] dV_x\, dV_y\, dV_z}{Z} \tag{3.63}$$

But, substituting Eq. 3.62 for Z, which was evaluated under the same assumptions as Eq. 3.63, this becomes

$$\frac{dN_{V_x,V_y,V_z}}{N} = \left(\frac{m}{2\pi kT}\right)^{3/2} \exp\left[-\frac{m}{2kT}(V_x^2 + V_y^2 + V_z^2)\right] dV_x\, dV_y\, dV_z \tag{3.64}$$

which is termed the Maxwell-Boltzmann distribution equation.

Instead of using the vector velocity components, it is often more convenient to use the scalar magnitude of velocity or speed of the particles, given by

$$V = |(V_x^2 + V_y^2 + V_z^2)^{1/2}| \tag{3.65}$$

It is helpful to think of the components V_x, V_y, V_z as the cartesian coordinates of a point on the surface of a sphere of radius V, as shown in Figure 3.4. Using spherical coordinates, the velocity components can be expressed by

$$\begin{aligned} V_x &= V \sin\theta \cos\phi \\ V_y &= V \sin\theta \sin\phi \\ V_z &= V \cos\theta \end{aligned} \tag{3.66}$$

It is important to keep in mind that, while the components V_x, V_y, V_z can have values from $-\infty$ to $+\infty$, the scalar V can have values only from 0 to $+\infty$. To write Eq. 3.64 in terms of V, we note that an integral over $dV_x\, dV_y\, dV_z$ in cartesian coordinates is replaced in spherical coordinates by the integral over $V^2 \sin\theta\, d\theta\, d\phi\, dV$. Integrating over θ, ϕ (surface area of the sphere), this factor becomes $4\pi V^2\, dV$. Consequently, Eq. 3.64 can be written as the fraction of the particles having a velocity magnitude or speed in the range from V to $V + dV$ as

$$\frac{dN_V}{N} = 4\pi\left(\frac{m}{2\pi kT}\right)^{3/2} V^2 \exp\left(-\frac{mV^2}{2kT}\right) dV \tag{3.67}$$

which is an alternative form of the Maxwell-Boltzmann velocity distribution equation, written here in terms of the molecular speeds.

A number of useful results are found from integration of either of the forms for the velocity distribution, Eq. 3.64 or 3.67.

The fraction of particles having an x-component velocity between some V_x and $V_x + dV_x$ and any value of V_y, V_z can be found by integrating Eq. 3.64 over all possible values of V_y and V_z. This yields

$$\begin{aligned}\frac{dN_{V_x}}{N} &= \int_{V_y}\int_{V_z} \frac{dN_{V_xV_yV_z}}{N} \\ &= \left(\frac{m}{2\pi kT}\right)^{3/2}\left[\int_{-\infty}^{+\infty}\int_{-\infty}^{+\infty} \exp\left[-\frac{m}{2kT}(V_x^2 + V_y^2 + V_z^2)\right] dV_y\, dV_z\right] dV_x \\ &= \left(\frac{m}{2\pi kT}\right)^{1/2} \exp\left[-\frac{m}{2kT} V_x^2\right] dV_x \end{aligned} \tag{3.68}$$

A plot of this function is shown in dimensionless form in Figure 3.5. It is apparent that the component velocity is symmetrical in the $\pm x$-direction with a maximum at zero. The average value of V_x is given by

$$\overline{V_x} = \left(\frac{m}{2\pi kT}\right)^{1/2} \int_{-\infty}^{+\infty} V_x \exp\left(-\frac{m}{2kT} V_x^2\right) dV_x = 0 \tag{3.69}$$

The integral is zero because of the symmetry displayed in Figure 3.5. This symmetry of component velocity and the result of Eq. 3.69 are in agreement with the assumption of Section 3.4 that the gas is isotropic, as stated by Eq. 3.46.

Let us return now to the velocity distribution as expressed by Eq. 3.67. The most probable speed V_{mp} is that corresponding to the maximum of that function found from

$$\frac{d}{dV}\left(\frac{dN_V}{N\, dV}\right) = 4\pi\left(\frac{m}{2\pi kT}\right)^{3/2} V \exp\left(-\frac{mV^2}{2kT}\right)\left[2 - \frac{mV^2}{kT}\right] = 0 \tag{3.70}$$

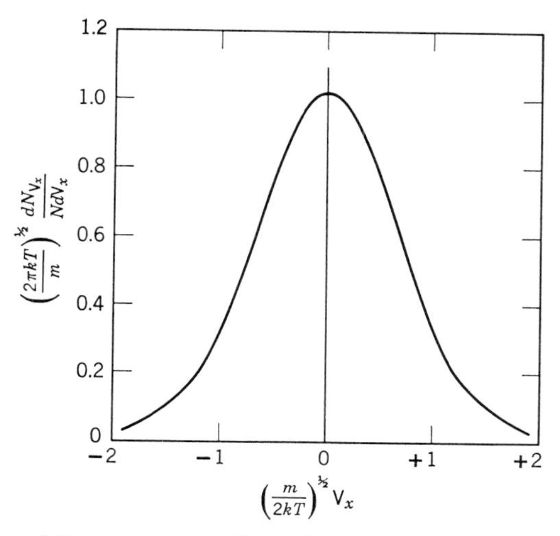

Figure 3.5 *x*-component Maxwell-Boltzmann velocity distribution.

so that

$$V_{mp} = \left(\frac{2kT}{m}\right)^{1/2} \tag{3.71}$$

On the other hand, the mean speed is given by

$$\begin{aligned} \bar{V} &= \int_0^\infty V\left(\frac{dN_V}{N\,dV}\right)dV = 4\pi\left(\frac{m}{2\pi kT}\right)^{3/2}\int_0^\infty V^3 \exp\left(-\frac{mV^2}{2kT}\right)dV \\ &= 4\pi\left(\frac{m}{2\pi kT}\right)^{3/2}\left[2\left(\frac{kT}{m}\right)^2\right] = \left(\frac{8kT}{\pi m}\right)^{1/2} \end{aligned} \tag{3.72}$$

or

$$\bar{V} = 1.1284 V_{mp} \tag{3.73}$$

Similarly, the mean-square speed is

$$\begin{aligned} \overline{V^2} &= \int_0^\infty V^2\left(\frac{dN_V}{N\,dV}\right)dV = 4\pi\left(\frac{m}{2\pi kT}\right)^{3/2}\int_0^\infty V^4 \exp\left(-\frac{mV^2}{2kT}\right)dV \\ &= 4\pi\left(\frac{m}{2\pi kT}\right)^{3/2}\left[\frac{3\sqrt{\pi}}{8}\left(\frac{2kT}{m}\right)^{5/2}\right] = \frac{3kT}{m} \end{aligned} \tag{3.74}$$

Thus the mean kinetic energy and, consequently, internal energy are in agreement with that found from Eq. 3.53. Using Eqs. 3.74 and 3.71, the root-mean-square speed is

$$(\overline{V^2})^{1/2} = 1.2248 V_{mp} \tag{3.75}$$

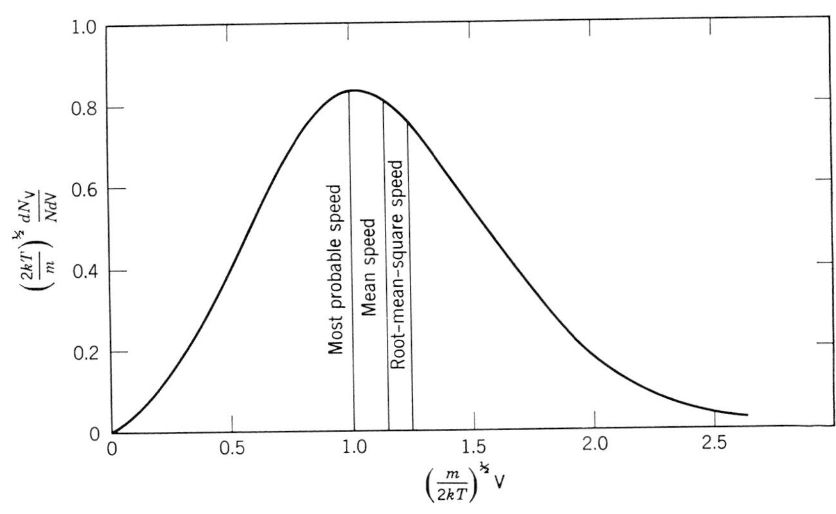

Figure 3.6 Maxwell-Boltzmann speed distribution.

If we now plot the speed distribution of Eq. 3.67, as in Figure 3.6, we see that the distribution is asymmetric, tending to zero for large V. The values given in Eqs. 3.71, 3.73, 3.75 are also pointed out in this figure.

Example 3.2

(*a*) Calculate the most probable speed, the mean speed, and the root-mean-square speed for helium at 0 C. (*b*) Plot the speed distribution for helium, $(dN_V/N\,dV)$, versus V at 0 C and 200 C.

(*a*) From Table A.2, the mass of the helium atom is

$$m = 1.66 \times 10^{-24} \times (4.003) = 6.645 \times 10^{-24}\ \text{gm}$$

From Eq. 3.71, the most probable speed at 0 C is

$$V_{mp} = \left(\frac{2kT}{m}\right)^{1/2} = \left[\frac{2(1.38044 \times 10^{-16})273.15}{6.645 \times 10^{-24}}\right]^{1/2}$$
$$= 10.64 \times 10^{4}\ \text{cm/sec} = 1064\ \text{meters/sec}$$

From Eq. 3.73, the mean or average speed is

$$\overline{V} = 1.1284 \times (1064) = 1201\ \text{meters/sec}$$

and, from Eq. 3.75, the root-mean-square speed is

$$(\overline{V^2})^{1/2} = 1.2248 \times (1064) = 1303\ \text{meters/sec}$$

While these values may seem to be quite high, we must remember that an atom undergoes frequent collisions with other atoms, so that an atom's direction is constantly changing and it does not progress very rapidly in any direction.

(*b*) From Eqs. 3.67 and 3.71,

$$\frac{dN_V}{N\,dV} = \frac{4}{\sqrt{\pi}\,V_{mp}}\left(\frac{V}{V_{mp}}\right)^2 \exp\left(-\frac{V}{V_{mp}}\right)^2$$

At 0 C,

$$\frac{dN_V}{N\,dV} = 0.00212\left(\frac{V}{1064}\right)^2 \exp\left(-\frac{V}{1064}\right)^2$$

At 200 C,

$$V_{mp} = 1064 \times \left(\frac{473.15}{273.15}\right)^{1/2} = 1400 \text{ meters/sec}$$

and

$$\frac{dN_V}{N\,dV} = 0.00161\left(\frac{V}{1400}\right)^2 \exp\left(-\frac{V}{1400}\right)^2$$

Plots of these distributions are shown in Figure 3.7. We note that, as the temperature is increased, the distribution becomes broader, more atoms shifting to the higher energy levels (higher velocities). There is, of course, a resulting increase in internal energy of the gas.

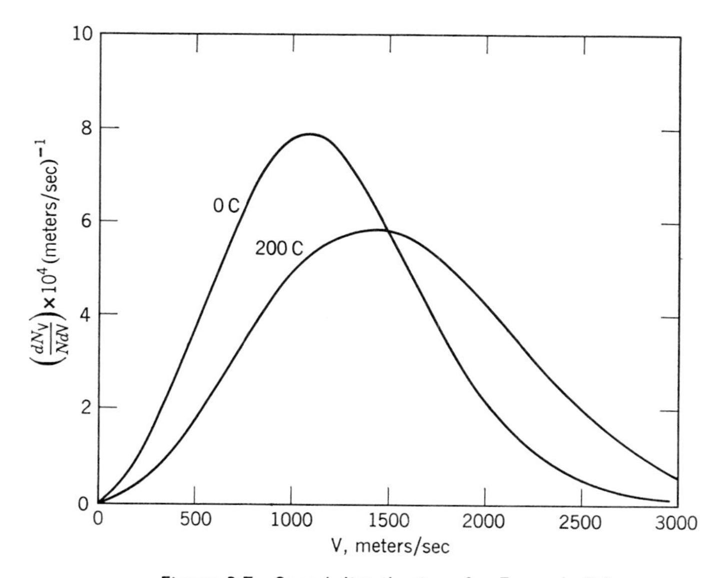

Figure 3.7 Speed distributions for Example 3.2.

3.7. The Equilibrium Distribution for the Components of a Mixture

In the final section of this chapter we consider the statistical mechanics for a mixture of two components, A and B, in which we assume a system of independent particles obeying corrected Boltzmann statistics. The development of the equilibrium distribution equation is analogous to the procedure followed in Section 3.3, but somewhat more involved.

Consider a system of volume V containing a mixture of a fixed number of particles, N_A, of species A and a fixed number, N_B, of species B. The mixture exists at temperature T and has a given internal energy U. Since we have assumed a system of independent particles, each species has its own set of energy levels, which for component A are

$$\epsilon_{A_1}, \epsilon_{A_2}, \ldots, \epsilon_{A_j}$$

with degeneracies

$$g_{A_1}, g_{A_2}, \ldots, g_{A_j}$$

and for component B the levels are

$$\epsilon_{B_1}, \epsilon_{B_2}, \ldots, \epsilon_{B_j}$$

with degeneracies

$$g_{B_1}, g_{B_2}, \ldots, g_{B_j}$$

At any instant of time, the N_A particles of A are distributed in some manner among the energy levels for A, and the N_B particles of B are distributed in some manner among the energy levels for B. Any given distribution is specified in terms of the number of particles of each species that are found in each of its levels, that is, by the sets

$$N_{A_1}, N_{A_2}, \ldots, N_{A_j}$$
$$N_{B_1}, N_{B_2}, \ldots, N_{B_j}$$

The possible distributions are, of course, subject to the restrictions

$$N_A = \sum_{A_j} N_{A_j} \tag{3.76}$$

$$N_B = \sum_{B_j} N_{B_j} \tag{3.77}$$

and also, since the system energy U is given,

$$U = U_A + U_B = \sum_{A_j} N_{A_j}\epsilon_{A_j} + \sum_{B_j} N_{B_j}\epsilon_{B_j} \tag{3.78}$$

The thermodynamic probabilities w_A, w_B for the components are multiplicative, since any of the microstates of component A can exist in combination with any of those for component B. Thus the thermodynamic probability w for the mixture is, using the corrected Boltzmann expression Eq. 3.34 for A and B,

$$w = w_A w_B = \left(\prod_{A_j} \frac{g_{A_j}^{N_{A_j}}}{N_{A_j}!}\right)\left(\prod_{B_j} \frac{g_{B_j}^{N_{B_j}}}{N_{B_j}!}\right) \tag{3.79}$$

If we rewrite Eq. 3.79 in logarithmic form and expand each term by Stirling's formula, the expression for w becomes

$$\ln w = N_A + \sum_{A_j} N_{A_j} \ln \frac{g_{A_j}}{N_{A_j}} + N_B + \sum_{B_j} N_{B_j} \ln \frac{g_{B_j}}{N_{B_j}} \tag{3.80}$$

We seek the most probable distribution of the sets of particles N_A, N_B, among their respective sets of energy levels; that is, the distribution having the maximum value of w. If we differentiate Eq. 3.80 and set $d \ln w = 0$ as in the previous developments, we find that

$$\sum_{A_j} \ln \frac{g_{A_j}}{N_{A_j}} \, dN_{A_j} + \sum_{B_j} \ln \frac{g_{B_j}}{N_{B_j}} \, dN_{B_j} = 0 \tag{3.81}$$

This relation specifies the most probable distribution, subject to the constraints

$$dN_A = \sum_{A_j} dN_{A_j} = 0 \tag{3.82}$$

$$dN_B = \sum_{B_j} dN_{B_j} = 0 \tag{3.83}$$

and

$$dU = \sum_{A_j} \epsilon_{A_j} \, dN_{A_j} + \sum_{B_j} \epsilon_{B_j} \, dN_{B_j} = 0 \tag{3.84}$$

Using the method of undetermined multipliers, we multiply Eq. 3.82 by α_A, Eq. 3.83 by α_B, and Eq. 3.84 by β. Adding the three resulting expressions to the negative of Eq. 3.81, we have

$$\sum_{A_j} \left(\ln \frac{N_{A_j}}{g_{A_j}} + \alpha_A + \beta \epsilon_{A_j}\right) dN_{A_j} + \sum_{B_j} \left(\ln \frac{N_{B_j}}{g_{B_j}} + \alpha_B + \beta \epsilon_{B_j}\right) dN_{B_j} = 0 \tag{3.85}$$

Since components A and B are independent, we conclude from Eq. 3.85 that, for the most probable distribution,

$$N_{A_j} = g_{A_j} e^{-\alpha_A} e^{-\epsilon_{Aj}/kT} \tag{3.86}$$

for all values of A_j, and

$$N_{B_j} = g_{B_j} e^{-\alpha_B} e^{-\epsilon_{Bj}/kT} \tag{3.87}$$

for all values of B_j. In both expressions the relation $\beta = 1/kT$ has been assumed, as in earlier developments.

If we now sum Eq. 3.86 over all A_j's,

$$N_A = e^{-\alpha_A} \sum_{A_j} g_{A_j} e^{-\epsilon_{A_j}/kT} = e^{-\alpha_A} Z_A \tag{3.88}$$

where, according to Eq. 3.55, Z_A is the partition function for A,

$$Z_A = \sum_{A_j} g_{A_j} e^{-\epsilon_{A_j}/kT} \tag{3.89}$$

Similarly, for component B, from Eq. 3.87,

$$N_B = e^{-\alpha_B} \sum_{B_j} g_{B_j} e^{-\epsilon_{B_j}/kT} = e^{-\alpha_B} Z_B \tag{3.90}$$

where

$$Z_B = \sum_{B_j} g_{B_j} e^{-\epsilon_{B_j}/kT} \tag{3.91}$$

Thus, eliminating the multipliers α_A, α_B, the equilibrium distribution for the binary mixture whose components obey corrected Boltzmann statistics can be expressed as

$$N_{A_j} = \frac{N_A g_{A_j} e^{-\epsilon_{A_j}/kT}}{Z_A} \quad \text{for all } A_j \tag{3.92}$$

$$N_{B_j} = \frac{N_B g_{B_j} e^{-\epsilon_{B_j}/kT}}{Z_B} \quad \text{for all } B_j \tag{3.93}$$

It is important to note that the component partition functions Z_A, Z_B, are to be evaluated for A and B, respectively, each at the temperature and volume of the system. Thus we conclude that each of the components of the mixture behaves as though it existed alone in the total volume V at the temperature of the mixture.

PROBLEMS

3.1 Consider a simple system having the same non-degenerate energy levels 0, 1, 2, 3 as the example in Section 3.1. This system, however, contains 8 particles and has a total energy of 6. Determine the thermodynamic probability for every distribution consistent with these constraints, assuming Boltzmann statistics.

3.2 Repeat Problem 3.1 for Boltzmann, Bose-Einstein, and Fermi-Dirac statistics, assuming that each of the energy levels has a degeneracy of 6.

3.3 A system contains 3000 particles and has an energy of 4100 units. There are three non-degenerate energy levels, of energies 1, 2, and 3 units, respectively.

(*a*) One distribution consistent with the given constraints is $N_1 = 2000$, $N_2 = 900$, $N_3 = 100$. Shifts of one particle from level 2 to 1 and one from 2 to 3 or shifts of one particle from level 1 to 2 and one from 3 to 2 represent small changes in both directions from the given distribution. Calculate the ratio of thermodynamic probabilities for each of these small changes. Is the stated distribution the equilibrium distribution consistent with the constraints?

(*b*) Determine analytically the equilibrium distribution for this system. Then use the particle-shifting procedure of part (*a*) to demonstrate that it is the equilibrium distribution.

3.4 A monatomic ideal gas is contained in a cylinder fitted with a piston, as shown in Figure 3.8. The piston is then moved very slowly toward the right at velocity V_p. An atom with velocity V striking the moving piston at angle θ rebounds with velocity V' and at angle θ'. Determine the relations between these quantities. Then show that the decrease in energy of the particle can be expressed as $(2mV \cos \theta)V_p$, and that for the system as $P(dV/dt)$.

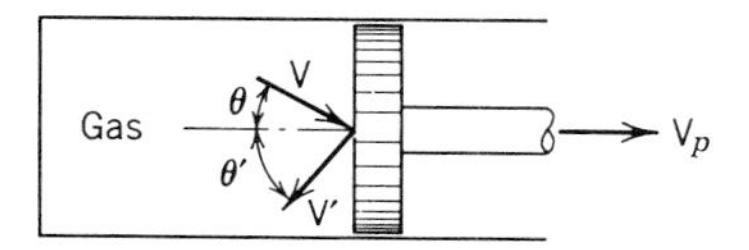

Figure 3.8 Sketch for Problem 3.4.

3.5 A two-dimensional ideal gas is a useful model for certain applications (for example, in analyzing the behavior of atoms adsorbed on a surface but still free to move in two directions). Determine the "pressure", force per unit length, for such a two-dimensional gas in terms of the energy and area.

3.6 Helium behaves according to Bose-Einstein statistics. As an indication of the applicability of the "corrected" Boltzmann model at different temperatures, evaluate the quantity e^{α} for helium at 1 atm pressure and 5, 50, and 500 K, assuming classical mechanics. What conclusions can be drawn from the results?

3.7 Assuming a Maxwell-Boltzmann velocity distribution, compare the average speeds of helium and xenon at 300 K and at 3000 K.

3.8 Consider a monatomic ideal gas. Show that the average positive x-component of velocity is

$$\bar{V}_x = \left(\frac{kT}{2\pi m}\right)^{1/2} \qquad (V_x \geq 0 \text{ only})$$

and also that

$$\overline{V_x^2} = \frac{kT}{m}$$

3.9 Consider a monatomic gas having a Maxwell-Boltzmann velocity distribution.

(*a*) Show that the corresponding energy distribution is given by

$$\frac{dN_\epsilon}{N} = \frac{2}{\sqrt{\pi}\,(kT)^{3/2}}\,\epsilon^{1/2} e^{-\epsilon/kT}\,d\epsilon$$

where dN_ϵ is the number of particles having energy in the interval ϵ to $\epsilon + d\epsilon$.

(*b*) For a system having the energy distribution of part (*a*), calculate the most probable energy and the average energy.

3.10 (*a*) For a gas with a Maxwell-Boltzmann velocity distribution, show that the fraction of atoms in the system having a speed between 0 and V can be expressed as

$$\frac{N_{0 \text{ to } V}}{N} = \operatorname{erf}(x) - \frac{2}{\sqrt{\pi}} x e^{-x^2}$$

where $x = V/V_{mp}$.

(*b*) Calculate the fraction of atoms in argon gas at 300 K with speeds less than 500 meters/sec at any instant of time.

3.11 At what temperature will 60% of the atoms in a helium-filled tank have speeds greater than 2000 meters/sec?

3.12 Consider a wall of a tank containing a monatomic gas. Taking x, y coordinates in the plane of the wall and z as the inward-directed normal, show that the distribution of particles striking the surface can be expressed in spherical coordinates as

$$\frac{\text{number}}{\text{cm}^2\text{-sec}} = \frac{N}{V}\left(\frac{m}{2\pi kT}\right)^{3/2} V^3 e^{-(m/2kT)V^2} \cos\theta \sin\theta \, d\theta \, d\phi \, dV$$

3.13 (*a*) Consider a gas contained in volume V and having an arbitrary speed distribution dN_V (the number of particles in an interval V to V + dV). Show that the rate at which particles of speed V strike a unit area of wall surface is given by

$$\frac{1}{4}\frac{V}{V}\, dN_V$$

(*b*) Assuming a Maxwell-Boltzmann distribution, show that the total number of atoms striking a unit area of the wall per unit time is given in terms of the average speed $\bar{V}$ by

$$\frac{1}{4}\frac{N}{V}\bar{V}$$

3.14 A cubic one-liter tank contains neon at 25 C, 1 atm pressure.

(*a*) From the results of Problem 3.13, calculate the number of atoms striking a 1 cm² surface of wall per second.

(*b*) Compare the total number of collisions of atoms with the walls per second and the total number of atoms in the tank.

3.15 A very small pinhole develops in the wall of a vessel, and every atom of the gas inside that strikes the hole escapes from the tank. Show that the average energy per particle escaping is $2kT$. Why is this different from the average energy of the gas inside the tank?

3.16 A classical experiment for verifying the Maxwell-Boltzmann velocity distribution involves the apparatus shown schematically in Figure 3.9. A unidirectional molecular beam (produced by collimating a stream of atoms escaping from a small hole in a vessel) enters a slit in a drum of diameter d. The drum rotates at high speed, n revolutions/sec, and these atoms are distributed along the recording plate according to their speeds. Develop an expression for

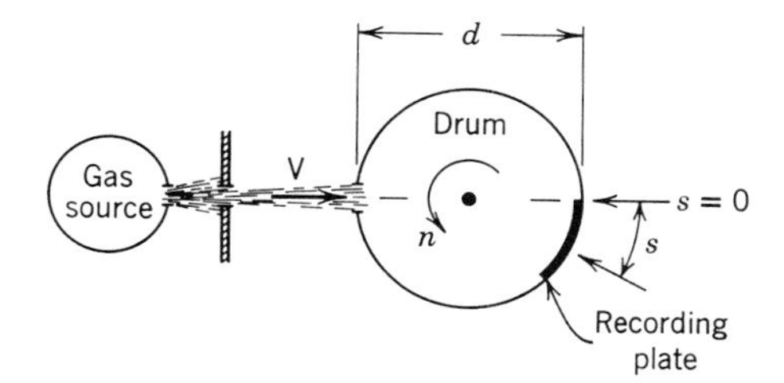

Figure 3.9 Sketch for Problem 3.16.

the intensity of deposition of atoms as a function of displacement s and other physical parameters. Sketch the relation between deposition intensity and displacement.

3.17 The equilibrium of a liquid with its vapor is of a dynamic character. At equilibrium, molecules of liquid evaporate at the same rate at which molecules of vapor condense, so that there is no net change in the amount of each phase present. Consider mercury at 0 C, at which temperature the vapor pressure is 1.85×10^{-4} mm Hg. If every vapor molecule striking the liquid surface condenses, what are the rates at which condensation and evaporation occur at equilibrium?

3.18 Consider a 1-liter vacuum container maintained at 1 micron (10^{-3} mm Hg) pressure. Air at ambient conditions surrounds the container. A pinhole of 10^{-4} mm diameter then develops in the wall.

(*a*) Find the rate at which molecules pass through the hole in each direction, assuming that every molecule arriving at the hole passes freely through. Treat air as a pure substance with $M = 28.97$.

(*b*) What is the pressure inside the tank after 30 minutes?

3.19 A "getter" is a substance used to remove gas molecules from a closed volume (by adsorption or chemical combination); it finds considerable application in high vacuum technology. As an example, consider a 100 cc volume containing air that is initially pumped to 1 mm Hg pressure and then sealed. Inside this volume is a "getter" of 1 cm^2 surface, which is assumed to remove 0.1% of the molecules striking the surface. Determine the pressure inside the volume as a function of time, assuming that the temperature remains constant at 25 C.

3.20 A "potential barrier" model is useful for many applications, especially in the field of electronics. Consider a monatomic ideal gas having Maxwell-Boltzmann velocity distribution and confined in an enclosure, one "wall" of which consists of a potential barrier φ (perhaps due to an electric field). Thus all particles that reach this wall with a normal component of kinetic energy greater than φ escape, while all those with energy component less than φ are reflected. Develop an expression for the flux of particles through this barrier in terms of the significant parameters.

3.21 A large tank contains a mixture of two gases A and B at temperature T and pressure P. The mole fractions are y_A and y_B, respectively. A very small hole is then opened in the wall of the container; thus every particle striking the hole may be assumed to escape. Does the escaping stream of particles have the same composition as the mixture in the tank? How do the relative proportions depend on molecular weights?

3.22 A 10-liter tank contains a mixture of 50% helium, 50% nitrogen at 30 C, 0.1 atm. One end of the tank contains a porous plug consisting of very fine holes, through which the gas is to be withdrawn. The total hole area is estimated to be 10^{-4} cm^2. Determine the pressure and composition inside the tank as a function of time.

4 Thermodynamics

In this chapter, we consider the first, second, and third laws of thermodynamics, as well as various thermodynamic properties, from a microscopic point of view. In so doing, we shall relate the definitions of the properties and statements of the laws that are given in classical thermodynamics to the concepts and definitions of statistical thermodynamics, and thus demonstrate the added insight into the nature of physical processes that a microscopic point of view provides. In addition to discussing these quantities, we shall develop expressions for the thermodynamic properties of interest in terms of the partition function, so that the subsequent problem of evaluating properties will be focused upon the determination of that quantity.

Since the developments of this chapter are based on the preceding chapter, for the present we must limit the discussion to systems of independent particles, and we shall concentrate primarily on systems following corrected Boltzmann statistics. However, the relevance of Bose-Einstein and Fermi-Dirac statistics at temperatures approaching absolute zero will also be demonstrated.

4.1. Internal Energy and Specific Heat

We found, from Eq. 3.58, that for corrected Boltzmann statistics the most probable distribution (the thermodynamic equilibrium state as discussed in Chapter 3) is given by the relation

$$N_j = \frac{N g_j e^{-\epsilon_j/kT}}{Z} \tag{4.1}$$

where the partition function Z is, by definition (Eq. 3.55),

$$Z = \sum_j g_j e^{-\epsilon_j/kT} \tag{4.2}$$

The internal energy corresponding to the most probable distribution was found to be (Eq. 3.59),

$$U = \sum_j N_j \epsilon_j = \frac{N}{Z} \sum_j g_j \epsilon_j e^{-\epsilon_j/kT} \tag{4.3}$$

The summation in this last expression is similar in form to the partition function. It would be very desirable to eliminate this summation in terms of Z to simplify the problem of evaluating the internal energy. It is possible to do this by differentiating Z with respect to temperature, but, in doing so, we must be careful to hold the various energies ϵ_j constant. In the developments of the previous chapter, we considered an energy level as specified by a value of the index j to be a group of quantum states all having the same, or very nearly the same, value of energy. The magnitude of the energy for each of the levels or groups of states is a function of the boundary parameters of the system. Thus, as the boundary parameters are varied, the energy of all the levels changes accordingly. (This point will be discussed again in Section 4.2 and more fully in Chapter 5.) For our applications, we shall be concerned primarily with a simple compressible substance, which by definition is one in which the only boundary parameter of importance is the system volume V. Other substances may have additional boundary parameters that are significant: length when tension in the system is of importance; area for surface tension; the position in an electric or magnetic field; etc. Thus, by specifying constant volume to hold the ϵ_j's constant, we imply fixing any other boundary parameter as necessary. Therefore

$$\left(\frac{\partial Z}{\partial T}\right)_V = \frac{1}{kT^2} \sum_j g_j \epsilon_j e^{-\epsilon_j/kT} \tag{4.4}$$

When we compare Eqs. 4.3 and 4.4, we can express the internal energy as

$$U = \frac{NkT^2}{Z}\left(\frac{\partial Z}{\partial T}\right)_V = NkT^2\left(\frac{\partial \ln Z}{\partial T}\right)_V \tag{4.5}$$

It is frequently convenient to have this expression for internal energy on a molal basis. Since k is the gas constant per molecule,

$$k = \frac{R}{N_0} \tag{4.6}$$

and the number of gram moles, n, is given by

$$n = \frac{N}{N_0} \tag{4.7}$$

The internal energy on a mole basis, u, is given by the relation

$$u = \frac{U}{n} = RT^2\left(\frac{\partial \ln Z}{\partial T}\right)_V \tag{4.8}$$

The constant-volume specific heat is defined in terms of the internal energy and temperature as

$$C_v = \left(\frac{\partial u}{\partial T}\right)_V$$

On differentiating Eq. 4.8, we have

$$C_v = \left(\frac{\partial u}{\partial T}\right)_V = RT^2\left(\frac{\partial^2 \ln Z}{\partial T^2}\right)_V + 2RT\left(\frac{\partial \ln Z}{\partial T}\right)_V \tag{4.9}$$

4.2. The First Law of Thermodynamics

Consider a thermodynamic system, which by definition consists of a fixed mass.† Neglecting changes in kinetic and potential energy of the system, the first law of thermodynamics is

$$dU = \delta Q - \delta W \tag{4.10}$$

where δQ is the heat transferred to the system and δW is the work done by the system on its surroundings during the process.

On differentiating Eq. 3.5,

$$dU = \sum_j \epsilon_j \, dN_j + \sum_j N_j \, d\epsilon_j \tag{4.11}$$

we see that a change in internal energy can be expressed as the sum of two contributions. The first in Eq. 4.11 is the change due to a net redistribution of particles among the energy levels, while the second is the change resulting from a shift in the magnitude of energy associated with each of the levels or groups of quantum states. As mentioned in the preceding section, such energy level shifts result from changes in the boundary parameters of the system.

Let us now try to associate these two contributions of Eq. 4.11 with the macroscopic quantities, work and heat transfer. For example, consider the work done when the system boundary is allowed to move. If the driving force is only infinitesimally greater than that resisting the movement, then the process will proceed very slowly, and the system will pass through a series of equilibrium states. In such a reversible process, a differential quantity of work is given by the product of the pressure and a

† This is the definition of a system when the concept of system and control volume is used; it corresponds to a closed system for the terminology closed and open system.

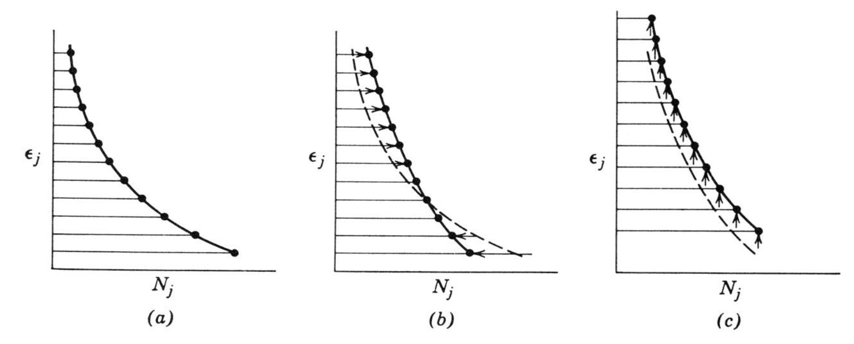

Figure 4.1 The microscopic viewpoint representation of the transfer of heat or work in a reversible process.

differential change in volume. Similarly, the work due to other thermodynamic potentials or driving forces in reversible processes is of the same form—the product of the potential and a change in the boundary parameter associated with that potential. We have previously stated that the energy levels are functions of the boundary parameters of the system, and we now see that the work for a reversible process is associated with changes in these parameters. Consequently, we associate the second term in Eq. 4.11 with the work done by a system in a reversible process, or

$$\delta W_{\text{rev}} = -\sum_j N_j \, d\epsilon_j \tag{4.12}$$

Comparing Eqs. 4.10 and 4.11, we identify the heat transfer for a reversible process with a net redistribution of particles among levels, so that

$$\delta Q_{\text{rev}} = \sum_j \epsilon_j \, dN_j \tag{4.13}$$

To aid in the visualization of the quantities work and heat transfer, let us represent the change in energy resulting from two processes as shown in Figure 4.1. Figure 4.1*a* represents the equilibrium distribution in a system of some particular energy U. Now consider a reversible heat transfer to the system during a process in which there is no work; that is, the various boundary parameters such as volume are held constant. Therefore the energies of the various levels remain fixed, and the increase in the energy of the system arises solely from a net shifting of particles from lower levels to higher levels, as shown in Figure 4.1*b*. Next, consider a reversible adiabatic process in which work is put into the system, as for example by boundary movement. From the first law, we conclude that the internal energy is increased. Taking a microscopic viewpoint, we see that the change in boundary parameters causes a shift in energy of the various levels, as shown in Figure 4.1*c*. The increase in energy of the system results in this case from the shifting of the energy levels, and not from a net redistribution of particles among levels.

Although our example is an extremely simple one, the ability to visualize a heat transfer or work on a microscopic scale should give us a further insight into the nature of these quantities, even if only for a reversible process. The nature of an irreversible process will be discussed in conjunction with the second law of thermodynamics, and the generalization to include energy changes due to mass transfer will be postponed until Chapter 10.

4.3. Entropy

Let us for the moment consider a thermodynamic system and proceed from a macroscopic or classical point of view. For example, if two metal blocks initially at different temperatures are brought into thermal communication, they will gradually come to the same temperature, some value intermediate to the two initial temperatures, by means of a spontaneous heat transfer between the two. We say that the blocks have reached a condition of thermal equilibrium. The change occurred because of the initial imbalance in temperature, the driving force, or thermodynamic potential. If we also include other thermodynamic potentials (pressure, chemical potential, etc.), we say that a condition of thermodynamic equilibrium exists in a system when the system is at equilibrium with respect to all potentials. That is, there is no imbalance in any driving force, and consequently no tendency for any spontaneous change to occur. From the methods of classical thermodynamics, the equilibrium state of a system of a given mass, volume, and internal energy is the state of maximum entropy. Mathematically, this is stated as

$$dS_{U,V,m} = 0 \qquad \text{for an infinitesimal change}$$

$$\Delta S_{U,V,m} < 0 \qquad \text{for a finite change}$$

Let us now view the system from a microscopic point of view. A particular distribution d of particles among the various energy levels can occur in w_d ways, the thermodynamic probability for that distribution. For an equilibrium state of the system of given V, N, U, all distributions consistent with the system constraints are accessible or available to the system. Thus the total number of quantum states available to the system, w_{tot}, is the summation of w_d over all possible distributions, or

$$w_{\text{tot}} = \sum_d w_d$$

On the other hand, we can also consider an inhibited state, one in which the system is restrained in some manner, at least temporarily, from reaching all possible distributions (and quantum states), so that w_{tot} for such a state is some smaller number than that given above.

Comparing this microscopic viewpoint of the equilibrium state at fixed V, N, U with the classical concept, we conclude that the entropy should be directly related to the total number of states available to the system. However, we must keep in mind that entropy is an extensive thermodynamic property; this statement means that, if two independent systems are combined, their entropies are additive. If the entropy is directly proportional to $\ln w_{tot}$ it will have this behavior, since the total thermodynamic probabilities of two systems are multiplicative when the systems are combined. Finally, the constant of proportionality must have the units of energy per degree.

With these considerations in mind, we define entropy from the microscopic viewpoint by the expression

$$S = k \ln w_{tot} \tag{4.14}$$

where k is the Boltzmann constant.

The definition of S according to Eq. 4.14 has the required characteristics of entropy discussed above. It remains to be shown that this definition is completely consistent with the classical definition of entropy from the second law of thermodynamics. Before proceeding to this demonstration, however, let us make a number of observations regarding Eq. 4.14. We note that any change from one macroscopic state to another that has a greater number of microscopic ways of occurring corresponds to an increase in entropy. In this sense, entropy can be considered as a measure of disorder, randomness, or lack of information about the microscopic configuration of the particles of which the system is comprised. A perfectly ordered system, with w_{tot} equal to unity, corresponds to zero entropy and implies a complete knowledge of the microscopic state of the system.

Our principal concern in this text, except in Chapter 12, is with equilibrium states. While all possible distributions consistent with the constraints are available to the system, the most probable distribution has a thermodynamic probability tremendously larger than all distributions differing from it by a distinguishable amount. Thus the thermodynamic equilibrium state may be represented for all practical purposes by the most probable distribution, that is,

$$\ln w_{mp} \approx \ln w_{tot}$$

so that the statistical mechanical entropy for an equilibrium state is given by

$$S = k \ln w_{mp} \tag{4.15}$$

In our consideration of equilibrium states, let us now express Eq. 4.15 in a more readily usable form. For corrected Boltzmann statistics, the thermodynamic probability for any given distribution is given by Eq. 3.34 which, in logarithmic form and with Stirling's formula applied, is

$$\ln w = \sum_j [N_j \ln g_j - N_j \ln N_j + N_j] = \sum_j N_j\left[\ln \frac{g_j}{N_j} + 1\right] \tag{4.16}$$

However, for the most probable distribution, the ratio g_j/N_j can be found from the equilibrium distribution equation, Eq. 4.1. Substituting this into Eq. 4.16, we have

$$\begin{aligned} \ln w_{\text{mp}} &= \sum_j N_j\left[\ln\left(\frac{Z}{N} e^{+\epsilon_j/kT}\right) + 1\right] \\ &= \ln \frac{Z}{N} \sum_j N_j + \frac{1}{kT} \sum_j N_j \epsilon_j + \sum_j N_j \\ &= N \ln \frac{Z}{N} + \frac{U}{kT} + N \end{aligned} \tag{4.17}$$

Using Stirling's formula for $N!$, we can also write this as

$$\ln w_{\text{mp}} = \ln \frac{(Z)^N}{N!} + \frac{U}{kT} \tag{4.18}$$

Substituting Eq. 4.18 into Eq. 4.15, we can express the entropy as

$$S = k \ln Z_N + \frac{U}{T} \tag{4.19}$$

where the quantity Z_N is defined for convenience as

$$Z_N = \frac{(Z)^N}{N!} \tag{4.20}$$

and is called the partition function for the system of N independent particles, as contrasted with the single-particle partition function Z.

We can use the equation for entropy in the form of Eq. 4.19 to show that our definition is in agreement with the definition of entropy from classical thermodynamics. Since all the quantities on the right side of Eq. 4.19 are point or state functions, S must also be a point function. Therefore, differentiating Eq. 4.19, we obtain

$$dS = k\, d \ln Z_N + d\left(\frac{U}{T}\right) \tag{4.21}$$

But, from Eq. 4.20,

$$d \ln Z_N = d \ln \frac{(Z)^N}{N!} = N\, d \ln Z = \frac{N}{Z}\, dZ \tag{4.22}$$

We find dZ from the definition of Z (Eq. 4.2), and Eq. 4.22 becomes

$$d \ln Z_N = \frac{N}{Z} \sum_j \left[-\frac{g_j}{kT} e^{-\epsilon_j/kT} \, d\epsilon_j + \frac{g_j \epsilon_j}{kT^2} e^{-\epsilon_j/kT} \, dT \right]$$

$$= -\frac{N}{ZkT} \sum_j g_j e^{-\epsilon_j/kT} \, d\epsilon_j + \frac{N}{ZkT^2} \sum_j g_j \epsilon_j e^{-\epsilon_j/kT} \, dT \quad (4.23)$$

Now, if the change is reversible, such that the system passes through a series of equilibrium states, then the equilibrium distribution equation (Eq. 4.1) must be valid at any point along the path. For this case, substituting Eq. 4.1 into Eq. 4.23, we have, for the reversible processes,

$$d \ln Z_N = \frac{-1}{kT} \sum_j N_j \, d\epsilon_j + \frac{1}{kT^2} \left(\sum_j N_j \epsilon_j \right) dT \quad (4.24)$$

Since Eq. 4.24 was derived for a reversible path, the summation in the first term on the right can be identified with the work done by the system according to Eq. 4.12. Furthermore, since the summation in the second term is equal to the internal energy, Eq. 4.24 can be rewritten:

$$d \ln Z_N = \frac{1}{kT} \, \delta W_{\text{rev}} + \frac{1}{kT^2} U \, dT \quad (4.25)$$

and, substituting Eq. 4.25 into Eq. 4.21, we have

$$dS = \frac{1}{T} \, \delta W_{\text{rev}} + \frac{U}{T^2} \, dT + d\left(\frac{U}{T} \right) = \frac{1}{T} (\delta W_{\text{rev}} + dU) = \frac{\delta Q_{\text{rev}}}{T} \quad (4.26)$$

which is the classical definition of entropy. Since this definition is made on a differential basis, it is still possible for the two values to differ by a constant of integration. This point will be resolved shortly through the third law of thermodynamics.

4.4. The Second Law of Thermodynamics

We have discussed the quantities heat and work and have also examined the first law of thermodynamics from a microscopic viewpoint. In doing so, we limited the discussion to reversible processes and thus were concerned with systems passing through a series of equilibrium states. We realize, of course, that macroscopic processes occurring in nature are irreversible. However, before proceeding to the subject of irreversible processes, let us first consider reversible processes from the standpoint of entropy and the second law of thermodynamics.

Suppose that a gas is contained in an insulated cylinder fitted with a

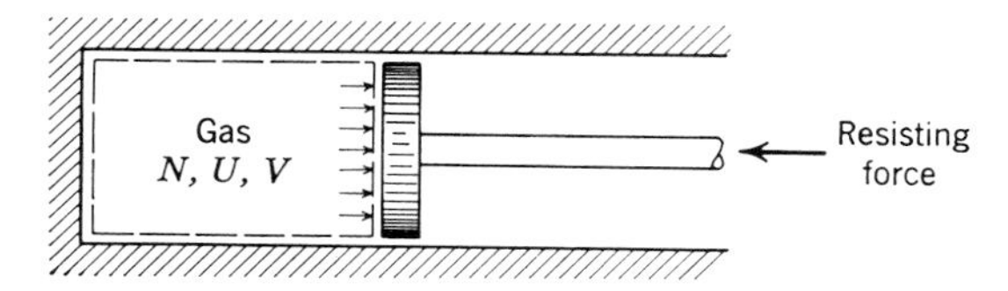

Figure 4.2 The reversible expansion of a gas.

frictionless piston, as shown in Figure 4.2. The system initially has some known number of particles N, internal energy U, and volume V, and is assumed to exist at a condition of equilibrium. The pressure of the gas on the face of the piston is exactly balanced by the proper resisting force (including the contribution of ambient pressure on the back face of the piston). Now let the resisting force be reduced by an infinitesimal amount, allowing the piston to move very slowly toward the right. The gas will expand accordingly, and, if the piston moves slowly enough, it may be assumed that the system passes through a series of equilibrium states, and undergoes a reversible expansion. Furthermore, since the cylinder is well insulated, there will be no heat transfer during the process.

From our discussion of the first law, we know that the energy change that takes place in this system during the process described is due solely to a shifting of the energy levels and not to any net redistribution of particles among the levels. We note from Eq. 4.16 that, for the succession of equilibrium states assumed by the system, the thermodynamic probability remains constant because there was no net redistribution of particles. Consequently, in accordance with our definition of entropy in Eq. 4.15, the entropy of the system remains constant during the process. This result is, of course, in agreement with our experience in classical thermodynamics, but the molecular point of view now gives us another way of visualizing the concept of an isentropic process.

Let us consider next the opposite extreme, an irreversible expansion in which no work is done, as represented in Figure 4.3. The tank is insulated, and a gas on the left side is restrained by a thin partition. The initial state of the gas is identical with that of the previous example: N particles, an internal energy U, and volume V. In this case, an equal volume on the other side of the partition is completely evacuated. If the partition is suddenly broken, the gas will expand to fill the entire volume. This expansion is irreversible and is certainly very different in nature from the reversible expansion examined previously.

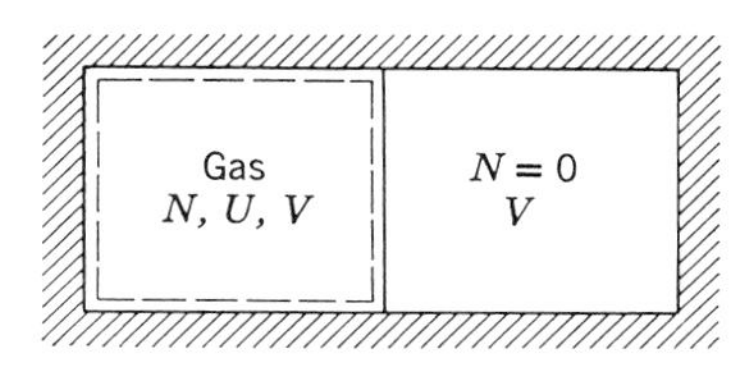

Figure 4.3 An irreversible expansion of a gas.

If we analyze this process from a classical viewpoint, we are limited to a description of the system before the process and after the gas has again come to equilibrium at the end of the process. We do not know exactly what happened between these two end states, since the changes occurred very rapidly and the system was not at equilibrium. However, as in the first example, there is no heat transfer during the process because the tank is well insulated. Furthermore, in this process no work is done, since there was no resistance to the expansion (assuming the hole in the partition to be reasonably large compared to molecular dimensions). We therefore conclude from the first law that there is no change in internal energy during the process.

When we examine this process from a microscopic viewpoint, we cannot describe the series of states that the system passes through; this is the same difficulty that is encountered in an analysis from the macroscopic point of view (the subject of non-equilibrium thermodynamics will be discussed in Chapter 12). We are, however, able to make some interesting observations about the process. There is a change in volume, and consequently there will be a shift in the energy levels. Since no work was done during the process, Eq. 4.12, which applies to a reversible process, cannot in general be valid. Also, since the energy of the system remained constant, we conclude from Eq. 4.11 that there must have been some net redistribution of particles among levels during the process. Because there was no heat transfer, we conclude also that Eq. 4.13 is not in general valid for an irreversible process. (The heat transfer is given, however, by Eq. 4.13 for a process in which all the boundary parameters are fixed.) Consequently, our description of an irreversible process is quite limited in scope. In our example there was a general shifting of levels and of particles among levels, and we did not know the path followed—only the end states. However, the microscopic viewpoint gives us another means of viewing an irreversible process, and in that respect is very useful.

The microscopic point of view is even more useful and informative when we consider the second law for this process. Since the system is isolated from its surroundings (no energy transfer), from a classical viewpoint the second law of thermodynamics can be written

$$\Delta S_{\substack{\text{isolated}\\ \text{system}}} \geq 0 \tag{4.27}$$

That is, the entropy change for an isolated system undergoing a spontaneous process can only increase, or, in the limit, for a completely reversible process, the entropy remains constant. We might add that Eq. 4.27 is a very general statement of the second law, because for any process it is possible to define an isolated system that includes everything affected by the process. This statement is also a most useful form of the second law, since it tells us the direction in which any process can proceed.

For an analysis from the microscopic viewpoint, let us imagine the system described in Figure 4.3 at the instant of time after the partition has been broken, before the gas has begun to expand. It is, at least in theory, possible that at the moment that the partition is ruptured, none of the molecules in the vicinity of the hole is moving toward the partition, and consequently all the gas will remain on the left side of the container for a period of time. If the hole is extremely small, of molecular dimensions, this might seem reasonable, but it certainly is unacceptable to us if the hole is large, as we assumed previously. That is, although there is a probability that all the gas will remain on the left side, that probability is so small that we do not expect to observe this behavior in nature. Thus it is overwhelmingly more probable that the gas will expand to fill the entire volume as this additional number of states becomes available to the system. Similarly, once the gas has filled the volume, there always exists the minute possibility that at some later time all the molecules will be found on the left side of the tank, but it is equally unlikely that this condition will exist for any finite period of time as it is that the gas would have remained on the left side when the partition ruptured, since the number of microscopic states resulting in such a macroscopic configuration is infinitesimal compared to the total number of states available.

Since the total thermodynamic probability for the state associated with the gas filling the volume is much greater than that for the initial state of the gas on the left side of the tank, we find from Eq. 4.15 that the spontaneous process occurring when the partition breaks corresponds to an increase in entropy, which is, of course, in agreement with the classical statement of the second law. The microscopic approach is also very informative, however, because it gives a different viewpoint from which to examine the second law of thermodynamics and, in particular, a more meaningful concept of the thermodynamic equilibrium state and the property entropy. For any change of state of an isolated system (during which there is an increase in entropy), a larger number of microscopic states is made available to the system, as for example by the removal of some internal inhibition or restriction on the system.

We should always bear in mind that the second law of thermodynamics is our attempt to describe systematically certain aspects of phenomena that we observe in nature. However, our description from the microscopic point of view involves a perspective quite different from that in our description from a macroscopic point of view, and provides very significant additional insight into the fundamental nature of the processes that give rise to the second law.

Before proceeding, we should mention that, although we have discussed only two specific examples here, a completely reversible expansion and an irreversible one with no work output, there are any number of degrees of irreversibility between these extremes. For example, in the first case, if the resisting force on the piston is reduced by a finite amount, then the process will proceed at a finite rate, and the system will not pass through a series of equilibrium states. The resulting process is irreversible to a certain extent, and the work done by the system on its surroundings will be some amount less than that for the reversible expansion. In such a case, the work and heat transfer are not in general given by Eqs. 4.12 and 4.13 since the process is not reversible, and the entropy will increase, although not as much as in the case of the completely irreversible expansion.

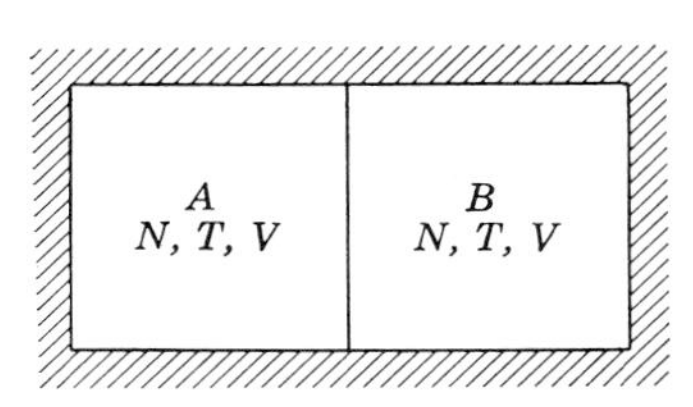

Figure 4.4 The mixing of two gases.

Another factor causing irreversibilities is the mixing of two different substances. Consider an insulated tank containing two ideal gases A and B, initially separated by a partition as shown in Figure 4.4. The number of molecules, N, of each gas is the same, and each gas initially occupies a volume V and is at temperature T. Gases A and B considered together constitute an isolated system, the entropy of which at the initial state 1 is

$$S_1 = S_{A_1} + S_{B_1} = k \ln w_{\mathrm{mp}_{A_1}} + k \ln w_{\mathrm{mp}_{B_1}} = k \ln (w_{\mathrm{mp}_{A_1}} \times w_{\mathrm{mp}_{B_1}}) \tag{4.28}$$

If the partition is now removed or broken, the two gases will gradually mix, and after a period of time the entire tank will be filled with a uniform mixture of A and B at a condition of equilibrium. The number of states consistent with the most probable distribution in the mixture at state 2 will be the product of those for the two components as they exist in the mixture. That is, as discussed in Section 3.7,

$$w_{\mathrm{mp}_{AB_2}} = w_{\mathrm{mp}_{A_2}} \times w_{\mathrm{mp}_B}$$

and the entropy of the mixture at state 2 is

$$S_2 = k \ln w_{\mathrm{mp}_{AB_2}} = k \ln (w_{\mathrm{mp}_{A_2}} \times w_{\mathrm{mp}_{B_2}}) \tag{4.29}$$

To determine the change in thermodynamic probability for A and B as a result of the mixing process, let us visualize the process in a manner similar to that used before. Immediately after the partition is removed, all the molecules of substance A are still on the left side of the tank and those of B on the right. Now there is certainly a possibility that A and B will remain for a period of time in this separated state, but we realize that, owing to the random motions of the molecules, the probability of the two gases mixing is much greater. That is,

$$\begin{aligned} w_{\text{mp}_{A_2}} &> w_{\text{mp}_{A_1}} \\ w_{\text{mp}_{B_2}} &> w_{\text{mp}_{B_1}} \end{aligned} \tag{4.30}$$

Using Eqs. 4.30 in comparing Eqs. 4.28 and 4.29, we find that the mixing process constitutes an increase in entropy, or

$$S_2 - S_1 > 0 \tag{4.31}$$

which is the result obtained from a classical analysis.

For this problem, it is important to realize that, if the same gas is initially on both sides of the partition, the preceding analysis is not valid. Once the partition is removed, we are no longer able to distinguish which molecules were initially on the left side and which were on the right. Consequently, the thermodynamic probability for the system after the partition is broken cannot be split into that associated with the molecules that started on the left and right sides, respectively. It equals the product of the initial values of w for the two parts, and there is no change in entropy.

In this section we have discussed irreversibility resulting from unrestrained expansion (boundary movement at a finite rate resulting from a finite pressure difference) and that due to mixing of different substances. In general, any change driven by a finite imbalance in a thermodynamic potential is irreversible. Since the force resisting the change is a finite amount less than the driving force, the process occurs at a finite rate. Consequently, the system does not pass through a series of equilibrium states and the process is not reversible. Other examples of irreversibility are processes involving friction, heat transfer through a finite temperature difference, and flow of electrical current through a finite resistance. A general study of irreversible processes will be left to Chapter 12.

4.5. The Third Law of Thermodynamics

We have noted in the last two sections that the classical definition of entropy is made only on a differential basis, or only in terms of a change between two states. It was shown that a change in the entropy as defined by Eq. 4.15 is consistent with the classical definition, but we realize that this would still be true if our definition contained an arbitrary constant of integration. Thus it would seem to be more general to define entropy by the expression

$$S = k \ln w_{\text{mp}} + S_0 \tag{4.32}$$

We should still keep in mind that, although it is more correct to use w_{tot} than w_{mp}, the two are essentially the same for our applications.

In an effort to determine the constant S_0, let us examine the behavior of a system at very low temperatures, values near absolute zero. As the temperature of a system is decreased, the system energy decreases accordingly as more and more of the particles drop into the lower energy levels. This is readily seen by examining the equilibrium distribution equation for any of the statistical models. Let us for the present assume corrected Boltzmann statistics and use the distribution equation to compare the number of particles N_0 in the ground energy level ϵ_0 to the number N_1 in the first higher level ϵ_1. From Eq. 4.1,

$$\frac{N_1}{N_0} = \frac{g_1}{g_0} e^{-(\epsilon_1 - \epsilon_0)/kT} \tag{4.33}$$

As the temperature approaches absolute zero, the ratio N_1/N_0 is seen to approach zero. Similarly, the ratio of the number of particles in any other level to that in the ground level approaches zero. That is, at a temperature of absolute zero, all the particles would be in the ground level of energy. Of course, the assumption of corrected Boltzmann statistics is at best only a crude approximation at extremely low temperature (the ratio of N_j/g_j could not be small for the heavily populated lower levels). In order to make an accurate analysis of the behavior of substances at low temperature, it is necessary to use the more general Bose-Einstein and Fermi-Dirac statistics, which will be discussed in Section 4.7.

To find the limiting value of entropy at absolute zero, we must determine the number of microstates consistent with the distribution in which all the particles are in the ground level; that is, the number of ways all the particles can be placed in the g_0 quantum states corresponding to the ground level. It is believed that the ground level is either non-degenerate ($g_0 = 1$) or has at most only a few states. That is, the thermodynamic probability w is either unity or very nearly so. Consequently, from Eq. 4.32,

$$\lim_{T \to 0} S = S_0 \tag{4.34}$$

It is of interest to compare this result with those of classical thermodynamics in an attempt to find a numerical value for S_0. First, we recall that, as a result of our definition of the absolute temperature scale, the absolute zero of temperature is not physically attainable. It is, instead, a limit. Our discussions have been in agreement with this. The ratio N_1/N_0, according to Eq. 4.33, is never identically zero; it would be so only in the limit as T goes to zero. The unattainability of absolute zero temperature is one way of stating the third law of thermodynamics. A far more useful form, however, concerns the nature of chemical reactions as the temperature approaches zero. It can be shown that the values of $(\Delta H/T)$ and $(\Delta G/T)$ for a reaction approach each other as the temperature becomes very small. From the definition of the Gibbs function, this requires that, for any chemical reaction,

$$\lim_{T\to 0} \Delta S = 0 \tag{4.35}$$

Therefore the entropies of all substances, ideally at least, must be the same at absolute zero. For convenience we choose the value†

$$S_0 = 0 \tag{4.36}$$

4.6. Thermodynamic Properties

In earlier sections of this chapter, we have developed expressions for the internal energy and entropy of a system of independent particles in terms of the partition function. For the internal energy, we have (Eq. 4.5)

$$U = NkT^2\left(\frac{\partial \ln Z}{\partial T}\right)_V$$

For the entropy, we have (Eq. 4.19)

$$S = k \ln Z_N + \frac{U}{T}$$

where, in accordance with Eq. 4.20,

$$Z_N = \frac{(Z)^N}{N!}$$

By using Stirling's formula for $N!$, Eq. 4.19 can also be written in the form

$$S = Nk\left[\ln \frac{Z}{N} + 1\right] + \frac{U}{T} \tag{4.37}$$

As mentioned at the beginning of this chapter, it is also of interest to express other thermodynamic properties in terms of the partition function. From the definition of the Helmholtz function and Eq. 4.37, we can write

$$A = U - TS = -NkT\left[\ln \frac{Z}{N} + 1\right] \tag{4.38}$$

Now suppose that we write the thermodynamic property relation for a pure substance in terms of the Helmholtz function. For the simple compressible substance (i.e., in the absence of tension, surface effects, electric

† The selection of this reference value for S_0 precludes a contribution due to nuclear spin. A special problem, for which such a contribution must be included, is discussed in Section 8.6.

and magnetic fields), this relation is

$$dA = -S\,dT - P\,dV + g\,dn \tag{4.39}$$

where g is the Gibbs function per mole. Consequently, we see from this expression that

$$g = \left(\frac{\partial A}{\partial n}\right)_{T,V} \tag{4.40}$$

But, from Eq. 4.7, since N_0 is a constant,

$$dn = n\frac{dN}{N} \tag{4.41}$$

Therefore

$$G = ng = N\left(\frac{\partial A}{\partial N}\right)_{T,V} = -NkT\left[\ln\frac{Z}{N} + 1\right] - N^2kT\left[\frac{N}{Z}\left(\frac{-Z}{N^2}\right)\right]$$

$$= -NkT\ln\frac{Z}{N} \tag{4.42}$$

From the definitions of the Gibbs and Helmholtz functions,

$$PV = G - A = NkT \tag{4.43}$$

which is the equation of state for an ideal gas.

Using the definition of enthalpy along with Eqs. 4.5 and 4.43, we can express that property as

$$H = U + PV = NkT\left[T\left(\frac{\partial \ln Z}{\partial T}\right)_V + 1\right] \tag{4.44}$$

There is another development of interest in regard to the equation of state. Suppose we consider a system in which the only significant boundary parameter is the volume. Then, for a small change occurring along a reversible path,

$$\delta W_{\text{rev}} = P\,dV = -\sum_j N_j\,d\epsilon_j \tag{4.45}$$

But, since the system passes through a series of equilibrium states, the equilibrium distribution equation, Eq. 4.1, may be substituted into Eq. 4.45, giving

$$P\,dV = -\frac{N}{Z}\sum_j g_j e^{-\epsilon_j/kT}\,d\epsilon_j \tag{4.46}$$

Note that the summation in this equation is the same as that obtained by differentiating Z at constant temperature. If we do this,

$$dZ = -\frac{1}{kT}\sum_j g_j e^{-\epsilon_j/kT}\,d\epsilon_j \tag{4.47}$$

we find that, at constant temperature,

$$P\,dV = \frac{NkT}{Z}\,dZ = NkT\,d\ln Z \tag{4.48}$$

Therefore, for a system at equilibrium,

$$P = NkT\left(\frac{\partial \ln Z}{\partial V}\right)_T \tag{4.49}$$

Since this development has been made for a system of independent particles, it follows that Eq. 4.49 must be identical with Eq. 4.43. A comparison of the two indicates that, for an ideal gas,

$$Z = V \times f(T) \tag{4.50}$$

This is, in fact, the result that was found in Eq. 3.62,

$$Z = V\left(\frac{2\pi mkT}{h^2}\right)^{3/2}$$

when the partition function was evaluated for the special case of a classical, monatomic ideal gas.

In practical applications, especially in tabulating the properties of a substance for use in thermodynamic calculations, it is convenient to write the equations for the various thermodynamic properties on a unit gram-mole basis as was done previously for the internal energy. The internal energy per mole was found to be (Eq. 4.8)

$$u = \frac{U}{n} = RT^2\left(\frac{\partial \ln Z}{\partial T}\right)_V$$

Using Eqs. 4.6 and 4.7, the equation of state can be written in the form

$$Pv = RT \tag{4.51}$$

where the specific volume is defined as

$$v = \frac{V}{n} \tag{4.52}$$

From Eq. 4.44, the enthalpy per mole is

$$h = \frac{H}{n} = RT\left[T\left(\frac{\partial \ln Z}{\partial T}\right)_V + 1\right] \tag{4.53}$$

Although this equation could be used to calculate values of enthalpy, it may prove more convenient in practice simply to calculate this quantity from the relation

$$h = u + Pv$$

after first determining the internal energy. The same remarks apply to the determination of specific heats. Equation 4.9 gives an expression for C_v in terms of the partition function. However, in many cases, it is more convenient to calculate C_v directly from its definition,

$$C_v = \left(\frac{\partial u}{\partial T}\right)_V$$

Similarly, the constant-pressure specific heat

$$C_p = \left(\frac{\partial h}{\partial T}\right)_P \tag{4.54}$$

is usually found from the enthalpy according to Eq. 4.54 rather than from an expression for C_p in terms of the partition function.

In the same manner, the equations for the other properties on a mole basis are found to be

$$s = \frac{S}{n} = R\left[\ln\frac{Z}{N} + 1\right] + \frac{u}{T} \tag{4.55}$$

$$a = -RT\left[\ln\frac{Z}{N} + 1\right] \tag{4.56}$$

and

$$g = -RT\ln\frac{Z}{N} \tag{4.57}$$

While the partition function Z is a function of volume, the internal energy and enthalpy are not, since the derivative of $\ln Z$ while V is held constant is independent of volume. However, in Eqs. 4.55 to 4.57, we must be careful that the values of Z and N used are both consistent with the volume of the system being examined. As is commonly the case, the system is considered to be one mole of gas. Consequently, the partition function is found by using the specific volume, and the N used in these equations is Avogadro's number.

While for purposes of calculation the properties are most conveniently expressed on a mole basis, we can also express these quantities in terms of the system partition function Z_N. From Eq. 4.20 at constant N (i.e., constant mass),

$$d\ln Z_N = N\,d\ln Z \tag{4.58}$$

so that the internal energy can be given in the form

$$U = kT^2\left(\frac{\partial \ln Z_N}{\partial T}\right)_V \tag{4.59}$$

Similarly, from Eq. 4.49,

$$P = kT\left(\frac{\partial \ln Z_N}{\partial V}\right)_T \tag{4.60}$$

Since the entropy has previously been given in terms of the function Z_N, the Helmholtz function is, from Eq. 4.19,

$$A = U - TS = -kT \ln Z_N \tag{4.61}$$

In expressing the Gibbs function and enthalpy, we could use for the product PV either the equation of state, Eq. 4.43, or the expression given by Eq. 4.60. In terms of the latter,

$$G = A + PV = -kT\left[\ln Z_N + V\left(\frac{\partial \ln Z_N}{\partial V}\right)_T\right] \tag{4.62}$$

and

$$H = U + PV = kT\left[T\left(\frac{\partial \ln Z_N}{\partial T}\right)_V + V\left(\frac{\partial \ln Z_N}{\partial V}\right)_T\right] \tag{4.63}$$

Now the interesting point regarding the development of properties in terms of the system partition function is that, while we have derived Eqs. 4.59 to 4.63 strictly under the assumption of ideal gas or independent particles, we shall find in Chapter 10 that these particular equations are in fact valid as well for systems of dependent particles, that is, for real substances. For real substances, however, the system partition function Z_N is not given by Eq. 4.20, and most unfortunately becomes a very difficult quantity to evaluate. The analysis of the behavior of real substances is the subject of Chapter 11.

4.7. Bose-Einstein and Fermi-Dirac Statistics

While considering the laws and functions of thermodynamics from a microscopic point of view, it is of interest to develop briefly the more general Bose-Einstein and Fermi-Dirac statistics. We shall not devote much time to these models here because we have found that, in the majority of cases of interest to us, each reduces to the much simpler common form that we have termed corrected Boltzmann statistics. However, since at very low temperatures the assumption that $g_j \gg N_j$ is not reasonable, it is important to determine the behavior of these more general models near absolute zero. We shall also find that the expressions for thermodynamic properties for these models are not easily simplified, as are those for corrected Boltzmann statistics.

For Bose-Einstein statistics, the equilibrium distribution equation obtained by maximizing the thermodynamic probability for given values of N, U, V was found to be (Eq. 3.24)

$$N_j = \frac{g_j}{e^{\alpha} e^{\beta \epsilon_j} - 1}$$

It was shown in Chapter 3 that in the special case of corrected Boltzmann statistics the multiplier β is equal to $1/kT$, and in the present chapter it was found that entropy defined in terms of w and using the relation $\beta = 1/kT$ is consistent with the entropy definition of classical thermodynamics from the second law. Therefore let us assume that this identification of β with the temperature according to Eq. 3.54 is valid in general, so that the Bose-Einstein distribution equation can be expressed as

$$N_j = \frac{g_j}{e^{\alpha} e^{\epsilon_j/kT} - 1} \tag{4.64}$$

To determine the entropy from Bose-Einstein statistics, we use the expression for thermodynamic probability given in Eq. 3.20:

$$\ln w = \sum_j [(g_j + N_j) \ln (g_j + N_j) - g_j \ln g_j - N_j \ln N_j]$$

It is more convenient for our purposes to write the equation in the form

$$\ln w = \sum_j \left[g_j \ln \left(1 + \frac{N_j}{g_j}\right) + N_j \ln \left(1 + \frac{g_j}{N_j}\right) \right] \tag{4.65}$$

For the most probable distribution, or macroscopic equilibrium state, we substitute Eq. 4.64 into Eq. 4.65 and obtain

$$\ln w_{\text{mp}} = \sum_j \left[g_j \ln (1 - e^{-\alpha} e^{-\epsilon_j/kT})^{-1} + N_j \left(\alpha + \frac{\epsilon_j}{kT}\right) \right]$$

$$= - \sum_j g_j \ln (1 - e^{-\alpha} e^{-\epsilon_j/kT}) + \alpha N + \frac{U}{kT} \tag{4.66}$$

Using Eq. 4.66, we find that the entropy is

$$S = k \ln w_{\text{mp}} = -k \sum_j g_j \ln (1 - e^{-\alpha} e^{-\epsilon_j/kT}) + \alpha N k + \frac{U}{T} \tag{4.67}$$

and, from the definition of the Helmholtz function,

$$A = U - TS = kT \sum_j g_j \ln (1 - e^{-\alpha} e^{-\epsilon_j/kT}) - \alpha N k T \tag{4.68}$$

We find that the expressions for these properties are not given in terms of the partition function as defined by Eq. 3.55. By using the logarithmic series, it is easy to show that, when the ratio N_j/g_j is much less than unity, the equations above reduce to those found previously for corrected Boltzmann statistics. This demonstration is left as an exercise for the student.

To find the Gibbs function, we approach the problem as before by making use of the thermodynamic property relation. From Eqs. 4.40 and 4.41,

$$G = N\left(\frac{\partial A}{\partial N}\right)_{T,V} \tag{4.69}$$

Before differentiating the equation for the Helmholtz function, however, we recall that in the special case it was found that α is given by Eq. 3.56, or

$$\alpha = \ln \frac{Z}{N} = f(N, T, V) \tag{4.70}$$

Therefore we assume that the multiplier α is some function of N, T, V in the general case also. Now, from Eqs. 4.68 and 4.69,

$$\begin{aligned} G &= N\left[kT \sum_j \frac{g_j}{e^{\alpha}e^{\epsilon_j/kT} - 1}\left(\frac{\partial \alpha}{\partial N}\right)_{T,V} - NkT\left(\frac{\partial \alpha}{\partial N}\right)_{T,V} - \alpha kT\right] \\ &= N\left[kT \sum_j N_j\left(\frac{\partial \alpha}{\partial N}\right)_{T,V} - NkT\left(\frac{\partial \alpha}{\partial N}\right)_{T,V} - \alpha kT\right] \\ &= -\alpha NkT \end{aligned} \tag{4.71}$$

Therefore

$$\alpha = -\frac{G}{NkT} \tag{4.72}$$

Previously, the Gibbs function for corrected Boltzmann statistics was found to be given by Eq. 4.42. From that equation and Eq. 4.70,

$$G = -NkT \ln \frac{Z}{N} = -NkT\alpha \tag{4.73}$$

which is identical with Eq. 4.72 for Bose-Einstein statistics.

Since it is desirable to eliminate the multiplier α from our distribution equation, Eq. 4.72 can be substituted into Eq. 4.64, and the Bose-Einstein distribution equation becomes

$$N_j = \frac{g_j}{e^{(\epsilon_j - G/N)/kT} - 1} \tag{4.74}$$

The development for the general Fermi-Dirac statistics will not be discussed here, because the method is entirely analogous to that for the

Bose-Einstein statistics. The details are left as an exercise. When the development is carried through, it is found that the entropy is expressed as

$$S = k \sum_j g_j \ln (1 + e^{-\alpha} e^{-\epsilon_j/kT}) + \alpha N k + \frac{U}{T} \tag{4.75}$$

and also that α is given by Eq. 4.72, as was found for Bose-Einstein statistics (and the corrected Boltzmann model as well). Consequently, the distribution equation for Fermi-Dirac statistics is conveniently written in the form

$$N_j = \frac{g_j}{e^{(\epsilon_j - G/N)/kT} + 1} \tag{4.76}$$

Equations for other thermodynamic properties are not presented here since, as for the entropy, they do not reduce to a convenient form. Evaluation of properties for the general Bose-Einstein and Fermi-Dirac statistical models is difficult for this reason.

As mentioned above, it is of interest to examine the two general models as the temperature approaches absolute zero. The third law was discussed in Section 4.5 considering only corrected Boltzmann statistics, which is not really a valid assumption in this region. For Bose-Einstein statistics, let us take $\epsilon_0 = 0$ and find the ratio of the number of particles in any level j to that in the ground level from the distribution equation in the form of Eq. 4.64. For any value of $j > 0$,

$$\frac{N_j}{N_0} = \frac{g_j}{g_0}\left[\frac{e^{\alpha} - 1}{e^{\alpha} e^{\epsilon_j/kT} - 1}\right] \tag{4.77}$$

If e^{α} approaches unity at low temperature, which is known to be the case for corrected Boltzmann statistics, then, since the ratio g_j/g_0 is finite, it is apparent from Eq. 4.77 that

$$\lim_{T \to 0} \left(\frac{N_j}{N_0}\right) = 0 \tag{4.78}$$

for any $j > 0$. That is, at the limiting temperature of absolute zero, all the particles would be in the ground level of energy, the same result as was found in Section 4.5. That e^{α} approaches unity as T approaches zero is in fact reasonable for Bose-Einstein statistics is seen by writing the distribution equation for the ground energy level $\epsilon_0 = 0$.

$$N_0 = \frac{g_0}{e^{\alpha} - 1} \tag{4.79}$$

As was discussed previously, the statistical weight of the ground level is known to be a small number (perhaps equal to unity). As the temperature is decreased, N_0 becomes increasingly larger, such that e^α must approach unity at very low temperature.

To determine the behavior of entropy for a Bose-Einstein system as the temperature goes to zero, let us examine the original expression for w for this model, Eq. 3.2. At the limit of T equal to zero, there is only one possible distribution among energy levels, namely that all the particles are in the ground level, and, for this distribution,

$$w = \frac{(g_0 + N - 1)!}{(g_0 - 1)!N!} \tag{4.80}$$

If the ground level is non-degenerate, then g_0 is equal to unity, w is identically one, and the entropy is exactly zero according to Eq. 4.15. Even if this is not exact, it is nevertheless true that g_0 is a small number, so that w is small, in which case the entropy is still very nearly equal to zero. This is in agreement with our earlier discussion of the third law.

For a system obeying Fermi-Dirac statistics, the behavior at temperatures approaching absolute zero is entirely different from the Bose-Einstein system, since in this case there can be no more than one particle per quantum state. The ground energy level will be completely filled by g_0 particles, the first level above that by g_1 particles, etc. Therefore, in this system at absolute zero, the particles would assume the lowest possible energy states, but could not all crowd into the ground level. This behavior can be seen by examining the distribution equation, Eq. 4.76, at low temperature. As the temperature approaches absolute zero, the Gibbs function per particle G/N, also called the chemical potential μ, approaches a zero-point value μ_0. For all the energy levels for which ϵ_j is less than μ_0,

$$\lim_{T \to 0} (e^{(\epsilon_j - \mu_0)/kT}) = e^{-\infty} = 0 \tag{4.81}$$

so, that for these levels,

$$\lim_{T \to 0} \left(\frac{N_j}{g_j}\right) = 1 \tag{4.82}$$

That is, all these levels are completely filled. On the other hand, for those levels for which ϵ_j is greater than μ_0,

$$\lim_{T \to 0} [e^{(\epsilon_j - \mu_0)/kT}] = e^{+\infty} \tag{4.83}$$

and

$$\lim_{T \to 0} \left(\frac{N_j}{g_j}\right) = 0 \tag{4.84}$$

That is, all energy levels above an energy equal to μ_0 are empty at absolute zero. This limiting distribution for Fermi-Dirac statistics at absolute zero

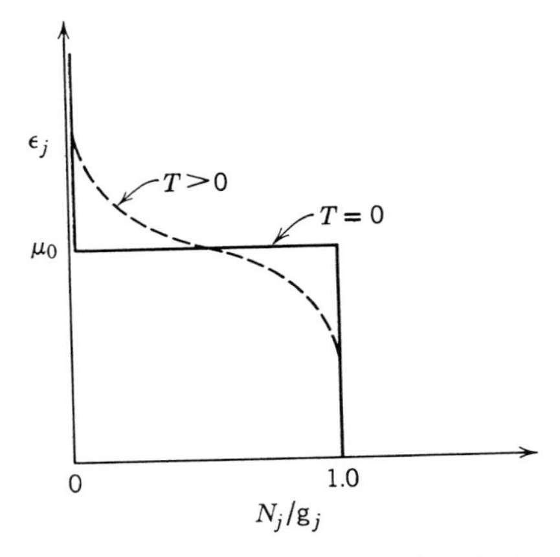

Figure 4.5 The Fermi-Dirac distribution.

temperature is depicted in Figure 4.5. The separating value of energy μ_0, below which all levels are filled and above which all levels are empty at absolute zero temperature, is known as the Fermi energy.

We also note that, as the temperature is increased, some particles shift into the higher energy levels, and the distribution changes to that shown by the broken line in Figure 4.5. As the temperature continues to increase, the upper portion of the distribution curve approaches the exponential function predicted by corrected Boltzmann statistics.

The internal energy and Gibbs function at absolute zero can be found,† since

$$\sum_j g_j \geq \sum_j N_j = N \tag{4.85}$$

where the summation extends over those levels containing particles. Since N and the various g_j's are known, Eq. 4.85 specifies the maximum value of j to be considered. Note that we have used the $>$ sign because it is possible that the uppermost level containing particles at zero temperature ($\epsilon_j = G/N$) will not be completely filled. This is certainly of minor importance, however, since the numbers become very large.

To evaluate the entropy at absolute zero, we use the equation for thermodynamic probability for Fermi-Dirac statistics, Eq. 3.3,

$$w = \prod_j \frac{g_j!}{(g_j - N_j)!N_j!}$$

† This problem will be considered in Section 7.4, in terms of the behavior of the conduction electrons in a metal.

For all levels of energy less than μ_0, N_j equals g_j and the corresponding term in Eq. 3.3 is unity. For energy levels of magnitude greater than μ_0, N_j equals zero, so that all these terms are also equal to one. Now, if the uppermost level containing particles is full, then w for the distribution at absolute zero is identically one and the entropy is zero as found before for the corrected Boltzmann case. Or, if there are not enough particles to fill the level $\epsilon_j = \mu_0$ completely, then for this level $N_j < g_j$, and the contribution to w for that level and consequently w for the entire distribution will be some number slightly greater than unity. However, the product of the logarithm of this number and the Boltzmann constant will be extremely small, so that the entropy even in this case is very nearly zero.

While the discussions of this section regarding the behavior of systems at temperatures approaching absolute zero are certainly more realistic than the discussion in Section 4.5 assuming corrected Boltzmann statistics, it must be kept in mind that we have assumed throughout a system of independent particles. Strictly speaking, then, our discussion should be limited to systems at very low pressure.

4.8. Entropy from Information Theory

In this section, we present the basic concepts of information theory, a subject of current interest in the field of communication. It is not our purpose here to present a thorough development of this subject, but only to discuss the concept of communication entropy and its relation to statistical thermodynamics. This should then provide us with still another means for viewing entropy, and further understanding of this property.

Any attempt to investigate the problem of the conveyance of information through the sending and receiving of messages will necessarily be intimately linked with the concepts of conditional probability, which has been discussed in some detail in Chapter 2. Suppose that we, as the potential receiver of some message, do not know the outcome of a particular experiment or the exact state of some system of interest to us. That is, we have a complete lack of knowledge about a certain event. Let us use as a simple example the result of a single toss of a die. If we have no knowledge of the result (or of the fairness of the die), we would assign a probability of 1/6 to each of the possible outcomes, 1, 2, . . . , 6. If we now receive a message concerning the outcome of the toss, that message may contain information that may enable us to eliminate some of the six possibilities, such that we could then reassign probabilities to those remaining possible outcomes. We must therefore define some rational scale of information that will enable us to compare the informational content of various messages. If all the possible outcomes have equal probabilities, then a logical choice for such a scale should be in terms of the ratio of the probability of an outcome after the message is received to the probability of an outcome before the message. Let us define information I_A conveyed by a message A concerning event j by the expression

$$I_A = K \ln \frac{P(j/A)}{P(j)} \tag{4.86}$$

where $P(j)$ is the probability of one of the equally likely outcomes before we receive the message, and $P(j/A)$ is the probability of one of the possible outcomes given the information of message A. The constant K and absolute magnitude of I_A are arbitrary and may be fixed in particular applications.

In defining a scale for information, we have used the logarithm for convenience, such that this quantity will have properties consistent with our feeling for the word information, as will be seen in this section. Of course, if message A contains no information that permits us to reassign probabilities, then

$$P(j/A) = P(j)$$

and the information conveyed is, from Eq. 4.86, equal to zero. We also realize that the maximum possible information is conveyed by a message if we are told the result, that is, if the system state is exactly specified. Given this information, we know that, for the only possible outcome,

$$P(j/A) = 1$$

and the information for such a message is

$$I_{\max} = K \ln \frac{1}{P(j)} = -K \ln P(j) \tag{4.87}$$

In the toss of a die, if we receive a message stating that the result is a 2, for example, the information conveyed by that message is the maximum for the experiment or event and is equal to

$$I_{\max} = -K \ln (1/6) = +K \ln 6$$

The information included in any other message that eliminates some, but not all, of the six possible results could be compared with this value for the purpose of determining the effectiveness of communication.

Since the amount of information conveyed by a message that exactly specifies the system state or outcome of an experiment is given by Eq. 4.87, that expression must also represent the uncertainty of the receiver regarding the outcome before the message was received. Let us then define uncertainty by this expression for the situation in which all possible outcomes are equally likely, or

$$\mathscr{U} = -K \ln P(j) \tag{4.88}$$

where $\mathscr{U}$ is associated with a lack of knowledge about an event or system. Of course, if we know the exact state, then $P(j) = 1$, and our uncertainty is zero. If we do not know the exact state, then $0 < P(j) < 1$, so that uncertainty is always greater than, or equal to, zero.

In general, a message gives us a certain amount of information but does not remove all doubt regarding the event in question. Although the message does in general reduce our uncertainty about the event, it does not necessarily reduce it to zero. In such a case, we see from Eqs. 4.86 and 4.88 that the information conveyed by a message is equal to the decrease in uncertainty of the receiver, or

$$I = \mathscr{U}_1 - \mathscr{U}_2 = -K \ln P(j)_1 + K \ln P(j)_2 = K \ln \frac{P(j)_2}{P(j)_1} \tag{4.89}$$

where states 1 and 2 are the initial and final states of the receiver. The probability $P(j)_2$ is merely a convenient way of expressing the quantity $P(j/A)$ used before.

Consider again the experiment of tossing a die. Given no information about the result, our uncertainty in the initial state 1 is

$$\mathscr{U}_1 = -K \ln (1/6) = K \ln 6$$

Suppose that we receive a message A which states "the result is a number less than 6." As a result of this message, we know that the outcome must be one of the numbers 1, 2, 3, 4, 5. We now assign each of these a probability of 1/5, and our uncertainty is now

$$\mathscr{U}_2 = -K \ln (1/5) = K \ln 5$$

which is less than that at the initial state. The information received from message A is

$$I_A = \mathscr{U}_1 - \mathscr{U}_2 = K \ln (6/5)$$

Now consider a second message B which states "the result is an even number." Using our prior knowledge of the event received from message A, we know that the result must be either a 2 or a 4, and we assign a probability of 1/2 to each of these possibilities. Our uncertainty has now been reduced from $K \ln 5$ to

$$\mathscr{U}_3 = -K \ln (1/2) = K \ln 2$$

and the information of message B, given A, is

$$I_{B/A} = \mathscr{U}_2 - \mathscr{U}_3 = K \ln (5/2)$$

The information conveyed by the succession A and B is

$$I_{AB} = \mathscr{U}_1 - \mathscr{U}_3 = K \ln 3$$

or

$$I_{AB} = I_A + I_{B/A} \tag{4.90}$$

On the other hand, suppose that message B had been received before A. The uncertainty in the receiver at state $2'$ after message B only would be

$$\mathscr{U}_{2'} = -K \ln (1/3) = K \ln 3$$

since the possible results are limited here to the numbers 2, 4, and 6. The information associated with the message is

$$I_B = \mathscr{U}_1 - \mathscr{U}_{2'} = K \ln 2$$

which is found to be different from its informational content if A were already known. Of the possible results 2, 4, 6, the number 6 is eliminated by reception of message A, after which

$$\mathscr{U}_{3'} = -K \ln (1/2) = K \ln 2$$

and we find

$$\mathscr{U}_{3'} = \mathscr{U}_3 \tag{4.91}$$

The information in message A when it follows B is

$$I_{A/B} = \mathscr{U}_{2'} - \mathscr{U}_{3'} = K \ln (3/2)$$

and that of the succession B and then A is

$$I_{BA} = \mathscr{U}_1 - \mathscr{U}_{3'} = K \ln 3$$

or

$$I_{BA} = I_B + I_{A/B} \tag{4.92}$$

We also note that

$$I_{BA} = I_{AB} \tag{4.93}$$

and we can conclude that the order in which the messages are received makes no difference in the total information conveyed. We also see from Eq. 4.91 that the order makes no difference in the uncertainty in the receiver at the final state.

Let us now consider the more general case in which the various probabilities are not necessarily all the same. It is convenient here to think in terms of messages using symbols of the English language. We are aware that, in normal usage, the various letters of the English alphabet do not appear with equal frequencies. The probabilities of these letters are given in Table 4.1. Before we receive any particular symbol in a transmitted message, we have some uncertainty about what that symbol will be (not considering our knowledge of those already received). As the symbol is received, it conveys to us a certain amount of information. If it happens to be an A, the information given us is

$$I = -K \ln 0.063 = 2.76K$$

whereas, if it had been an X, the information would have been greater:

$$I = -K \ln 0.002 = 6.21K$$

since the letter X does not normally appear so frequently as does the letter A. In a long message, where the probability of some symbol j is $P(j)$, the average information conveyed per symbol is, using the definition of a mean or average value from Eq. 2.31,

$$\frac{I}{N} = -K \sum_{j} P(j) \ln P(j) \tag{4.94}$$

which is known as Shannon's formula. This quantity is also seen to be the average uncertainty per symbol about the message before its reception.

TABLE 4.1

Relative Frequencies of Appearance of Letters in Printed English

Symbol	Probability	Symbol	Probability
Blank space	0.198	U	0.022
E	0.105	M	0.021
T	0.072	P	0.017
O	0.065	Y	0.012
A	0.063	W	0.012
N	0.059	G	0.011
I	0.055	B	0.010
R	0.054	V	0.008
S	0.052	K	0.003
H	0.047	X	0.002
D	0.035	J	0.001
L	0.029	Q	0.001
C	0.023	Z	0.001
F	0.022		

Naturally, in the field of communications, a problem of great importance is the maximization of informational content in a message of a given number of symbols. If the expression in Eq. 4.94 is maximized subject to the constraint of constant N, it is found that the maximum average information per symbol results when the given symbols appear with equal frequencies. This is certainly not the case with the English language, as seen from Table 4.1.

The feature of information theory that is of interest here is that the quantity represented by Eq. 4.94 is called the entropy of the signal or message source. In the development of the expression for thermodynamic entropy defined by Eq. 4.15, if we had expressed the thermodynamic probability for Boltzmann statistics in terms of the numbers of particles in each quantum state instead of evaluating over energy levels having statistical weights, then the value of w could be given by

$$w = \frac{N!}{\prod_i N_i!} \tag{4.95}$$

where the symbol Π_i refers to the product over all states, and not energy levels. In terms of the logarithm, Eq. 4.95 becomes

$$\ln w = \ln N! - \sum_i \ln N_i!$$

But, using Stirling's formula under the assumption that the N_i for those states contributing significantly to $\ln w$ are relatively large,

$$\ln w = N \ln N - N - \sum_i N_i \ln N_i + \sum_i N_i = - \sum_i N_i \ln \frac{N_i}{N} \tag{4.96}$$

It is found that

$$k \ln w = -Nk \sum_i P(i) \ln P(i) \tag{4.97}$$

where

$$P(i) = \frac{N_i}{N} \tag{4.98}$$

the fraction of the particles in the various states. Equation 4.97 is the thermodynamic entropy if the $P(i)$ are distributed according to the most probable distribution for the given number of particles and energy of the system. While we assumed Boltzmann statistics here, the same result, Eq. 4.97, will be found in Chapter 10 for the general case of the canonical ensemble when we consider systems of dependent particles, in which case the N_i represent small systems instead of individual particles.

We find, by comparing Eqs. 4.94 and 4.97, a great similarity between the entropy of thermodynamics and that of information theory. In addition to the identity in form of the mathematical expressions, we recall that communication entropy is a measure of the uncertainty about the message before it is received, while the thermodynamic entropy is a measure of microscopic disorder, or uncertainty about the microscopic state of the thermodynamic system. We now realize that entropy in terms of uncertainty is a general and basic concept in science, but it must always be kept in mind that its evaluation and subsequent utility depend upon the constraints placed on the system in the particular field of investigation.

4.9. The Properties of an Ideal Gas Mixture

In Section 3.7, we developed the expressions for the most probable or thermodynamic equilibrium distribution for a system consisting of a mixture of A and B, under the assumption of independent particles and corrected Boltzmann statistics. We found there that the equilibrium distribution equations are (Eq. 3.92 and 3.93)

$$N_{A_j} = \frac{N_A g_{A_j} e^{-\epsilon_{A_j}/kT}}{Z_A} \qquad \text{for all } A_j$$

$$N_{B_j} = \frac{N_B g_{B_j} e^{-\epsilon_{B_j}/kT}}{Z_B} \qquad \text{for all } B_j$$

where the partition functions Z_A, Z_B were defined by Eqs. 3.89 and 3.91,

$$Z_A = \sum_{Aj} g_{A_j} e^{-\epsilon_{A_j}/kT}$$

$$Z_B = \sum_{Bj} g_{B_j} e^{-\epsilon_{B_j}/kT}$$

each at the temperature and volume of the system. The internal energy of the mixture is (Eq. 3.78)

$$U = U_A + U_B = \sum_{Aj} N_{A_j}\epsilon_{Aj} + \sum_{Bj} N_{B_j}\epsilon_{B_j}$$

But, substituting Eqs. 3.92 and 3.93 for N_{A_j} and N_{B_j}, and by the same procedure as in Section 4.1, we have for U_A

$$U_A = \sum_{A_j} N_{A_j}\epsilon_{A_j} = \frac{N_A}{Z_A}\sum_{A_j} g_{A_j}\epsilon_{A_j} e^{-\epsilon_{A_j}/kT}$$

$$= \frac{N_A kT^2}{Z_A}\left(\frac{\partial Z_A}{\partial T}\right)_V = n_A RT^2\left(\frac{\partial \ln Z_A}{\partial T}\right)_V \tag{4.99}$$

and similarly for U_B,

$$U_B = \sum_{Bj} N_{B_j}\epsilon_{B_j} = n_B RT^2\left(\frac{\partial \ln Z_B}{\partial T}\right)_V \tag{4.100}$$

Substituting these relations into Eq. 3.78 and using Eq. 4.8, we find that the internal energy of the system is

$$U = n_A RT^2\left(\frac{\partial \ln Z_A}{\partial T}\right)_V + n_B RT^2\left(\frac{\partial \ln Z_B}{\partial T}\right)_V = n_A u_A + n_B u_B \tag{4.101}$$

where u_A is the molal internal energy for substance A as though it existed alone in volume V at temperature T, while u_B is the corresponding value for substance B. This conclusion is in accord with the corresponding analysis made for an ideal gas mixture from the viewpoint of classical thermodynamics.

The entropy of a system at equilibrium is defined by Eq. 4.15,

$$S = k \ln w_{\text{mp}}$$

Therefore, substituting Eqs. 3.92 and 3.93 for the ratios g_{A_j}/N_{A_j} and g_{B_j}/N_{B_j}, which correspond to the most probable distributions, into Eq. 3.80, we obtain

$$
\begin{aligned}
\ln w_{\mathrm{mp}} &= N_A + \sum_{Aj} N_{A_j} \ln\left(\frac{Z_A}{N_A}\, e^{+\epsilon_{A_j}/kT}\right) + N_B + \sum_{Bj} N_{B_j} \ln\left(\frac{Z_B}{N_B}\, e^{+\epsilon_{B_j}/kT}\right) \\
&= N_A + \ln\frac{Z_A}{N_A}\sum_{Aj} N_{A_j} + \frac{1}{kT}\sum_{Aj} N_{A_j}\epsilon_{A_j} \\
&\qquad + N_B + \ln\frac{Z_B}{N_B}\sum_{Bj} N_{B_j} + \frac{1}{kT}\sum_{Bj} N_{B_j}\epsilon_{B_j} \\
&= N_A + N_A \ln\frac{Z_A}{N_A} + \frac{U_A}{kT} + N_B + N_B\ln\frac{Z_B}{N_B} + \frac{U_B}{kT} \\
&= N_A\left[\ln\left(\frac{Z_A}{N_A}\right) + 1\right] + \frac{U_A}{kT} + N_B\left[\ln\left(\frac{Z_B}{N_B}\right) + 1\right] + \frac{U_B}{kT}
\end{aligned}
\tag{4.102}
$$

If we now substitute Eq. 4.102 into Eq. 4.15, we obtain for the entropy of the mixture

$$
\begin{aligned}
S &= N_A k\left[\ln\left(\frac{Z_A}{N_A}\right) + 1\right] + \frac{U_A}{T} + N_B k\left[\ln\left(\frac{Z_B}{N_B}\right) + 1\right] + \frac{U_B}{T} \\
&= n_A\left\{R\left[\ln\left(\frac{Z_A}{N_A}\right) + 1\right] + \frac{u_A}{T}\right\} + n_B\left\{R\left[\ln\left(\frac{Z_B}{N_B}\right) + 1\right] + \frac{u_B}{T}\right\}
\end{aligned}
\tag{4.103}
$$

or, using Eq. 4.55, we can put this expression in the form

$$s = n_A s_A + n_B s_B \tag{4.104}$$

In Eq. 4.104, s_A is the entropy of substance A evaluated as though it exists alone in the system volume V at temperature T, and s_B is the corresponding value for substance B. We find again that the result is in accord with an analysis of an ideal gas mixture made from the classical viewpoint.

Expressions for the other thermodynamic properties of interest for ideal gas mixtures, C_p, C_v, H, A, G, follow readily from their definitions and the relations for internal energy, entropy, and the ideal gas equation of state.

PROBLEMS

4.1 Using the expression for Z given by Eq. 3.62, calculate the constant-volume and constant-pressure specific heats for a classical, monatomic ideal gas Determine the ratio C_p/C_v.

4.2 The entropy of carbon dioxide at 25 C, 1 atm pressure, is 51.072 cal/mole-K. Estimate the thermodynamic probability for the most probable distribution.

4.3 In Chapter 3, it was demonstrated that $\beta = 1/kT$ is valid for a classical monatomic gas, after which the relation was presumed to be valid in general.

(*a*) Starting with the partition function and equilibrium distribution equation both in terms of β (instead of $1/kT$), derive an expression for internal energy as a function of β analogous to the development of Section 3.1.

(*b*) Carry out the development of Section 3.3 in terms of β to find an expression for entropy as a function of β. Without assuming a classical monatomic gas, show that, in order that $dS = (\delta Q/T)_{\text{rev}}$, β must be $1/kT$.

4.4 The most probable distribution in a thermodynamic system has been presumed to be extremely more likely than any distribution differing from it by a distinguishable amount. For a volume V containing N particles at energy U, consider a distribution d for which $N_j = N_{j_{\text{mp}}} + \delta N_j$, where $|\delta N_j| \ll N_{j_{\text{mp}}}$. Starting with Eq. 4.16, show that the entropy for such a distribution must be less than that corresponding to the most probable distribution.

4.5 Consider 1 mole of gas having a distribution d for which

$$\frac{|\delta N_j|}{N_{j_{\text{mp}}}} = 0.001\,\%$$

for every level j. Calculate the ratio of thermodynamic probabilities w_{mp}/w_d, and find the corresponding difference in entropies.

4.6 Consider a classical monatomic gas contained in one side of a tank as shown in Figure 4.3. The partition is broken, and the gas is allowed to fill the entire volume. Using the expression for Z given by Eq. 3.62, analyze this process from the microscopic viewpoint. Show that the expression for the net entropy change is of the same form as the classical result.

4.7 Develop the entropy equation for a Fermi-Dirac system, Eq. 4.75, and demonstrate that Eq. 4.72 is a valid relation for such a system.

4.8 Consider a Bose-Einstein system at very low temperature. Show that the limiting relation, Eq. 4.78, is valid without requiring that the ground level energy ϵ_0 is equal to zero.

4.9 Suppose that a Bose-Einstein system has a ground level degeneracy g_0 of 10. Is the entropy of this system equal to zero at absolute zero temperature? Calculate the value at zero temperature for a mole of this substance.

4.10 An experiment consists in rolling a pair of dice and recording the result. This experiment is repeated many times, after which the results are announced one by one. What is the average information conveyed per symbol?

4.11 What is the average information per character (including blanks) of the first four words in this sentence? If every letter "e" is deleted, how does this change the result?

4.12 Maximize Eq. 4.97 for $k \ln w$ at fixed N. What is the nature of the resulting distribution? Is the corresponding entropy identical with thermodynamic entropy?

4.13 Two dissimilar ideal gases A and B are contained in an insulated tank separated by a wall. There are N_A molecules of A in volume V_A and N_B mplecules of B in volume V_B. Both are at the same pressure and temperature. The wall is then broken and the gases are allowed to mix.

(*a*) Use the partition functions to show that the entropy change due to mixing is given by

$$-R \sum_i n_i \ln y_i$$

(*b*) If gases A and B are identical, show that there is no change in entropy when the wall is broken.

4.14 (*a*) The system partition function Z_N is defined for a pure substance by Eq. 4.20. Find the corresponding definition of Z_N for a mixture of A and B in terms of Z_A, Z_B, N_A, N_B such that the entropy equation, Eq. 4.19, is satisfied.

(*b*) For this definition of Z_N, show that Eqs. 4.59 and 4.99 are consistent.

(*c*) Show that Eq. 4.60 reduces to Dalton's law of partial pressures.

5 Quantum Mechanics

In Chapter 1, we discussed briefly some of the background and hypotheses of the subject of quantum mechanics. Succeeding chapters have led to the expression of thermodynamic properties in terms of the partition function, evaluation of which depends upon a knowledge of the energy levels as evaluated from quantum mechanics. We must therefore develop the mathematical aspects of quantum mechanics, at least to the extent required for the determination of the energy levels, and this is the purpose of this chapter.

5.1. The Bohr Theory of the Atom

As an introduction to the subject of quantum mechanics, we describe the Bohr theory of the atom, which was presented originally by Niels Bohr in 1913 and was frequently modified during the succeeding decade until its general replacement by the new quantum mechanics.

Bohr's theory is fundamentally a combination of classical mechanics and Planck's quantum theory of radiation.† The atomic model consists of electrons orbiting around a central nucleus. We consider here the simplest case, namely, the hydrogen atom, which consists of a single electron revolving about a nucleus comprised of one proton. The force exerted on the electron by the nucleus is the Coulomb potential $C\mathscr{Z}e^2/r^2$, where $\mathscr{Z}$ is the atomic number of the substance (i.e., the charge on the nucleus, which is equal to unity for hydrogen), e is the electron charge, and r is the radius of the orbit. Bohr postulated that only discrete orbits are permissible, these being specified by the requirement that the angular momentum of the electron be equal to $\mathbf{n}(h/2\pi)$, where $\mathbf{n}$ is the principal quantum number, which may have only integral values 1, 2, 3, ... For

† The topic of thermal radiation will be treated in some detail in Section 6.5 as an example of a Bose-Einstein system.

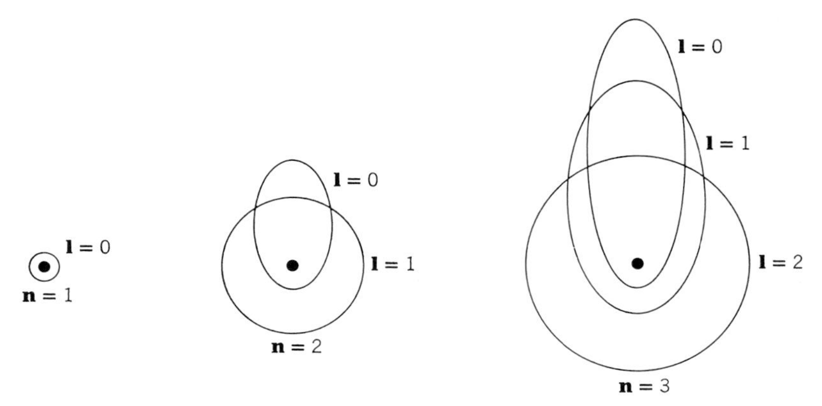

Figure 5.1 Electron orbits for the Bohr theory of the hydrogen atom.

an electron to jump from one orbit to a higher one, it must receive a quantum of radiation (energy according to Planck's theory) of exactly the appropriate frequency. This absorption of energy results in a discrete line in the spectrum observed for hydrogen. The various wavelengths at which hydrogen and other atoms absorb or emit energy will be discussed in Section 5.9. With Bohr's hypothesis and using the Coulomb potential, it is found from Newton's second law of motion that the radii of these allowed orbits are directly proportional to $\mathbf{n}^2$.

This basic theory is far too simple to explain the observed spectral lines for even the hydrogen atom, much less those for more complex atoms. Modifications to the theory allow for elliptical orbits as well as circular, as specified by the azimuthal quantum number **l**, which may take on only the values $\mathbf{l} = 0, 1, 2, \ldots (\mathbf{n} - 1)$. Figure 5.1 shows the Bohr orbits for the first few values of the quantum numbers **n** and **l**. The energy is dependent primarily upon the principal quantum number **n**, and energy differences for different values of **l** are very small for the hydrogen atom. The magnetic quantum number $\mathbf{m}_l$, with allowed values $\mathbf{m}_l = 0, \pm 1, \pm 2, \ldots, \pm \mathbf{l}$, was introduced to account for the observed splitting of spectral lines in a magnetic field, and the electron spin quantum number $\mathbf{m}_s = \pm\frac{1}{2}$ was introduced in an attempt to account for the behavior of helium and other more complex substances.

In spite of its growing complexity, the Bohr theory was still unable to account for or explain many of the observed phenomena, and it was therefore discarded in favor of the new quantum mechanics. The model is still useful today, however, as a tool for many calculations and especially as a convenient pictorial representation of the atom.

5.2. Wave Characteristics of Electrons and the Heisenberg Uncertainty Principle

The introduction of Planck's quantum theory of radiation began a great controversy about the fundamental nature of light and electromagnetic radiation in general. Radiation having a corpuscular nature could not easily explain the problems of dispersion, reflection, interference, diffraction, and others handled so readily by the wave theory. On the other hand, using the quantum theory viewpoint, one could rather simply explain the problem of energy exchange—the photoelectric effect in particular—and other difficulties that had plagued the wave theory for years. Furthermore, it seemed for a time that Bohr's atomic model, which incorporated the quantum theory, would provide a basis for the explanation of spectra and atomic behavior in general.

Let us consider a photon of electromagnetic radiation. From Planck's equation for energy exchange,

$$\epsilon = h\nu \tag{5.1}$$

where ν is the frequency, and from Einstein's equation,

$$\epsilon = mc^2 \tag{5.2}$$

we find that, although a photon or quantum of radiation has no mass, it does have associated with it a momentum (mc),

$$(mc) = \frac{h\nu}{c} = \frac{h}{\lambda} \tag{5.3}$$

in which the wavelength λ is given by

$$\lambda = \frac{c}{\nu} \tag{5.4}$$

We view radiation, then, as having a dualistic nature—the wave properties frequency and wavelength and the particle or corpuscular property of momentum.

In 1923, when the controversy over the nature of radiation had reached essentially an impasse with both viewpoints capable of only partial success, de Broglie postulated that matter may also possess a similar dualistic nature. His hypothesis that a wavelength given by the relation

$$\lambda = \frac{h}{m\mathrm{V}} \tag{5.5}$$

can be associated with an electron of mass m and velocity V was first confirmed by Davisson and Germer in 1925, and later by others. It was found that passing a narrow beam of electrons through a thin metal foil produces a set of concentric diffraction rings, similar to the optical rings of light diffraction, a phenomenon explained by wave interference. The electron ring spacings could also be predicted according to Eq. 5.5. From the magnitude of Planck's constant h, it is clear that the de Broglie matter wavelengths for large macroscopic bodies are so small as to result in a totally negligible influence, but for electrons and other atomic-scale considerations, a clear distinction between particle and wave is not always possible.

The Heisenberg uncertainty principle can be demonstrated by the following thought experiment. Suppose that we wish to observe an electron through some powerful microscope. It is impossible, even in theory, to determine the exact position of the electron at any instant of time, inasmuch as some source of radiation must be used to illuminate the electron. Consequently, the minimum uncertainty Δx in the position of the electron is of the order of magnitude of the wavelength of the illuminating radiation. That is,

$$\Delta x \sim \lambda$$

We conclude that, in order to minimize the uncertainty in position of the electron, it would be desirable to have a microscope using radiation of a very short wavelength. However, in making this measurement or observation, at least one photon of radiation must strike the electron and be reflected to the observer. This photon possesses a momentum h/λ, at least part of which is transferred to the electron as they collide. Consequently, there is a resulting uncertainty about the momentum of the electron of the order of magnitude h/λ, or

$$\Delta(m\mathsf{V}_x) \sim \frac{h}{\lambda}$$

with the opposite dependency on λ. We find that radiation having a short wavelength results in a small uncertainty in position of the electron but a large uncertainty in its momentum. Conversely, by using radiation of long wavelength we could determine the momentum of an electron quite accurately, but we would then be very uncertain about its position. The product of these effects,

$$\Delta x \times \Delta(m\mathsf{V}_x) \sim h \tag{5.6}$$

is an approximate statement of Heisenberg's uncertainty principle, one of the fundamental equations of quantum mechanics. As pointed out in

Chapter 1, this basic uncertainty principle is not observed or of consequence in ordinary macroscopic measurements, because of the extremely small magnitude of Planck's constant. However, it is of prime importance on the atomic scale, where we are concerned with small masses and distances.

5.3. The Schrödinger Wave Equation

As a useful preliminary to the Schrödinger wave equation for matter waves, we first develop and solve the differential equation representing a vibrating string. Consider a small segment of the string shown in Figure 5.2. The variable y represents the amplitude of vibration at time t for any distance x from the fixed end point, and the mass of the string per unit length is designated by the symbol ρ.

At any time t, the net force in the y-direction is, for small amplitude,

$$F_{y_{\text{net}}} = F\left[\left(\frac{\partial y}{\partial x}\right)_{x+\Delta x} - \left(\frac{\partial y}{\partial x}\right)_{x}\right] \tag{5.7}$$

However, from Newton's second law,

$$F_{y\text{net}} = \rho\,\Delta x\left(\frac{\partial^2 y}{\partial t^2}\right) \tag{5.8}$$

Therefore

$$\left[\frac{(\partial y/\partial x)_{x+\Delta x} - (\partial y/\partial x)_x}{\Delta x}\right] = \frac{\rho}{F}\left(\frac{\partial^2 y}{\partial t^2}\right) \tag{5.9}$$

and in the limit, as $\Delta x \to 0$,

$$\left(\frac{\partial^2 y}{\partial x^2}\right) = \frac{\rho}{F}\left(\frac{\partial^2 y}{\partial t^2}\right) \tag{5.10}$$

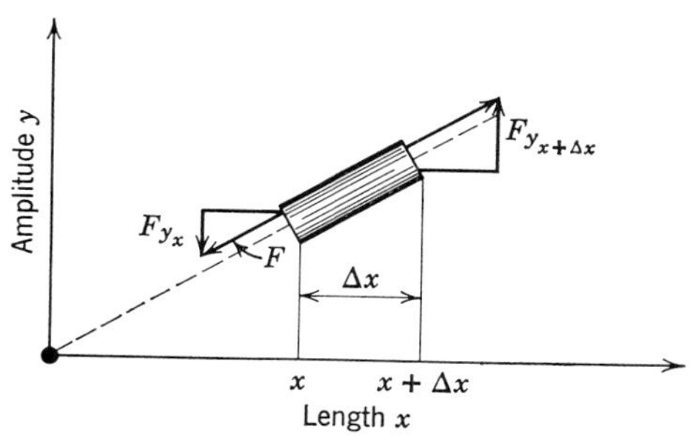

Figure 5.2 An element of a vibrating string.

Since the ratio F/ρ has the units (length/time)^2, it is convenient to write Eq. 5.10 as

$$\left(\frac{\partial^2 y}{\partial x^2}\right) = \frac{1}{V_p^2}\left(\frac{\partial^2 y}{\partial t^2}\right) \tag{5.11}$$

where

$$V_p = \left(\frac{F}{\rho}\right)^{1/2} \tag{5.12}$$

and is called the phase velocity. Equation 5.11 is then the standard partial differential equation for a string vibrating in one direction.

In general, the amplitude y is a function of both x and t. Let us assume that the dependency is separable, so that y can be expressed as a product

$$y = f(x) \times g(t) \tag{5.13}$$

where f is a function only of distance and g is a function only of time. The partial derivatives required in Eq. 5.11 can then be expressed as the ordinary derivatives of Eq. 5.13. Since g is not a function of x, it follows that

$$\frac{\partial^2 y}{\partial x^2} = g \times \left(\frac{d^2 f}{dx^2}\right) \tag{5.14}$$

Similarly, since f is not a function of t,

$$\left(\frac{\partial^2 y}{\partial t^2}\right) = f \times \left(\frac{d^2 g}{dt^2}\right) \tag{5.15}$$

Substituting Eqs. 5.14 and 5.15 into Eq. 5.11 and rearranging, we obtain

$$\frac{1}{f}\left(\frac{d^2 f}{dx^2}\right) = \frac{1}{gV_p^2}\left(\frac{d^2 g}{dt^2}\right) = -\alpha^2 \tag{5.16}$$

We note that one side of this equation is a function only of x, and the other a function only of t. Inasmuch as x and t are independent, Eq. 5.16 can be valid only if both sides are constant, which for convenience is set equal to $-\alpha^2$.

We can write the second equality of Eq. 5.16 as

$$\frac{d^2 g}{dt^2} + \alpha^2 V_p^2 g = 0 \tag{5.17}$$

The solution of this differential equation is

$$g = C_1 \cos(\alpha V_p t) + C_2 \sin(\alpha V_p t) = C_3 \sin(\alpha V_p t + \delta) \tag{5.18}$$

where

$$\delta = \tan^{-1}(C_1/C_2)$$
$$C_3 = \sqrt{C_1^2 + C_2^2} \tag{5.19}$$

From Eq. 5.18 the variable g is seen to be a periodic function of time, having the frequency

$$\nu = \frac{\alpha V_p}{2\pi} \tag{5.20}$$

Therefore the constant α can be expressed as

$$\alpha = \frac{2\pi\nu}{V_p} = \frac{2\pi}{\lambda} \tag{5.21}$$

where λ is the wavelength.

Using Eq. 5.21 and the equality involving f and α in Eq. 5.16, it follows that

$$\frac{d^2 f}{dx^2} + \frac{4\pi^2}{\lambda^2} f = 0 \tag{5.22}$$

The function f then represents the amplitude of the standing wave at any position x for a vibration of associated wavelength λ.

In 1926, Schrödinger applied the differential wave equation to the matter waves of de Broglie. Using the symbol Ψ_x for the x-direction amplitude or wave function (the physical interpretation of which will be discussed shortly), and the de Broglie equation (Eq. 5.5), the differential equation is

$$\frac{d^2\Psi_x}{dx^2} + \frac{4\pi^2(mV_x)^2}{h^2}\Psi_x = 0 \tag{5.23}$$

It is more convenient to use the total energy ϵ_x, where

$$\epsilon_x = (\text{K.E.})_x + (\text{P.E.})_x = \tfrac{1}{2}mV_x^2 + \Phi_x \tag{5.24}$$

where Φ_x is the potential energy.

The one-dimensional wave equation then becomes

$$\frac{d^2\Psi_x}{dx^2} + \frac{8\pi^2 m}{h^2}(\epsilon_x - \Phi_x)\Psi_x = 0 \tag{5.25}$$

In order to generalize the equation to three dimensions, we define a wave function Ψ as

$$\Psi = \Psi_x\Psi_y\Psi_z \tag{5.26}$$

and recognize that

$$\epsilon = \epsilon_x + \epsilon_y + \epsilon_z \tag{5.27}$$

and

$$\Phi = \Phi_x + \Phi_y + \Phi_z \tag{5.28}$$

so that, in three dimensions,

$$\nabla^2\Psi + \frac{8\pi^2 m}{h^2}(\epsilon - \Phi)\Psi = 0 \tag{5.29}$$

where the operator ∇^2 is

$$\nabla^2 = \frac{\partial^2}{\partial x^2} + \frac{\partial^2}{\partial y^2} + \frac{\partial^2}{\partial z^2} \tag{5.30}$$

Equation 5.29 is known as the time-independent Schrödinger wave equation. In general, as for the vibrating string, the wave function Ψ is dependent upon time as well as position and may even be a complex function having real and imaginary parts. Consequently, there is a corresponding general time-dependent Schrödinger wave equation, but for stationary states of the system (standing waves), the case of interest to us here, we need consider only the special form, Eq. 5.29.

A basic hypothesis of wave mechanics is that the wave function Ψ has the characteristic of a probability function. Since Ψ may in general be complex while a probability must be real, the product of Ψ and its complex conjugate Ψ^* is taken as representing probability. Furthermore, a probability function must be continuous and single-valued and must approach zero for large values of distance. It is found that a Ψ function having the required behavior is obtained from solutions of the Schrödinger wave equation (Eq. 5.29) only for discrete energy states of the system. When the function Ψ is real, the product $\Psi\Psi^*$ reduces to Ψ^2, so that for stationary states, where the particle being represented is presumed to have a definite energy ϵ_i, the quantity

$$\Psi_i^2 \, dx \, dy \, dz$$

becomes the probability of finding the particle of energy ϵ_i in the element of space x to $x + dx$, y to $y + dy$, z to $z + dz$.

5.4. Translation

In applying the Schrödinger wave equation to the molecules of a gas, we assume that the various energy modes are separable. We consider first the translational energy, which would constitute the entire energy of a monatomic ideal gas with all the atoms in the electronic ground state.

The gas is confined in a box of dimensions X, Y, Z, but to simplify the problem let us first consider a one-dimensional case, in which the particles move only in the x-direction, between 0 and X. Applying the Schrödinger equation to a single particle, for the x-direction only, yields

$$\frac{d^2\Psi_x}{dx^2} + \frac{8\pi^2 m}{h^2}(\epsilon_x - \Phi_x)\Psi_x = 0 \tag{5.31}$$

However, since the particle must remain inside the box, the probability, and consequently the wave function Ψ_x, must be zero for all x outside this range. This physical requirement is satisfied by making the potential energy $\Phi_x = \infty$ for $x < 0$ and $x > X$. We also note that, for an ideal gas, $\Phi_x = 0$ inside the box. Therefore, for the range

$$0 < x < X \tag{5.32}$$

the Schrödinger wave equation is

$$\frac{d^2\Psi_x}{dx^2} = \frac{8\pi^2 m}{h^2}\epsilon_x\Psi_x = 0 \tag{5.33}$$

with the boundary conditions

$$\Psi_x(0) = 0 \tag{5.34}$$

$$\Psi_x(X) = 0 \tag{5.35}$$

The solution to Eq. 5.33 can be written in the form

$$\Psi_x = C_1 \sin\left[x\sqrt{\frac{8\pi^2 m\epsilon_x}{h^2}}\right] + C_2 \cos\left[x\sqrt{\frac{8\pi^2 m\epsilon_x}{h^2}}\right] \tag{5.36}$$

From the boundary condition at $x = 0$ (Eq. 5.34), it follows that

$$0 = C_1(0) + C_2(1)$$

and we conclude that

$$C_2 = 0 \tag{5.37}$$

From the boundary condition at $x = X$ (Eq. 5.35),

$$0 = C_1 \sin\left[X\sqrt{\frac{8\pi^2 m\epsilon_x}{h^2}}\right] \tag{5.38}$$

Now the possibility that $C_1 = 0$ would result in the physically impossible solution that $\Psi_x = 0$ for all values of x. The only alternative is that

$$\sin\left[X\sqrt{\frac{8\pi^2 m\epsilon_n}{h^2}}\right] = 0$$

or

$$X\sqrt{\frac{8\pi^2 m\epsilon_x}{h^2}} = \mathbf{k}_x\pi \tag{5.39}$$

with

$$\mathbf{k}_x = 1, 2, 3, \ldots$$

The value $\mathbf{k}_x = 0$ results in a trivial solution ($\Psi_x = 0$ for all x) and is therefore not permissible.

It is seen from Eq. 5.39 that only discrete energy states are allowed, these states being given in terms of the x-direction translational quantum number $\mathbf{k}_x$ as

$$\epsilon_{\mathbf{k}_x} = \frac{h^2}{8mX^2}\mathbf{k}_x{}^2 \qquad \mathbf{k}_x = 1, 2, 3, \ldots \tag{5.40}$$

For each energy state $\epsilon_{\mathbf{k}_x}$ as specified by the quantum number $\mathbf{k}_x$, there is a corresponding x-direction wave function $\Psi_{\mathbf{k}_x}$, which from Eqs. 5.36, 5.37 and 5.39 can be expressed in the form

$$\Psi_{\mathbf{k}_x}(x) = C_1 \sin\left(\mathbf{k}_x \pi \frac{x}{X}\right) \tag{5.41}$$

The constant C_1 in this equation can be evaluated by normalization, because the square of the wave function represents the probability. Therefore

$$\int_{x=0}^{X} \Psi_{\mathbf{k}_x}{}^2(x)\, dx = 1 \tag{5.42}$$

since, for each $\mathbf{k}_x$, a particle having the corresponding energy $\epsilon_{\mathbf{k}_x}$ must be found somewhere in the box, according to Eq. 5.32. If we substitute Eq. 5.41 into Eq. 5.42,

$$C_1{}^2 \int_0^X \sin^2\left(\mathbf{k}_x \pi \frac{x}{X}\right) dx = 1 \tag{5.43}$$

and integrate, we find

$$C_1 = \sqrt{\frac{2}{X}} \tag{5.44}$$

Therefore each quantum state as specified by $\mathbf{k}_x$ has associated with it the wave function

$$\Psi_{\mathbf{k}_x}(x) = \sqrt{\frac{2}{X}} \sin\left(\mathbf{k}_x \pi \frac{x}{X}\right) \tag{5.45}$$

The diagrams of Figure 5.3 show the wave functions and square of the wave functions for the first few quantum states for the particle in the one-dimensional box.

To this point we have discussed only the one-dimensional case, where the particle was restricted to the range from $x = 0$ to $x = X$. If we now generalize the problem to three dimensions, so that $0 < x < X$, $0 < y < Y$, $0 < z < Z$, then, since the x-, y-, z-directions are assumed to be independent, it follows that

$$\epsilon = \epsilon_x + \epsilon_y + \epsilon_z \tag{5.46}$$

From the Schrödinger time-independent equation, Eq. 5.29, and the boundary conditions, it is found that only discrete energy states are

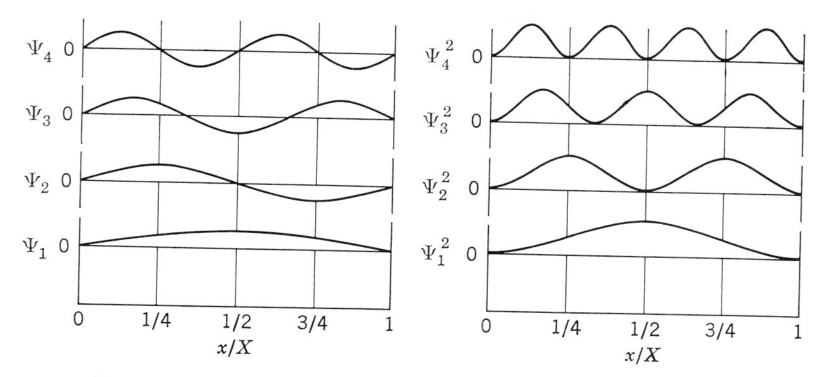

Figure 5.3 The functions $\Psi_{\mathbf{k}_x}$ and $\Psi_{\mathbf{k}_x}{}^2$ for x-directional translation.

permissible; these energies are given in terms of three quantum numbers for translation by the expression

$$\epsilon_{\mathbf{k}_x,\mathbf{k}_y,\mathbf{k}_z} = \frac{h^2}{8m}\left(\frac{\mathbf{k}_x{}^2}{X^2} + \frac{\mathbf{k}_y{}^2}{Y^2} + \frac{\mathbf{k}_z{}^2}{Z^2}\right) \tag{5.47}$$

in which each of the quantum numbers $\mathbf{k}_x$, $\mathbf{k}_y$, $\mathbf{k}_z$ may have any integral value 1, 2, 3, ... A particular quantum state is then specified by the values of each of these numbers. It is important to realize that the states (5, 2, 3), (5, 3, 2), (2, 5, 3), etc., are all different quantum states, although all have exactly the same magnitude of energy. This point should help to clarify the distinction between energy (or quantum) states and energy levels. Each of the energy states $\epsilon_{\mathbf{k}_x,\mathbf{k}_y,\mathbf{k}_z}$, has associated with it a wave function

$$\Psi_{\mathbf{k}_x,\mathbf{k}_y,\mathbf{k}_z}(x, y, z) = \sqrt{\frac{8}{XYZ}} \sin\left(\mathbf{k}_x\pi\frac{x}{X}\right) \sin\left(\mathbf{k}_y\pi\frac{y}{Y}\right) \sin\left(\mathbf{k}_z\pi\frac{z}{Z}\right) \tag{5.48}$$

where the constant has been evaluated by normalizing the wave function, as in the one-dimensional problem.

Very commonly, we are concerned with a system for which

$$X = Y = Z = V^{1/3} \tag{5.49}$$

so that Eq. 5.47 can be simplified to the form

$$\epsilon_{\mathbf{k}_x,\mathbf{k}_y,\mathbf{k}_z} = \frac{h^2}{8mV^{2/3}}(\mathbf{k}_x{}^2 + \mathbf{k}_y{}^2 + \mathbf{k}_z{}^2) \tag{5.50}$$

with $\mathbf{k}_x$, $\mathbf{k}_y$, $\mathbf{k}_z = 1, 2, 3, \ldots$. In this form, we see that the energy associated with a given quantum state is dependent upon the volume of the system, as stated earlier in Chapter 4; this statement was especially important in relating changes in energy to volume changes and ultimately to boundary movement work for a reversible process involving a simple compressible substance.

5.5. Application of the Wave Equation to Molecules

In the preceding section, we applied the Schrödinger wave equation to the problem of translation, the only energy mode for a monatomic ideal gas with all the atoms in the electronic ground level. In proceeding to the more general case of a molecule comprised of more than a single atom, the problem becomes relatively more complex because there may be several modes of energy storage. Let us consider a system of independent particles (an ideal gas), so that there is no system energy resulting from intermolecular forces and we may speak of the energy per molecule.

We assume first that the energy of the molecule can be separated into two parts: the translational energy of the molecule as a whole, expressed in terms of the motion of the center of mass of the molecule; and the energy internal to the molecule itself, that is, energy with respect to a coordinate system having its origin at the center of mass of the molecule. For the energy of the molecule, we may write

$$\epsilon = \epsilon_t + \epsilon_{\text{int}} \tag{5.51}$$

and, for the wave function,

$$\Psi = \Psi_t \Psi_{\text{int}} \tag{5.52}$$

We note also that, since we are concerned with independent particles, any potential energy is associated with the internal modes of energy. The Schrödinger wave equation can then be separated into two parts, one dependent solely upon the motion of the center of mass, and the other only upon the energy of the component atoms with respect to the center of mass. The first of these

$$\left[\frac{\partial^2 \Psi_t}{\partial x_{\text{C.M.}}^2} + \frac{\partial^2 \Psi_t}{\partial y_{\text{C.M.}}^2} + \frac{\partial^2 \Psi_t}{\partial z_{\text{C.M.}}^2}\right] + \frac{8\pi^2 m}{h^2} \epsilon_t \Psi_t = 0 \tag{5.53}$$

has been evaluated for translational energies and wave functions in the previous section for a monatomic ideal gas. The same results apply for the translation of a molecule consisting of more than one atom, provided that the x, y, z in Eq. 5.53 are the coordinates of the center of mass of the molecule (relative to an external coordinate frame) and that m is the mass of the molecule, i.e., the sum of the masses of the component atoms.

The second equation, dependent only upon the internal modes, becomes

$$\left(\frac{\partial^2\Psi_{\text{int}}}{\partial x^2} + \frac{\partial^2\Psi_{\text{int}}}{\partial y^2} + \frac{\partial^2\Psi_{\text{int}}}{\partial z^2}\right) + \frac{8\pi^2 m}{h^2}(\epsilon_{\text{int}} - \Phi_{\text{int}})\Psi_{\text{int}} = 0 \quad (5.54)$$

In this equation, the coordinate system x, y, z is taken with respect to the center of mass of the molecule. The energy may include contributions from rotational, vibrational, and electronic states, as discussed in Chapter 1. These energy modes are not strictly independent. However, for most substances, the energy of the first excited electronic level is so high that essentially all the molecules will be found in the ground level. In addition, the vibrational energy levels will be found to be relatively large (compared to those for rotation), so that, at moderate temperature, most molecules in a system will be found in the lowest few vibrational levels at any time. Under such conditions, there will not be an appreciable interaction between rotation and vibration, and the vibration can reasonably be assumed to be harmonic.

Our simple model for internal energy modes consists therefore of a rigid rotator, harmonic oscillator, and ground electronic level. For the energy contribution of the internal modes, we can write

$$\epsilon_{\text{int}} = \epsilon_r + \epsilon_v + \epsilon_{e0} \quad (5.55)$$

and, for the internal wave function,

$$\Psi_{\text{int}} = \Psi_r \Psi_v \Psi_{e0} \quad (5.56)$$

We find that Eqs. 5.55 and 5.56 permit a separation of the wave equation, Eq. 5.54, into two parts, one for rotation and the other for vibration (ϵ_{e_0} is a constant). In the next section, we consider that portion of the wave equation for a rigid rotator, and in Section 5.7 that for a harmonic oscillator.

At high temperatures, this simple internal model becomes inaccurate, and corrections for rotational centrifugal stretching, anharmonicity of the vibration, and rotation-vibration coupling must be included. For substances in which excited electronic levels are significant (O_2 and NO, for example), the rotation and vibration must be considered for each electronic level, and evaluation becomes extremely tedious. These considerations are taken up after our discussions of the rigid rotator and harmonic oscillator.

5.6. Rigid Rotator

In the simplified model for internal modes of energy, interaction between rotation and vibration is assumed to be negligible, and we

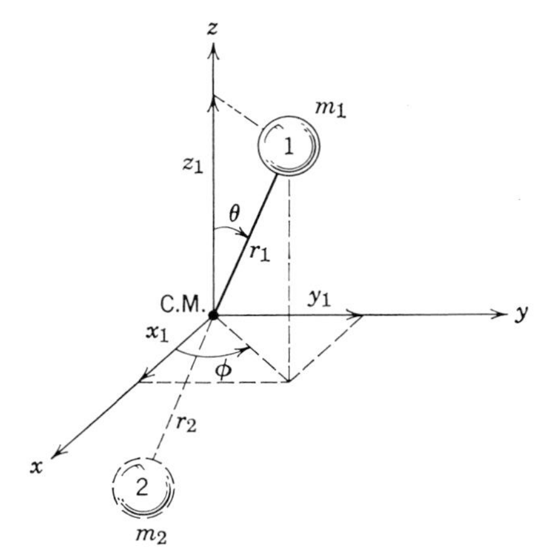

Figure 5.4 The spherical coordinate system for a diatomic molecule.

consider the molecule to be rigid insofar as rotation is concerned. In evaluating the rotational contribution to energy, we find it convenient to work in spherical polar coordinates, taking the origin at the center of mass of the molecule, as shown for a diatomic molecule in Figure 5.4. The molecule is comprised of two atoms, 1 and 2, of masses m_1 and m_2, respectively. The position of atom 1 with respect to the center of mass is specified by the radius r_1, the angle with the z-axis, θ, and the angle of the projection of r_1 in the xy-plane with the x-axis, ϕ. Angle θ is restricted to $0 \leq \theta \leq \pi$, while, for ϕ, $0 \leq \phi \leq 2\pi$.

From Figure 5.4, it is seen that the relations between the two coordinate systems are

$$\begin{aligned} x_1 &= r_1 \sin\theta \cos\phi \\ y_1 &= r_1 \sin\theta \sin\phi \\ z_1 &= r_1 \cos\theta \end{aligned} \tag{5.57}$$

Therefore the kinetic energy for atom 1 at constant distance r_1 can be expressed as

$$\begin{aligned} \text{K.E.}_{m_1} &= \frac{m_1}{2}\left[\left(\frac{\partial x_1}{\partial t}\right)^2 + \left(\frac{\partial y_1}{\partial t}\right)^2 + \left(\frac{\partial z_1}{\partial t}\right)^2\right] \\ &= \frac{m_1 r_1{}^2}{2}\left[\left(\frac{\partial \theta}{\partial t}\right)^2 + \sin^2\theta\left(\frac{\partial \phi}{\partial t}\right)^2\right] \end{aligned} \tag{5.58}$$

For atom 2, the angle θ is replaced by $(\pi - \theta)$, and the angle ϕ by $(\pi + \phi)$. The expression for the kinetic energy of atom 2 in spherical coordinates is then found to be identical with Eq. 5.58, with $m_1 r_1{}^2$ replaced by $m_2 r_2{}^2$, so that the total kinetic energy of the molecule is given by

$$\text{K.E.}_{m_1+m_2} = \frac{I_\varepsilon}{2}\left[\left(\frac{\partial \theta}{\partial t}\right)^2 + \sin^2\theta\left(\frac{\partial \phi}{\partial t}\right)^2\right] \tag{5.59}$$

where the moment of inertia I_ε of the rigid molecule is defined by

$$I_\varepsilon = \sum_i m_i r_i^2 = m_1 r_1^2 + m_2 r_2^2 \tag{5.60}$$

We find it convenient to use the total distance between atoms (centers) making up the molecule, instead of the distances r_1 and r_2. As shown in Figure 5.5, the equilibrium distance r_ε is defined as

$$r_\varepsilon = r_1 + r_2 \tag{5.61}$$

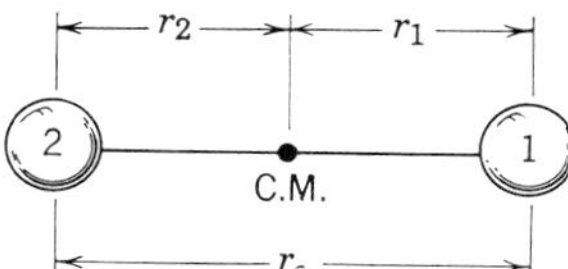

Figure 5.5 The coordinates of a diatomic rigid rotator.

From a balance of moments taken about the center of mass,

$$m_1 r_1 = m_2 r_2 \tag{5.62}$$

so that Eq. 5.60 is conveniently written in the form

$$I_\varepsilon = \left(\frac{m_1 m_2}{m_1 + m_2}\right) r_\varepsilon^2 = m_r r_\varepsilon^2 \tag{5.63}$$

where the quantity m_r is defined as

$$m_r = \frac{m_1 m_2}{m_1 + m_2} \tag{5.64}$$

and is called the reduced mass of the molecule.

We find that the total kinetic energy of rotation is equivalent to that of a single particle of mass m_r moving on a sphere of radius r_ε. Thus, to determine the energy levels and wave functions for a rigid diatomic rotator, we apply the Schrödinger wave equation to such a particle. The wave equation for internal modes, Eq. 5.54, is separated according to the expressions given by Eqs. 5.55 and 5.56. The potential energy Φ is assumed to be a function of r only, and as such will be considered in the vibrational energy evaluation. After separation, the partial differential equation for rotation as applied to a particle of mass m_r and constant dimension r_ε is

$$\nabla^2 \Psi_r + \frac{8\pi^2 m_r}{h^2}\epsilon_r \Psi_r = 0 \tag{5.65}$$

The Laplacian operator ∇^2 (Eq. 5.30), written in spherical polar coordinates, becomes

$$\nabla^2 = \frac{1}{r^2}\frac{\partial}{\partial r}\left(r^2 \frac{\partial}{\partial r}\right) + \frac{1}{r^2 \sin\theta}\frac{\partial}{\partial \theta}\left(\sin\theta \frac{\partial}{\partial \theta}\right) + \frac{1}{r^2 \sin^2\theta}\frac{\partial^2}{\partial \phi^2} \tag{5.66}$$

However, for our application, $r = r_\varepsilon =$ constant, so that the first term in Eq. 5.66 equals zero. Substituting this relation into Eq. 5.65, and noting from Eq. 5.63 that $I_\varepsilon = m_r r_\varepsilon^2$, the wave equation for the rigid rotator is

$$\frac{1}{\sin\theta}\frac{\partial}{\partial \theta}\left(\sin\theta \frac{\partial \Psi_r}{\partial \theta}\right) + \frac{1}{\sin^2\theta}\frac{\partial^2 \Psi_r}{\partial \phi^2} + \frac{8\pi^2 I_\varepsilon \epsilon_r}{h^2}\Psi_r = 0 \tag{5.67}$$

Even for our simplified model of rigid rotation, the wave equation, Eq. 5.67, is a relatively complex partial differential equation in θ and ϕ and does not readily yield a solution. Suppose that we assume a product solution, similar to the procedure used in solving the vibrating string equation in Section 5.3. We assume in this case that the wave function for rotation can be expressed as a product of two functions, one dependent only upon θ and the other only upon ϕ, or

$$\Psi_r = f(\theta) \times g(\phi) \tag{5.68}$$

The partial derivatives required in Eq. 5.67 now become the ordinary derivatives of f and g according to Eq. 5.68.

$$\frac{\partial \Psi_r}{\partial \theta} = g\left(\frac{df}{d\theta}\right) \tag{5.69}$$

$$\frac{\partial^2 \Psi_r}{\partial \phi^2} = f\left(\frac{d^2 g}{d\phi^2}\right) \tag{5.70}$$

Substituting Eqs. 5.68 to 5.70 into the wave equation, Eq. 5.67, the result is easily rearranged to the form

$$\frac{\sin\theta}{f}\frac{d}{d\theta}\left[\sin\theta\left(\frac{df}{d\theta}\right)\right] + \frac{8\pi^2 I_\varepsilon \epsilon_r}{h^2}\sin^2\theta = -\frac{1}{g}\frac{d^2 g}{d\phi^2} = \lambda^2 \tag{5.71}$$

where λ^2 is a constant. The reasoning in Eq. 5.71 is analogous to that in the solution of the vibrating string equation. The left side of Eq. 5.71 is a function only of θ, and the right side only of ϕ. Therefore, since θ and ϕ can be varied arbitrarily and independently of each other, the equality can be valid only if both sides are constant. This constant is taken to be λ^2, and we now have two ordinary differential equations. From the second equality of Eq. 5.71, it follows that

$$\frac{d^2 g}{d\phi^2} + \lambda^2 g = 0 \tag{5.72}$$

the solution of which is

$$g = C_1 \cos \lambda\phi + C_2 \sin \lambda\phi \tag{5.73}$$

If Ψ_r^2 is to have the necessary characteristics of a probability density, then Ψ_r (and f and g as well) must be finite and single-valued. It is seen from Eq. 5.73 that, for g to be single-valued, it is required that, as ϕ increases by 2π, the function given by Eq. 5.73 must return to the same value. That is,

$$\sin \lambda\phi = \sin \lambda(\phi + 2\pi) = \sin(\lambda\phi)\cos(2\pi\lambda) + \cos(\lambda\phi)\sin(2\pi\lambda) \tag{5.74}$$

and

$$\cos \lambda\phi = \cos \lambda(\phi + 2\pi) = \cos(\lambda\phi)\cos(2\pi\lambda) - \sin(\lambda\phi)\sin(2\pi\lambda) \tag{5.75}$$

Now, Eqs. 5.74 and 5.75 are seen to be correct only if

$$\cos 2\pi\lambda = 1 \tag{5.76}$$

$$\sin 2\pi\lambda = 0 \tag{5.77}$$

and we draw the important conclusion that λ must be integral, or

$$\lambda = 0, \pm 1, \pm 2, \pm 3, \ldots \tag{5.78}$$

We now turn to the equality involving f and λ in Eq. 5.71 with the knowledge that λ is restricted according to Eq. 5.78. Expanding the derivative and rearranging, we have

$$\frac{d^2 f}{d\theta^2} + \frac{\cos\theta}{\sin\theta}\left(\frac{df}{d\theta}\right) + \left[\frac{8\pi^2 I_\varepsilon \epsilon_r}{h^2} - \frac{\lambda^2}{\sin^2\theta}\right] f = 0 \tag{5.79}$$

Let us now perform a change of independent variable in order to convert this differential equation to a standard form. Letting

$$x = \cos\theta \tag{5.80}$$

the differential equation, Eq. 5.79, becomes

$$(1 - x^2)\frac{d^2 f}{dx^2} - 2x\frac{df}{dx} + \left[\frac{8\pi^2 I_\varepsilon \epsilon_r}{h^2} - \frac{\lambda^2}{1 - x^2}\right] f = 0 \tag{5.81}$$

with λ given by Eq. 5.78.

At this point, we must digress for a moment to discuss the standard differential equation of the same form, the associated Legendre equation,

$$(1 - z^2)\frac{d^2 y}{dz^2} - 2z\frac{dy}{dz} + \left[\mathbf{j}(\mathbf{j} + 1) - \frac{m^2}{1 - z^2}\right] y = 0 \tag{5.82}$$

When $\mathbf{j}$ and m are non-negative integers, a solution of this equation is given by

$$y = P_{\mathbf{j}}^{m}(z) = \frac{(1 - z^2)^{m/2}}{2^{\mathbf{j}}\,\mathbf{j}!} \frac{d^{m+\mathbf{j}}}{dz^{m+\mathbf{j}}} [(z^2 - 1)^{\mathbf{j}}] \tag{5.83}$$

The functions given by Eq. 5.83 are called associated Legendre functions. For any given value of $\mathbf{j}$ (0, 1, 2, . . .) it is seen that the maximum power of z inside the brackets is $2\mathbf{j}$, so that the maximum non-zero derivative is the $(2\mathbf{j})$th derivative. All higher derivatives must be zero. We find, therefore, that the maximum value of m yielding a non-zero solution is m equal to $\mathbf{j}$. It is also of interest to note that m equal to zero is permissible, in which case the solution is the ordinary Legendre polynomial. The allowed values of m for any $\mathbf{j}$ are then

$$m = 0, 1, 2, \ldots, (\mathbf{j} - 1), \mathbf{j} \tag{5.84}$$

We realize also that the solution of Eq. 5.83 is not a general solution of Eq. 5.82, since that equation is a second-order differential equation. However, the other independent solution to this equation includes an infinite series and consequently is not physically permissible for our particular application, as it would not permit Ψ_r^2 to have the required characteristics.

Returning now to our problem, if we compare our differential equation, Eq. 5.81, with the associated Legendre equation, Eq. 5.82, we find that

$$\frac{8\pi^2 I_\varepsilon \epsilon_r}{h^2} = \mathbf{j}(\mathbf{j} + 1) \qquad \mathbf{j} = 0, 1, 2, \ldots \tag{5.85}$$

and

$$|\lambda| = m \qquad m = 0, 1, 2, \ldots, \mathbf{j} \tag{5.86}$$

Consequently, only discrete energy states are allowed for the rigid rotator, with the magnitudes of these energies given in terms of the rotational quantum number $\mathbf{j}$ by the relation

$$\epsilon_r = \frac{h^2}{8\pi^2 I_\varepsilon} \mathbf{j}(\mathbf{j} + 1) \qquad \mathbf{j} = 0, 1, 2, 3, \ldots \tag{5.87}$$

and the rotational wave functions are

$$\Psi_r(\theta, \phi) = [C_1 \sin \lambda\phi + C_2 \cos \lambda\phi] \times P_{\mathbf{j}}^{|\lambda|}(\cos \theta) \tag{5.88}$$

with

$$\lambda = 0, \pm 1, \pm 2, \ldots, \pm(\mathbf{j} - 1), \pm \mathbf{j} \tag{5.89}$$

The constants of integration, C_1, C_2, are eliminated by normalization in the θ- and ϕ-directions, a rather complex problem that does not concern us here.

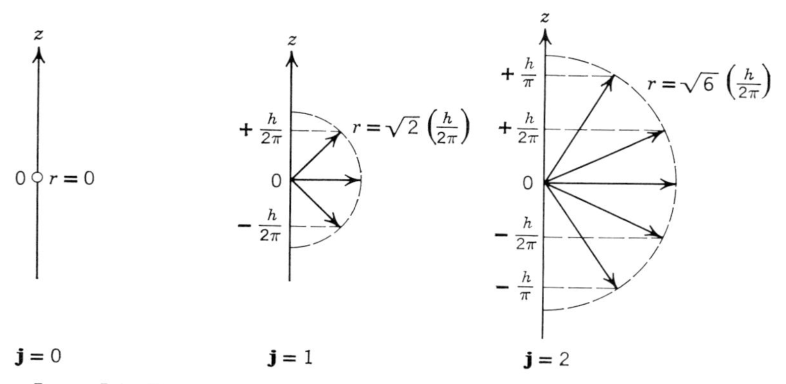

Figure 5.6 The total angular momentum vector for a rigid rotator.

One very important result is apparent from Eqs. 5.87, 5.88, and 5.89. From the last two of these, it is seen that for each value of **j** there are (2**j** + 1) values of λ and therefore (2**j** + 1) different and distinct wave functions (or discrete states), because the associated Legendre functions are not the same for different values of λ. However, since there is only one value of energy ϵ_r for each **j**, we conclude that the rotational energy levels given by Eq. 5.87 are degenerate, the degeneracy being given by

$$g_r = (2\mathbf{j} + 1) \tag{5.90}$$

The significance of the rotational degeneracy can be seen from an examination of the angular momentum of rotation, $H_{\mathbf{j}}$, given by

$$\begin{aligned} H &= m_r \mathsf{V} r_\varepsilon = (2m_r r_\varepsilon^{\,2} \times \tfrac{1}{2} m_r \mathsf{V}^2)^{1/2} = (2I_\varepsilon \epsilon_r)^{1/2} \\ &= [\mathbf{j}(\mathbf{j}+1)]^{1/2} \times \frac{h}{2\pi} \end{aligned} \tag{5.91}$$

From this relation, it is evident that the basic unit of total angular momentum is $h/2\pi$. The degeneracy is then explained in the following manner: Corresponding to each rotational quantum number **j** there is a total angular momentum $H_{\mathbf{j}}$, which can be represented by a vector. The projection of this total angular momentum vector along the z-axis is quantized so that the z-component can have only the values $(h/2\pi) \times \lambda$, with λ given by Eq. 5.89. The representation shown in Figure 5.6 is for the first three values of **j**, and shows the (2**j** + 1) orientations of the vector with respect to the z-axis for each value of **j**.

5.7. Harmonic Oscillator

In evaluating the energy and wave function for our simple vibrational model, the harmonic oscillator, we consider again a diatomic molecule, as shown in Figure 5.7. At some instant of time, atoms 1 and 2 are at distances r_1 and r_2, respectively, from the center of mass, with a total distance of separation

$$r = r_1 + r_2 \tag{5.92}$$

The distance r is, in general, different from the equilibrium distance r_ε (the distance at which the force between atoms changes sign) used for the rigid rotator.

We assume in our model that the vibration is harmonic; this means that the restoring force is directly proportional to $(r - r_\varepsilon)$, the displacement of the atoms from their equilibrium separation. This assumption is reasonable if the displacement $(r - r_\varepsilon)$ remains relatively small. Writing Newton's second law for atom 1,

$$m_1 \frac{d^2 r_1}{dt^2} + K(r - r_\varepsilon) = 0 \tag{5.93}$$

where K is the proportionality or spring constant. From a moment balance,

$$m_1 r_1 = m_2 r_2 \tag{5.94}$$

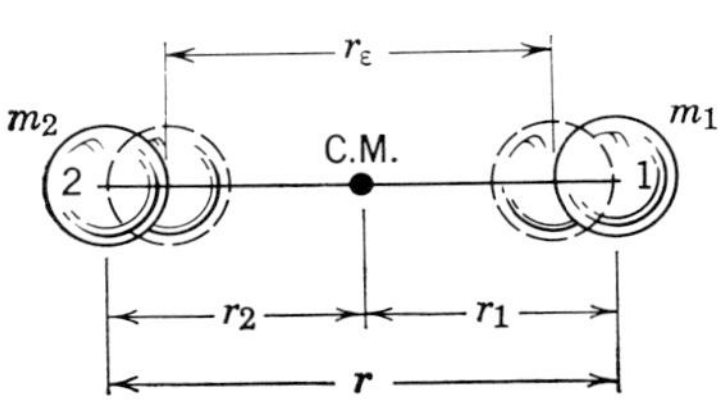

Figure 5.7 The coordinates of a harmonic oscillator.

and from Eqs. 5.92 and 5.94 we find the relation

$$r_1 = \left(\frac{m_2}{m_1 + m_2}\right) r \tag{5.95}$$

This expression can be differentiated and used to eliminate r_1 from Eq. 5.93, which then becomes

$$\left(\frac{m_1 m_2}{m_1 + m_2}\right) \frac{d^2 r}{dt^2} + K(r - r_\varepsilon) = 0 \tag{5.96}$$

The coefficient of the derivative in this expression has previously been termed the reduced mass m_r according to Eq. 5.64. If we now perform a change of variable by defining

$$x = r - r_\varepsilon \tag{5.97}$$

then Eq. 5.96 can be written as

$$m_r \frac{d^2 x}{dt^2} + Kx = 0 \tag{5.98}$$

The same result is obtained by writing Newton's second law for atom 2 and proceeding as above.

The solution to this differential equation has previously been found to be

$$x = C_1 \cos\sqrt{\frac{K}{m_r}}\,t + C_2 \sin\sqrt{\frac{K}{m_r}}\,t = C_3 \sin\left(\sqrt{\frac{K}{m_r}}\,t + \delta\right) \quad (5.99)$$

where C_1, C_2, C_3, and δ are constants. We conclude that x is a periodic function of time, with a frequency ν_ε equal to

$$\nu_\varepsilon = \frac{1}{2\pi}\sqrt{\frac{K}{m_r}} \quad (5.100)$$

Thus the spring constant K can be eliminated in terms of the frequency of vibration according to the relation

$$K = 4\pi^2\nu_\varepsilon^2 m_r \quad (5.101)$$

In order to apply the Schrödinger wave equation to the problem of vibration, it is necessary to know the potential energy resulting from the force between atoms. For any value x, the potential energy Φ is, using Eq. 5.101,

$$\Phi(x) = -\int_0^x F\,dx = \int_0^x Kx\,dx = \tfrac{1}{2}Kx^2 = 2\pi^2\nu_\varepsilon^2 m_r x^2 \quad (5.102)$$

which is seen to be parabolic in shape about the equilibrium point separation.

For our simple internal model, separation of the wave equation for internal modes (Eq. 5.54) into rotational and vibrational components has been discussed in Section 5.6. Therefore, we write the vibrational portion of Eq. 5.54 for a hypothetical particle of mass m_r which moves in the single direction x (equal to $r - r_\varepsilon$), and substitute Eq. 5.102 for the potential energy, $\Phi(x)$. The resulting wave equation for the harmonic oscillator is

$$\frac{d^2\Psi_v}{dx^2} + \frac{8\pi^2 m_r}{h^2}(\epsilon_v - 2\pi^2\nu_\varepsilon^2 m_r x^2)\Psi_v = 0 \quad (5.103)$$

In an attempt to convert this differential equation into a standard form as was done for the rigid rotator, let us first simplify the equation through a change of independent variable from x to z, where z is defined by the relation

$$z = \sqrt{\frac{4\pi^2\nu_\varepsilon m_r}{h}}\,x \quad (5.104)$$

Using the variable z, Eq. 5.103 reduces to the form

$$\frac{d^2\Psi_v}{dz^2} + (\alpha - z^2)\Psi_v = 0 \quad (5.105)$$

where, for convenience,

$$\alpha = \frac{2\epsilon_v}{h\nu_\varepsilon} \tag{5.106}$$

Next we transform the dependent variable Ψ_v to w by the definition

$$\Psi_v = e^{-z^2/2}w \tag{5.107}$$

Substituting Eq. 5.107 and the second derivative with respect to z into Eq. 5.105, we obtain the second-order differential equation

$$\frac{d^2w}{dz^2} - 2z\frac{dw}{dz} + (\alpha - 1)w = 0 \tag{5.108}$$

which is a standard form of differential equation.

Hermite's differential equation, using x and y as variables, is

$$\frac{d^2y}{dx^2} - 2x\frac{dy}{dx} + 2\mathbf{v}y = 0 \tag{5.109}$$

The solution of this differential equation for $\mathbf{v}$ equal to any non-negative integer is

$$y = H_\mathbf{v}(x) = (-1)^\mathbf{v} e^{x^2} \frac{d^\mathbf{v}}{dx^\mathbf{v}}[e^{-x^2}] \tag{5.110}$$

where the expressions of the form given by Eq. 5.110 are known as Hermite polynomials. An independent solution to the second-order differential (Eq. 5.109) includes an infinite series and is therefore not of interest to us because of our interpretation of the function Ψ_v^2.

Comparing our differential equation, Eq. 5.108, and the Hermite differential equation, Eq. 5.109, we find acceptable solutions of the wave equation only for

$$(\alpha - 1) = 2\mathbf{v} \qquad \mathbf{v} = 0, 1, 2, \ldots \tag{5.111}$$

From the definition of α, this specifies the allowed vibrational energy states in terms of the vibrational quantum number $\mathbf{v}$, and we find

$$\epsilon_v = h\nu_\varepsilon(\mathbf{v} + \tfrac{1}{2}) \qquad \mathbf{v} = 0, 1, 2, 3, \ldots \tag{5.112}$$

Frequently, it is more convenient to use the vibrational wave number ω_ε, defined by

$$\omega_\varepsilon = \frac{\nu_\varepsilon}{c} \tag{5.113}$$

where c is the velocity of light. In this form, the vibrational energy is written

$$\epsilon_v = hc\omega_\varepsilon(\mathbf{v} + \tfrac{1}{2}) \qquad \mathbf{v} = 0, 1, 2, 3, \ldots \tag{5.114}$$

It is also of interest to note here that for the ground state $\mathbf{v} = 0$ there is a residual or zero-point energy contribution of $\frac{1}{2}h\nu_\varepsilon$ or $\frac{1}{2}hc\omega_\varepsilon$.

The vibrational wave function Ψ_v is now found to be

$$\Psi_v(z) = Ce^{-z^2/2}H_v(z) \tag{5.115}$$

where

$$z = \sqrt{\frac{4\pi^2 c\omega_\varepsilon m_r}{h}}\,(r - r_\varepsilon) \tag{5.116}$$

Unlike the result for the rigid rotator, we find here but one wave function corresponding to each value of energy, and therefore we conclude that the vibrational levels for a diatomic molecule have a statistical weight of unity.

The constant in Eq. 5.115 is determined by normalization such that $\Psi_v^2(z)$ represents the probability of finding a molecule of energy ϵ_v at position z. The details of carrying out this process need not concern us here, but the results present a very informative picture, as shown in Figure 5.8. For each energy level, the probability distribution as represented by Ψ_v^2 has been superimposed on the diagram of energy versus distance from the equilibrium point.

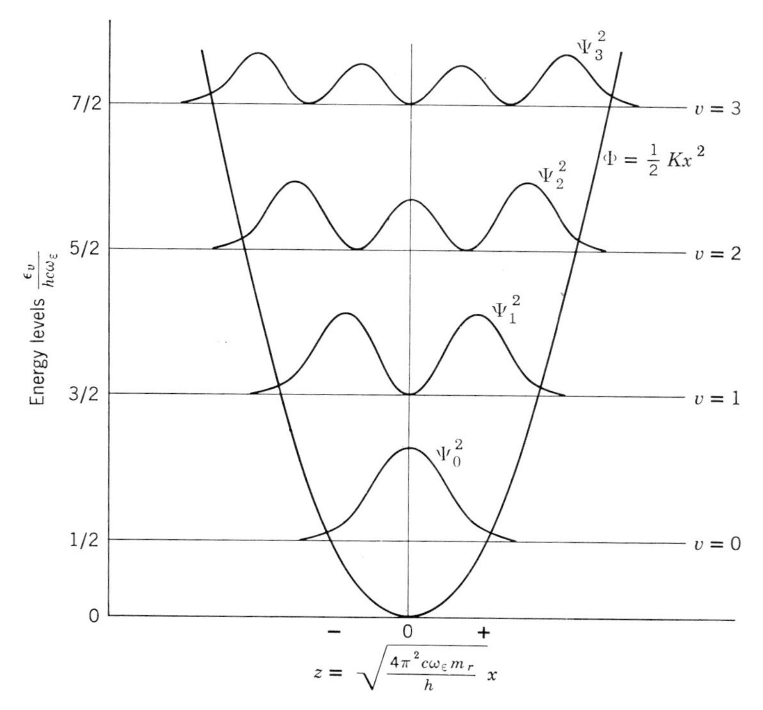

Figure 5.8 The probability function and energy levels of a harmonic oscillator.

5.8. General Rotation-Vibration

In the three preceding sections, we have considered the so-called simple internal model for a diatomic molecule. This model consists of a rigid rotator, harmonic oscillator, and electronic ground level. For the rigid rotator, we found that the rotational energy ϵ_r is given in terms of the rotational quantum number **j** by the relation (Eq. 5.87)

$$\epsilon_r = \frac{h^2}{8\pi^2 I_\varepsilon} \mathbf{j}(\mathbf{j} + 1) \qquad \mathbf{j} = 0, 1, 2, 3, \ldots$$

and it was also found (Eq. 5.90) that each of the rotational levels has a degeneracy g_r given by

$$g_r = (2\mathbf{j} + 1)$$

Similarly, for the harmonic oscillator the energy ϵ_v is given in terms of the vibrational quantum number **v** by Eq. 5.114 as

$$\epsilon_v = hc\omega_\varepsilon(\mathbf{v} + \tfrac{1}{2}) \qquad \mathbf{v} = 0, 1, 2, 3, \ldots$$

where each vibrational level has a degeneracy of unity.

We have stated in Section 5.5 that the simple internal model is a reasonable approximation under most conditions, but that it becomes increasingly inaccurate with increasing temperature because of rotational centrifugal stretching of the molecule, anharmonicity of the vibration, and interaction or coupling between rotation and vibration. Let us now examine each of these effects in terms of a correction to the results of the simple model, so that we may obtain a suitable internal model of the diatomic molecule for high temperature application.

It is convenient to express the energies in wave number units (ϵ/hc), for which the rigid rotator energy given by Eq. 5.87 becomes

$$\frac{\epsilon_r}{hc} = B_\varepsilon \mathbf{j}(\mathbf{j} + 1) \tag{5.117}$$

in which the constant B_ε has been defined as

$$B_\varepsilon = \frac{h}{8\pi^2 I_\varepsilon c} \tag{5.118}$$

This result is found to be very reasonable at moderate temperatures but at high temperatures where the rotational energy becomes very large, the molecule stretches due to centrifugal force, with a resulting increase in the moment of inertia above the value I_ε. A correction for this centrifugal stretching can be included by expressing rotational energy as

$$\frac{\epsilon_r}{hc} = B_\varepsilon \mathbf{j}(\mathbf{j}+1) - D_\varepsilon \mathbf{j}^2(\mathbf{j}+1)^2 + \cdots \tag{5.119}$$

where D_ε is called the centrifugal or rotational stretching constant. Other correction terms might also be considered, but are not included here.

In order to examine the effects of anharmonicity, we refer to the harmonic oscillator results. It was found in Section 5.7 that the vibrational energy levels for a harmonic oscillator are equally spaced according to Eq. 5.112 or Eq. 5.114 and as indicated in Figure 5.8. In general, the potential energy function is of a more complex nature, as shown by the solid curve in Figure 5.9. The resulting energy levels for vibration are also shown there. For the anharmonic oscillator, the potential energy curve tends toward a limiting value as the distance of separation becomes very large, that is, as the molecule dissociates into two separate atoms. This value is commonly taken as the zero of energy, such that the energy at the minimum point of the curve is $-D$, which is called the total dissociation energy of the molecule. However, we realize that this minimum point is not accessible to the molecule, which in its ground state still possesses a zero-point energy corresponding to $\mathbf{v} = 0$. Thus the observed dissociation energy of the molecule will be that corresponding to the ground state $\mathbf{v} = 0$, with an energy $-D_0$ relative to that of the dissociated molecule.

It is important to note two features of Figure 5.9. The harmonic oscillator curve, shown by the broken line, is seen to become badly in

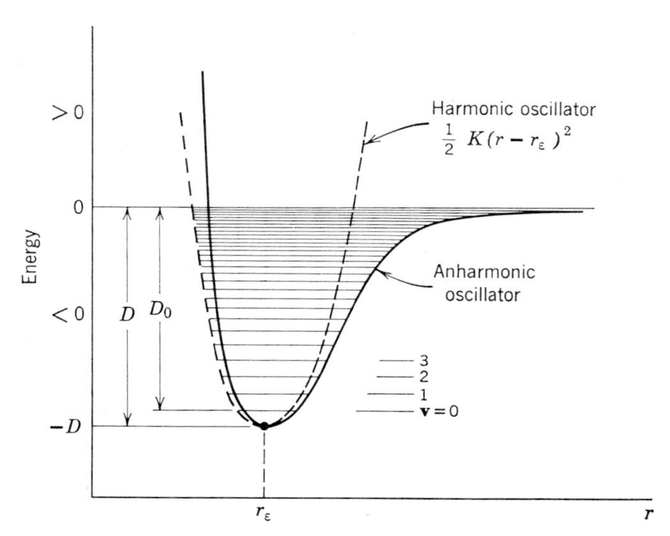

Figure 5.9 The potential energy function of an anharmonic oscillator.

error for $r \gg r_\varepsilon$. In addition, the vibrational energy level spacings are approximately equal only for the lower levels. We must conclude that the harmonic oscillator model for vibration will become erroneous at high temperature (say 3000–4000 K) where an appreciable fraction of the molecules are excited into the higher vibrational levels.

For the anharmonic oscillator, we express the energy as that for the harmonic oscillator, Eq. 5.114, plus correction terms, or

$$\frac{\epsilon_v}{hc} = \omega_\varepsilon(\mathbf{v} + \tfrac{1}{2}) - x_\varepsilon\omega_\varepsilon(\mathbf{v} + \tfrac{1}{2})^2 + y_\varepsilon\omega_\varepsilon(\mathbf{v} + \tfrac{1}{2})^3 + \cdots \qquad (5.120)$$

where x_ε and y_ε are called anharmonicity constants, which are generally small. Thus the correction terms of Eq. 5.119 become significant only for large values of the vibrational quantum number $\mathbf{v}$.

The vibrational energy as given by Eq. 5.120 is the energy with respect to the minimum point on the potential curve. We have noted, however, that the molecule never reaches this point. For the ground state of the molecule ($\mathbf{v} = 0$), the energy is an amount above the minimum point,

$$\epsilon_{v_0} = hc(\tfrac{1}{2}\omega_\varepsilon - \tfrac{1}{4}x_\varepsilon\omega_\varepsilon + \tfrac{1}{8}y_\varepsilon\omega_\varepsilon + \cdots) = \tfrac{1}{2}h\nu_\varepsilon(1 - \tfrac{1}{2}x_\varepsilon + \tfrac{1}{4}y_\varepsilon + \cdots) \qquad (5.121)$$

which is the correct difference between the D_0 and the D of Figure 5.9, instead of the $\tfrac{1}{2}h\nu_\varepsilon$ zero-point value resulting from the harmonic oscillator model.

There is still another significant problem at high temperature, that of interaction between rotation and vibration. Since the mean distance of separation between atoms increases as the vibrational level increases, the distance to be used in calculating the moment of inertia for Eq. 5.119 will be a function of the vibrational state. This effect will, of course, be negligible at moderate temperatures, where nearly all the molecules are in the lowest few vibrational states. However, at high temperature, Eq. 5.119 must be rewritten as

$$\frac{\epsilon_r}{hc} = B_v\mathbf{j}(\mathbf{j} + 1) - D_v\mathbf{j}^2(\mathbf{j} + 1)^2 + \cdots \qquad (5.122)$$

where

$$B_v = B_\varepsilon - \alpha(\mathbf{v} + \tfrac{1}{2}) + \cdots \qquad (5.123)$$

$$D_v = D_\varepsilon + \beta(\mathbf{v} + \tfrac{1}{2}) + \cdots \qquad (5.124)$$

The constants α and β are called coupling or vibrational stretching constants.

Including the three high temperature effects, centrifugal stretching, anharmonicity, and rotation-vibration coupling, the combined rotational-vibrational energy for a diatomic molecule in its electronic ground level can be written as the sum of Eqs. 5.120 and 5.122, which is

$$\frac{\epsilon_{rv}}{hc} = \omega_\varepsilon(\mathbf{v} + \tfrac{1}{2}) - x_\varepsilon\omega_\varepsilon(\mathbf{v} + \tfrac{1}{2})^2 + B_\varepsilon\mathbf{j}(\mathbf{j} + 1) - D_\varepsilon\mathbf{j}^2(\mathbf{j} + 1)^2 - \alpha(\mathbf{v} + \tfrac{1}{2})\mathbf{j}(\mathbf{j} + 1) + \cdots \quad (5.125)$$

The constants y_ε, β, and other higher order corrections are included when necessary for very precise calculations.

Equation 5.125, which is a general expression for rotational-vibrational energy, has been developed here strictly as an empirical equation. It consists of terms that give the rotational and vibrational energies for the simple internal model plus a number of correction terms, which take into account the deviations of actual behavior from that of the simple model. These correction constants, x_ε, D_ε, α, . . . , which, with ω_ε, I_ε best fit the data according to Eq. 5.125, can be determined from spectroscopic data for the energy levels of the molecule.

There is another approach to the problem, that is, to determine the energy levels for general rotation-vibration analytically from the unseparated Schrödinger wave equation, Eq. 5.54. This approach requires an expression for the potential energy of the shape shown in Figure 5.9. One such analytical function used with considerable success is the Morse potential,

$$\Phi(r) = D[e^{-2b(r-r_\varepsilon)} - 2e^{-b(r-r_\varepsilon)}] \quad (5.126)$$

For this potential, $\Phi \rightarrow -D$ as $r \rightarrow r_\varepsilon$, so that D is the total dissociation energy discussed previously. We also note the correct behavior of this function at large r, where $\Phi \rightarrow 0$. The constant b in the Morse potential is assigned a value to give the curve the proper shape in the vicinity of the equilibrium point, where the vibration is known to be essentially harmonic. To determine this value, we expand the two exponentials of Eq. 5.126 in series of $(r - r_\varepsilon)$ and, after collecting terms, identify the first term in the series with the potential energy of a harmonic oscillator. The result is

$$b = \omega_\varepsilon\left(\frac{2\pi^2c^2m_r}{D}\right)^{1/2} \quad (5.127)$$

Even for the Morse potential (Eq. 5.126), the wave equation, Eq. 5.54, cannot be solved exactly to determine rotational-vibrational energy levels because of the complexity of the differential equation. However, by means of a double series expansion and comparison of the first terms of the result with those of Eq. 5.125, the correction constants can be identified with known quantities. By this technique, we find the anharmonicity constant given by

$$x_\varepsilon = \frac{hc\omega_\varepsilon}{4D} \quad (5.128)$$

the centrifugal stretching constant given by the expression

$$D_\varepsilon = \frac{4B_\varepsilon^{\ 3}}{\omega_\varepsilon^{\ 2}} \tag{5.129}$$

and the coupling constant given by

$$\alpha = \frac{6B_\varepsilon^{\ 2}}{\omega_\varepsilon}\left[\left(\frac{x_\varepsilon \omega_\varepsilon}{B_\varepsilon}\right)^{1/2} - 1\right] \tag{5.130}$$

The set of relations, Eqs. 5.128 to 5.130, for the correction constants is especially useful for substances for which only minimal data are available, namely, the values for atomic masses, ω_ε, r_ε, and D (from observed values of D_0). However, if sufficient spectroscopic data are known, it would be somewhat more precise to determine these and additional constants empirically to best fit Eq. 5.125, because of the approximations made in developing Eqs. 5.128 to 5.130.

5.9. Electronic States of Atoms and Molecules

In the application of the Schrödinger wave equation to internal energy modes of a molecule, we have already analyzed the simple internal model—rigid rotator, harmonic oscillator, ground electronic level—and have also discussed the more general combined rotation-vibration of a molecule. We have not as yet discussed the electronic states for atoms or molecules. For our purposes, a detailed investigation of this subject is not necessary, because for most substances in the temperature range of interest the electronic energy levels are so widely spaced that essentially all the particles are found in the ground state at any instant of time. Nevertheless, we should be familiar enough with the subject to be able to recognize the exceptions to this statement. A second and still more important reason for at least a qualitative discussion of electronic states is that the ground level of an atom or molecule may be degenerate, resulting in a contribution to the entropy of the substance.

Let us consider the most elementary case, the hydrogen atom, as we did at the beginning of the chapter in our discussion of Bohr's theory. We apply the Schrödinger wave equation, Eq. 5.29, to the atom's single electron, which is under the influence of the nucleus according to the Coulomb potential. It is again convenient to work in spherical polar coordinates, so that the Laplacian operator is given by Eq. 5.66, which was used for the molecular rigid rotator. In direct contrast to that application, however, the distance r in this case is not constant, and the first term of Eq. 5.66 is not equal to zero. To achieve a solution of the differential equation, we assume that the wave function can be expressed as the product

$$\Psi(r, \theta, \phi) = \Psi_r(r)\Psi_\theta(\theta)\Psi_\phi(\phi) \tag{5.131}$$

where each of the three functions is dependent upon only one of the coordinates. Substitution of Eq. 5.131 into the wave equation results in a separation of the differential equation into three terms, each in terms of ordinary derivatives. Since the variables r, θ, ϕ are independent, we conclude that each term must be a constant. Solution of the θ and ϕ terms then follows directly in exactly the same manner as for the molecular rigid rotator. From this solution we obtain the pair of quantum numbers **l**, a non-negative integer analogous to the **j** for a molecule, and $\mathbf{m}_l$, where

$$\mathbf{m}_l = 0, \pm 1, \pm 2, \ldots, \pm \mathbf{l} \tag{5.132}$$

and is analogous to the λ for a molecule.

The remaining second-order differential equation in r then includes the potential energy resulting from the Coulomb force and also the term in $\mathbf{l}(\mathbf{l}+1)$. We find that a solution of this difficult equation is valid only if the energy quantized according to the relation

$$\epsilon = \left(\frac{2\pi^2 C^2 \mathscr{L}^2 e^4 m}{h^2}\right)\frac{1}{\mathbf{n}^2} \tag{5.133}$$

where the quantum number **n**, called the principal quantum number, can assume only positive integral values,

$$\mathbf{n} = 1, 2, 3, \ldots \tag{5.134}$$

For an acceptable solution it is also required that the azimuthal number **l** be restricted to the values

$$\mathbf{l} = 0, 1, 2, \ldots, (\mathbf{n} - 1) \tag{5.135}$$

while the values of $\mathbf{m}_l$, the magnetic quantum number, are as given by Eq. 5.132. The orbital angular momentum is found to be

$$\sqrt{\mathbf{l}(\mathbf{l}+1)}\left(\frac{h}{2\pi}\right)$$

in a manner similar to that for a molecule in Eq. 5.91. The $(2\mathbf{l}+1)$ orientations of the corresponding vector with respect to the z-axis are given by the z-component $\mathbf{m}_l\ (h/2\pi)$, similar to that shown for a molecule in Figure 5.6.

From our solution, we find that the energy levels for atomic hydrogen are dependent only upon the principal quantum number **n**, as given by

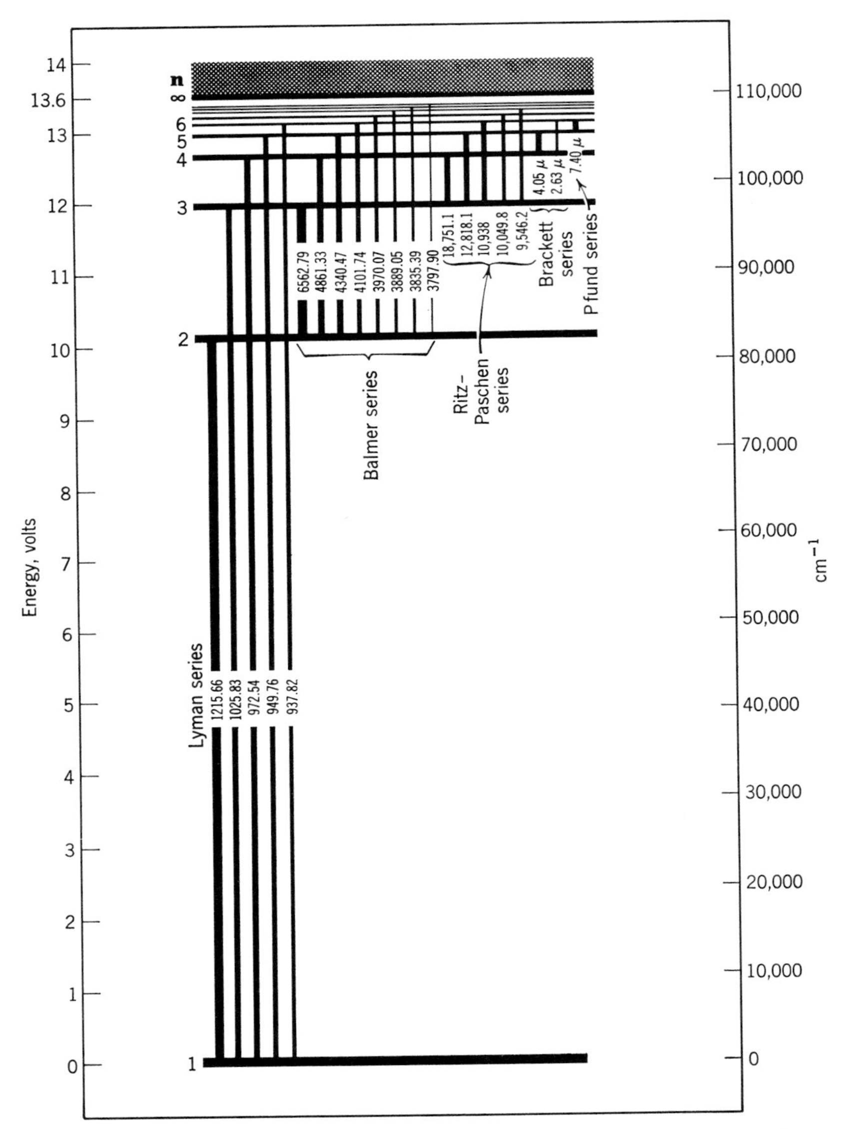

Figure 5.10 Electronic energy levels of the hydrogen atom. From G. Herzberg, *Atomic Spectra and Atomic Structure*, Dover Publications, New York, 1944, p. 24.

Eq. 5.133 and shown in Figure 5.10. Energies are given in units of electron volts and also reciprocal centimeters (ϵ/hc) with respect to the ground state. The limiting value for very large **n** is 13.595 volts (this is called the ionization potential), under which condition the electron leaves the influence of the nucleus. The vertical lines in the diagram show some of the transitions observed in the spectra of the hydrogen atom; the length of the line between any two levels is proportional to the wave number of the observed spectral line and the width is proportional to its intensity. Since the energy levels depend only upon **n**, the levels must be degenerate according to the number of allowed values of the quantum numbers **l** and $\mathbf{m}_l$. Actually, from a more general solution which includes the effects of relativity and electron spin, it is found that there are very slight differences in energy for different values of **l** and $\mathbf{m}_l$; this is in agreement with very precise spectral observations. The energy levels for atomic hydrogen, including the azimuthal number **l**, is shown in Figure 5.11. The dashed lines indicate so-called forbidden transitions, which ordinarily are not observed in the spectra.

The results discussed here for the hydrogen atom are the same as those found from Bohr's theory. We find very striking differences between the theories, however, even for hydrogen. In the new quantum or wave mechanics, the results follow directly from the sole assumption of the validity of the wave equation. Another important point that should be emphasized is that, although the energy is sharply defined according to Eq. 5.133, there has been no mention of the nature of orbits as assumed by Bohr. Indeed, if we evaluate the wave function of Eq. 5.131 in terms of r, θ, ϕ and normalize in the three directions, the square of the result then designates the probability of locating the electron in a certain position with respect to the nucleus. Let us consider here the radial direction r only. For given values of **n** (and energy) and **l**, the quantity Ψ'^2 represents the probability of locating the electron within an element of volume. Since the volume element is represented by $4\pi r^2\,dr$, the probability that the electron will be found between r and $r + dr$ and with any values of θ, ϕ is

$$4\pi\Psi'^2 r^2\,dr$$

The nature of the function $\Psi'^2 r^2$ is depicted in Figure 5.12 for the first few values of **n** and **l**. It is of particular interest to compare the probability functions of Figure 5.12 with the corresponding orbits of the Bohr theory shown in Figure 5.1. The marks on the r-axes in Figure 5.12 indicate the radii of the corresponding circular Bohr orbits, which are seen to coincide with the maxima of the probability function whenever $\mathbf{l} = \mathbf{n} - 1$. Whenever $\mathbf{l} \neq \mathbf{n} - 1$, there is more than one peak in the probability function, which corresponds to the elliptical orbits of the old

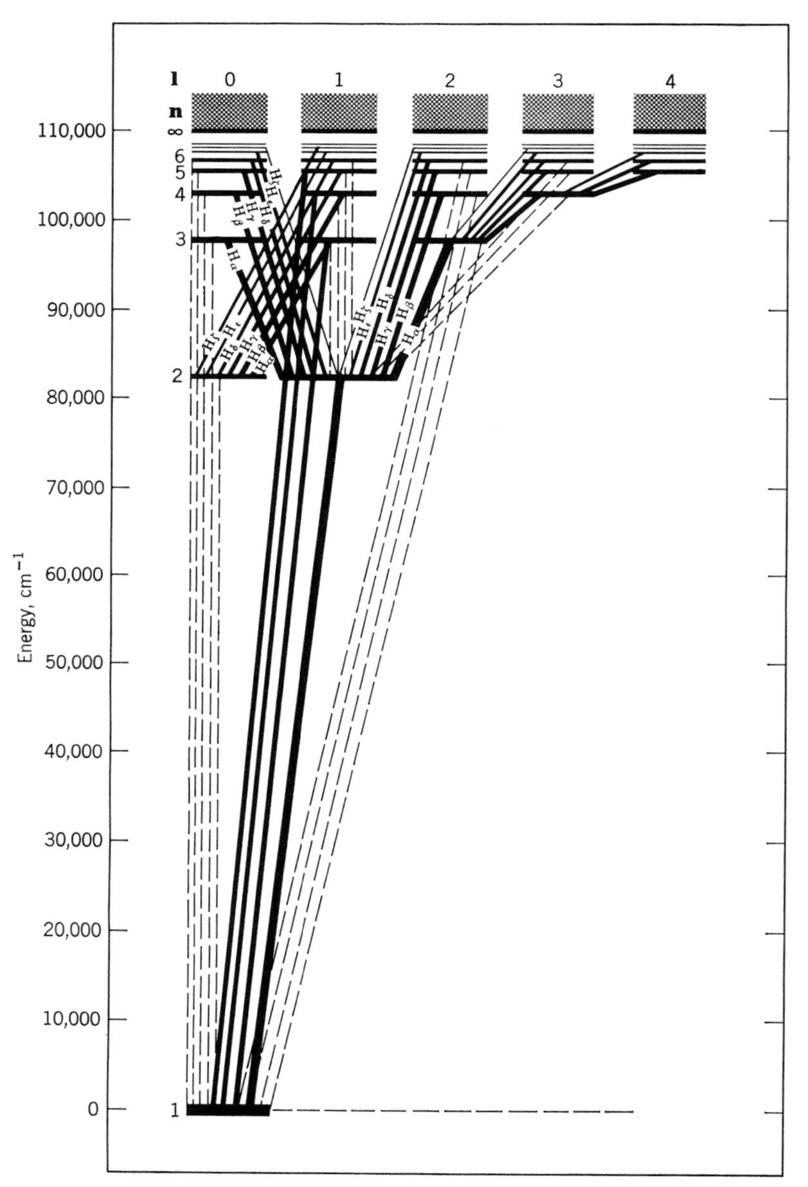

Figure 5.11 Electronic energy levels of the hydrogen atom. From G. Herzberg, *Atomic Spectra and Atomic Structure*, Dover Publications, New York, 1944, p. 26.

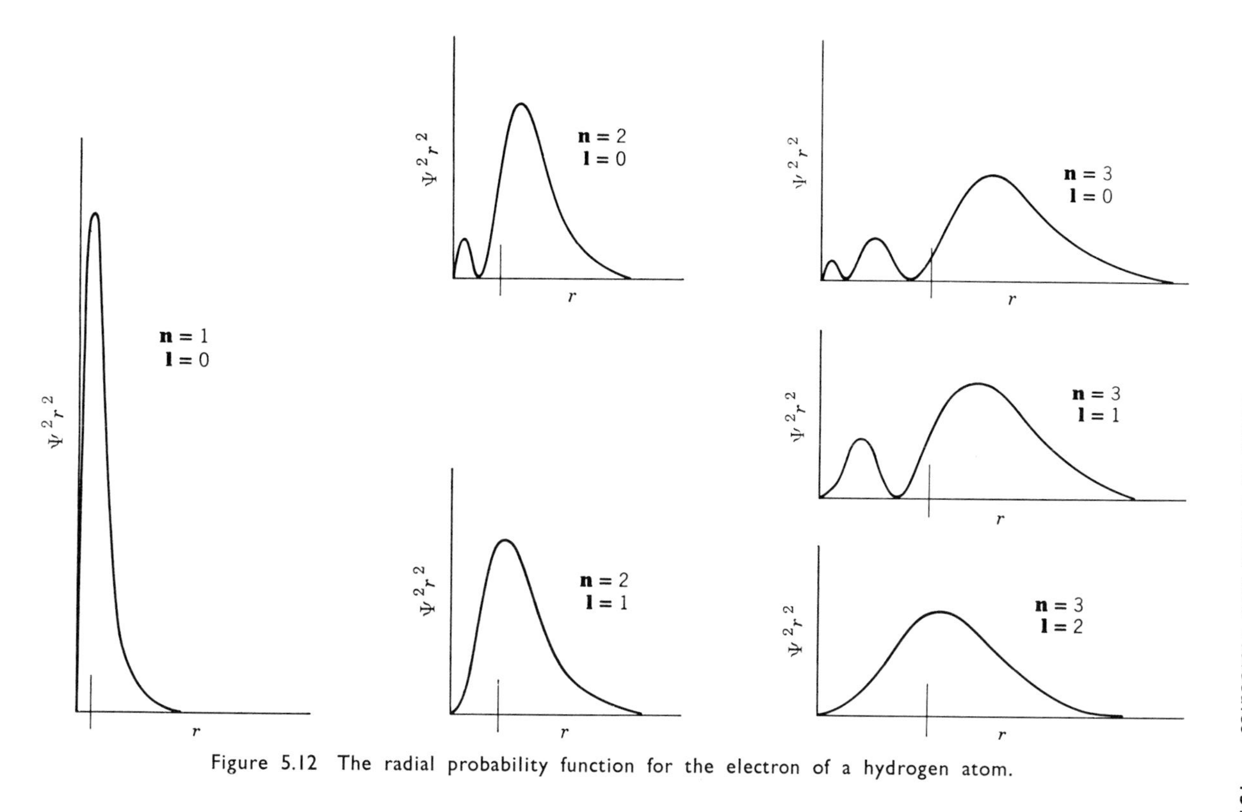

Figure 5.12 The radial probability function for the electron of a hydrogen atom.

theory. From this comparison, we can understand why Bohr's theory was able to predict certain behavior as well as it did although it failed in other cases.

It should also be emphasized that the values of the azimuthal and magnetic quantum numbers $\mathbf{l}$ and $\mathbf{m}_l$ give a shape to the atom as they influence the wave function and consequently the probability in the θ and ϕ directions.

TABLE 5.1

Electron State	$\mathbf{n}$	$\mathbf{l}$	$\mathbf{m}_l$	$\mathbf{m}_s$
1*s*	1	0	0	$\pm\frac{1}{2}$
2*s*	2	0	0	$\pm\frac{1}{2}$
2*p*		1	-1	$\pm\frac{1}{2}$
			0	$\pm\frac{1}{2}$
			$+1$	$\pm\frac{1}{2}$
3*s*	3	0	0	$\pm\frac{1}{2}$
3*p*		1	-1	$\pm\frac{1}{2}$
			0	$\pm\frac{1}{2}$
			$+1$	$\pm\frac{1}{2}$
3*d*		2	-2	$\pm\frac{1}{2}$
			-1	$\pm\frac{1}{2}$
			0	$\pm\frac{1}{2}$
			$+1$	$\pm\frac{1}{2}$
			$+2$	$\pm\frac{1}{2}$

The fourth quantum number of the atom, the electron spin number $\mathbf{m}_s$, did not appear in our solution of the wave equation, but it does result logically in the more general solution. We can justify the existence of such a number here in the same way as it was done before the development of wave mechanics. It is observed that the spectral lines of certain atoms split into even-numbered multiples under the influence of a magnetic field, while the splitting of the states of different $\mathbf{m}_l$ can only result in odd numbers 1, 3, 5, ... Therefore there must be a fourth quantum number to account for this observed phenomenon. Considering an electron to behave as a tiny spinning magnet with $\mathbf{m}_s = \pm\frac{1}{2}$ (indicating the orientation of the spin angular momentum vector) resolves this problem. Table 5.1 lists the first few quantum states for the single electron of the hydrogen atom and shows the restrictions on the four quantum numbers. For the hydrogen atom in the absence of a magnetic field, the energy is primarily a function of the principal quantum number. Thus there are two states of the same energy for the ground level ($\mathbf{n} = 1$), eight states of approximately the same energy for the first excited level ($\mathbf{n} = 2$), eighteen for the second excited level ($\mathbf{n} = 3$), etc. Since the energy actually does depend slightly upon $\mathbf{l}$ and $\mathbf{m}_l$ even for hydrogen, we must be careful not to say that these states have exactly the same energy.

The spectroscopic notation used to designate electron configuration is also listed in Table 5.1. In this notation, the number refers to the value of $\mathbf{n}$ and the letter following denotes the azimuthal number $\mathbf{l}$. Values of $\mathbf{l}$ equal to 0, 1, 2, 3, . . . are designated by the letters *s*, *p*, *d*, *f*, . . . , in this notation. For atoms having more than one electron, a superscript after the letter gives the number of electrons with those values of $\mathbf{n}$ and $\mathbf{l}$. For example, the symbol $2p^4$ means that four electrons have the numbers $\mathbf{n} = 2, \mathbf{l} = 1$.

So far we have discussed only the hydrogen atom, because of its simplicity. The first logical extension of the theory would be to the He^+ ion, which has a nucleus of two protons and two neutrons and a single electron. Evaluation of this problem is the same as for the H atom, except for a different Coulomb potential between the nucleus and electron. The complexity grows rapidly when we consider the helium atom, however, because now there are two electrons instead of a single one. There will thus be a potential between each electron and the nucleus, and another between the pair of electrons. However, the latter potential is relatively small, and this is neglected in order to achieve a solution to the problem. The wave equation is then written for each electron, and after combination it is found that valid solutions exist only for sets of the quantum numbers discussed previously and as given in Table 5.1. There is an additional consideration now, though, that of the Pauli exclusion principle, which will be discussed in the following section. For the present, let us say that one statement of the exclusion principle is that no two electrons in a single atom may simultaneously have an identical set of quantum numbers. Therefore, again referring to Table 5.1, we conclude that, for the ground state of the helium atom, one electron has the numbers 1, 0, 0, $+\frac{1}{2}$, and the other has 1, 0, 0, $-\frac{1}{2}$. That is, if the numbers $\mathbf{n}$, $\mathbf{l}$, $\mathbf{m}_l$ are all the same for two electrons, then their spins must be in opposite directions.

For more complex atoms with greater numbers of electrons, we find that, according to Pauli's exclusion principle, additional electrons must fall in successively higher states because the lower ones are already filled. As an example, the ground level of the oxygen atom has the electron configuration $1s^2 2s^2 2p^4$.

The periodic table of the elements is also readily understood as a result of this theory. We will consider two examples here, the inert gases and the halogens. The inert gases—helium (2 electrons), neon (10), argon (18), krypton (36), xenon (54)—all have closed or filled outer shells or subshells and are therefore very stable and chemically inactive (for Xe, there are no $4f$ electrons, which would have a higher energy than $5s$ and $5p$ electrons; hence the closed outer shell is $5p^6$). On the other hand, the halogens—fluorine (9 electrons), chlorine (17), bromine (35), iodine (53)—all lack a single electron to complete a shell or subshell and are consequently very active chemically (the situation with I is like that for Xe, the outer shell being $5p^5$). Each of the halogens has a low-lying electronic

TABLE 5.2
Atomic Energy Levels

Substance	Electron Configuration	Term Symbol	Energy, cm^{-1}
H (1 electron)	$1s$	$^2S_{1/2}$	0.00
	$2p$	$^2P_{1/2}$	82258.907
	$2s$	$^2S_{1/2}$	82258.942
	$2p$	$^2P_{3/2}$	82259.272
	$3p$	$^2P_{1/2}$	97492.198
He (2)	$1s^2$	$^1S_0$	0
	$1s2s$	$^3S_1$	159850.318
		$^1S_0$	166271.70
N (7)	$1s^22s^22p^3$	$^4S_{3/2}$	0
		$^2D_{5/2}$	19223.9
		$^2D_{3/2}$	19233.1
O (8)	$1s^22s^22p^4$	$^3P_2$	0
		$^3P_1$	158.5
		$^3P_0$	226.5
		$^1D_2$	15867.7
F (9)	$1s^22s^22p^5$	$^2P_{3/2}$	0
		$^2P_{1/2}$	404.0
	$1s^22s^22p^43s$	$^4P_{5/2}$	102406.5

From C. E. Moore, "Atomic Energy Levels," *National Bureau of Standards Circular* **467**, Vols. I–III (1949, 1952, 1958).

state. The first few electronic states observed for several simple atoms are given in Table 5.2.

For the majority of substances the energy of the first excited electronic level is so high that at moderate temperature all but a negligible fraction of the atoms are in the ground level at any time. Even for such substances, it is necessary that the degeneracy of the ground level be known, and, as this can be found from the term symbol for the level, we should be familar with this notation. As an example, let us consider the ground level of the helium atom with the term symbol $^1S_0$. The letter indicates the value of the orbital angular momentum quantum number **L** for the atom, as did the number **l** for a single electron. Thus the states S, P, D, F, . . . represent

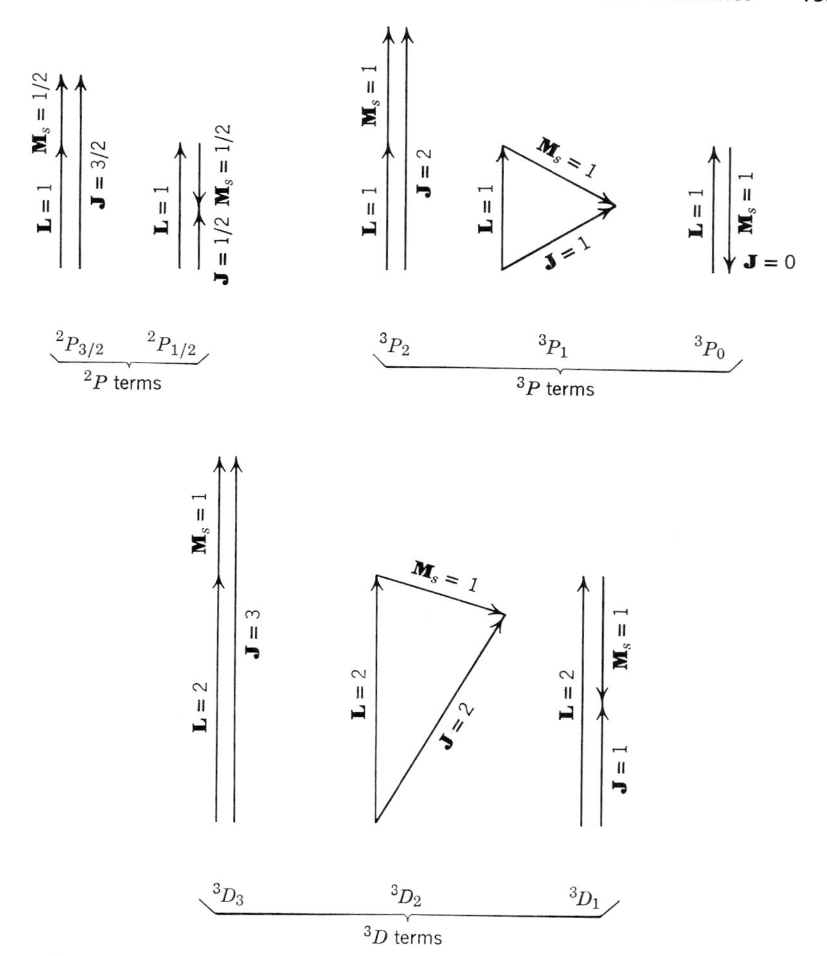

Figure 5.13 Vector representation of the electronic states of an atom.

values of **L** of 0, 1, 2, 3, . . . , respectively. The superscript at the left indicates the multiplicity of the state and equals $2\mathbf{M}_s + 1$, where $\mathbf{M}_s$ is the electron spin number of the atom. Since a single electron can have only the spin numbers $\pm\frac{1}{2}$, the spin number for an atom must necessarily be an integer, 0, 1, 2, 3, . . . , for an atom having an even number of electrons, and a half-integer, $\frac{1}{2}$, $\frac{3}{2}$, $\frac{5}{2}$, $\frac{7}{2}$, . . . , for one having an odd number of electrons. The subscript on the right of the term symbol is the quantity of particular interest to us here. This subscript gives the total angular momentum number **J** of the atom, the resultant of the vector addition of the quantities **L** and $\mathbf{M}_s$, a few examples of which are shown in Figure 5.13.

For each term there are $2\mathbf{J} + 1$ components of the total angular momentum $\sqrt{\mathbf{J}(\mathbf{J} + 1)}(h/2\pi)$ along a chosen axis, each with essentially the same energy in the absence of a magnetic field. Therefore there are $(2\mathbf{J} + 1)$ wave functions for each $\mathbf{J}$, and the degeneracy of each term is given by that quantity $(2\mathbf{J} + 1)$. For example, the ground electronic level of monatomic oxygen is $^3P_2$, and the first two excited levels are $^3P_1$, $^3P_0$. Consequently, the corresponding degeneracies are 5, 3, 1, respectively. The energy of the next excited level is relatively high and would not

TABLE 5.3
Molecular Electronic Levels

Molecule	Ground Term	Energy (ϵ/hc) of Next Term, cm^{-1}
H_2	$^1\Sigma_g^+$	90,171
NO	$^2\Pi$	121
O_2	$^3\Sigma_g^-$	7,882
Cl_2	$^1\Sigma_g^+$	18,147
I_2	$^1\Sigma_g^+$	11,803

From G. Herzberg, *Molecular Spectra and Molecular Structure, I, Spectra of Diatomic Molecules*, 2nd edition, D. Van Nostrand Co., Princeton, N.J., 1950.

contribute significantly to thermodynamic properties at moderate temperature. This term, $^1D_2$ ($\mathbf{M}_s = 0$, $\mathbf{L} = 2$, $\mathbf{J} = 2$), also has a degeneracy of 5. For hydrogen we find that only the ground level is important at moderate temperature since the first excited level is so high, but we draw the most important conclusion that the ground level degeneracy is equal to 2. We note also that the degeneracies of the next three terms are 2, 2, 4, for a total of eight states all having very nearly the same observed energy, which of course is in agreement with our previous discussion and Table 5.1, where there were eight states of $\mathbf{n} = 2$.

A discussion of the electronic states for molecules is necessarily even more complex than for atoms. Fortunately, for most molecules even the first excited level is sufficiently high in energy that it is satisfactory to assume that all molecules are in the ground level at any given time. This means that our only problem is that of determining the degeneracy of the electronic ground level, which can be found from the term symbol for the molecule. To illustrate the procedure, let us consider diatomic hydrogen, with the ground level term symbol $^1\Sigma_g^+$. The Greek symbol indicates the value of the molecule's orbital angular momentum number $\mathbf{\Lambda}$, just as do the English letters for an atom. Thus terms corresponding to values of $\mathbf{\Lambda}$

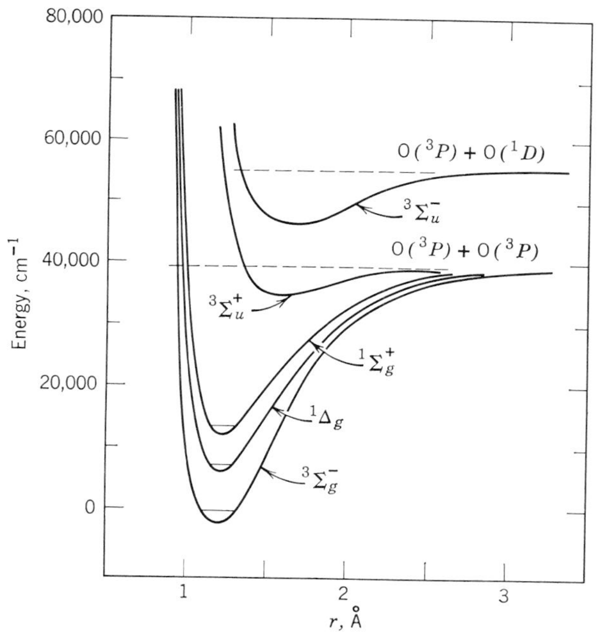

Figure 5.14 The potential energy functions for diatomic oxygen. From G. Herzberg, *Molecular Spectra and Molecular Structure, I, Spectra of Diatomic Molecules*, 2nd edition, D. Van Nostrand Co., Princeton, N.J., 1950, p. 446.

of 0, 1, 2, 3, . . . are represented by Σ, Π, Δ, Φ, . . . , respectively. The left superscript is $2\mathbf{M}_s + 1$, where $\mathbf{M}_s$ is the electron spin number. The right subscript and superscript on the term symbol refer to symmetry characteristics of the wave function; they do not concern us here.

For a diatomic molecule, the degeneracy of an electronic level is found to be $2(2\mathbf{M}_s + 1)$, except for a Σ term, which has a degeneracy of $(2\mathbf{M}_s + 1)$. Very briefly, the reason is that the orbital angular momentum vector is quantized along the axis joining the nuclei and can have only the components $\pm\mathbf{\Lambda}(h/2\pi)$, or a degeneracy of 2 when $\mathbf{\Lambda} \neq 0$. The spin vector can have $(2\mathbf{M}_s + 1)$ orientations; hence the combination of the two results in the values given above.

The ground level term symbol and energy of the next term for several diatomic molecules are listed in Table 5.3. It is seen that the NO molecule has an extremely low-lying electronic level, and that O_2 has one low enough to contribute to properties at moderate temperature. Actually, diatomic oxygen has several terms of not extremely different energies, as shown in Figure 5.14. For molecules with such low-lying terms, the problem of evaluation of the partition function becomes extremely tedious, because the rotational and vibrational constants are in general somewhat different for the different electronic levels. Thus each electronic level must be evaluated separately, as each has its own rotation and vibration.

5.10. The Pauli Exclusion Principle

In this section we discuss the symmetry nature of wave functions, the Pauli exclusion principle, and the very important point—whether Fermi-Dirac statistics or Bose-Einstein statistics represent the behavior of a given system of particles.

For simplicity, let us consider two quantum states i and j, having energies ϵ_i and ϵ_j, respectively. Now suppose that there are two identical particles 1 and 2, one in each of the states i, j. This pair may be two electrons, two protons, etc. The total energy of the system, then, is

$$\epsilon = \epsilon_i + \epsilon_j$$

and we have a situation referred to as exchange degeneracy, since the same energy results if particle 1 is in state i and particle 2 is in state j, or vice versa. The system wave functions describing these two possibilities can be written as the product of the single-particle wave functions, neglecting interactions, or

$$\Psi_i(1)\Psi_j(2)$$

or

$$\Psi_i(2)\Psi_j(1)$$

The first of these is the product of the single-particle functions (including spin coordinates) for particle 1 in state i, particle 2 in state j. The second is the product of the functions for particle 2 in state i, particle 1 in state j. Obviously, both must satisfy the combined or system wave equation describing the two-particle system. We have a problem, though, in that either of these system functions considers that the two particles can be distinguished from one another, such that we can tell which particle is in which quantum state. Such a function would, of course, be unacceptable, as a basic hypothesis of quantum mechanics is that the particles are identical and indistinguishable. This difficulty is resolved when we realize that linear combinations of these functions must also satisfy the wave equation. We have two choices, then, for a system wave function, namely a symmetric function

$$\Psi_S = \Psi_i(1)\Psi_j(2) + \Psi_i(2)\Psi_j(1) \tag{5.136}$$

or an antisymmetric function

$$\Psi_A = \Psi_i(1)\Psi_j(2) - \Psi_i(2)\Psi_j(1) \tag{5.137}$$

We note that the sign of the symmetric system wave function remains the same upon interchange of the coordinates of the two identical particles, but the antisymmetric system function changes sign if the particles are interchanged.

For any system of a pair of identical particles, either Eq. 5.136 or Eq. 5.137 must be valid (neglecting interaction). The important point is that both cannot be valid because then their linear combination would also necessarily be valid, and we would be back to our original, unacceptable functions.

From the experimentally observed behavior, it is found that particles with half-integral spin—electrons, protons, nuclei of odd mass number, for example—have only antisymmetric states and wave functions. Particles with integral spin—photons, nuclei of even mass number—have only symmetric states and wave functions.

It is not difficult to construct symmetric or antisymmetric system functions from single-particle functions for a system of more than two quantum states and more than two identical particles. We realize immediately that it is impossible to construct an antisymmetric function from a collection of single-particle functions if any two of the single-particle functions (i.e., quantum states) are identical. In such a case those two states could be exchanged with no effect on the system function. We recognize this statement as one form of the Pauli exclusion principle presented earlier: no two electrons in an atom can simultaneously have an identical set of quantum numbers. Thus we now find that this statement is equivalent to the wave mechanical statement that the system wave function is antisymmetric. This, then, is the Pauli exclusion principle. For the electrons in an atom, it is commonly stated as: the wave function for any atom must be antisymmetric in all the electrons, since interchange of the coordinates of any pair must result in a sign change of the system function. We also note that the Pauli exclusion principle says nothing about any two particles that are not identical, as, for example, a proton and an electron, or about particles having symmetric wave functions.

We are now in a position to make the necessary connection between quantum mechanics and the types of statistics developed earlier. We find that the particles with antisymmetric wave functions discussed above are subject to the exclusion principle and consequently can have no two identical single-particle states occupied. These particles therefore obey the Fermi-Dirac statistics, developed under the assumption of no more than one particle per state. Conversely, for particles having symmetric wave functions there is no such restriction; hence these particles obey the Bose-Einstein statistics, for which it is possible to have more than one particle per state.

5.11. Band Theory of Solids

In Section 5.9, we discussed the electronic states of atoms, which were assumed to be independent of one another, that is, at infinite separation. In this, the final section of this chapter, we consider, in a qualitative sense, the effects on the energy levels of bringing such atoms into close proximity with one another—which is the situation in the lattice structure of a solid. The result of this interaction is shown schematically in Figure 5.15, where, at infinite separation ($1/r = 0$), the energy levels are those for the individual atoms. As the distance between atoms is progressively decreased,

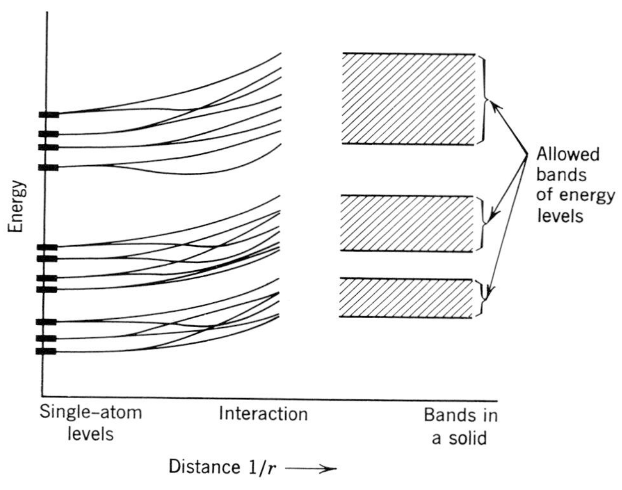

Figure 5.15 Energy bands in a solid.

these levels split and change as indicated on the diagram. A solid is comprised of a large number of atoms packed in a crystal lattice, for which the original single-atom levels have blended into energy bands, each comprised of a large, but finite, number of electron levels. Between these allowed energy bands are forbidden ranges in which there are no states available to the electrons.

In Section 5.10 we determined from the exclusion principle that electrons behave according to Fermi-Dirac statistics. From the discussion of Fermi-Dirac statistics in Section 4.7, we conclude that, at absolute zero temperature, electrons fill the lower energy levels (and therefore bands) up to an energy μ_0, the Fermi level, above which the energy levels are empty. For an electrical insulator, this Fermi level is such that there is a filled energy band above which there is a large energy gap (5–10 ev) to the next allowed band, which is empty. This situation is depicted in Figure 5.16*a*. The highest filled band is termed the valence band, and the next band above this (empty) is called the conduction band. For a metal, an electrical conductor, the situation at absolute zero temperature is different; the Fermi level occurs in the middle of an allowed band, as shown in Figure 5.16*b*. Thus there are electrons in the conduction band at this condition.

As the temperature is increased, to room temperature for example, because of the large energy gap only a very few electrons in an insulator have received sufficient energy to move into the conduction band. Consequently, the electrons in an insulator are rather immobile, and the

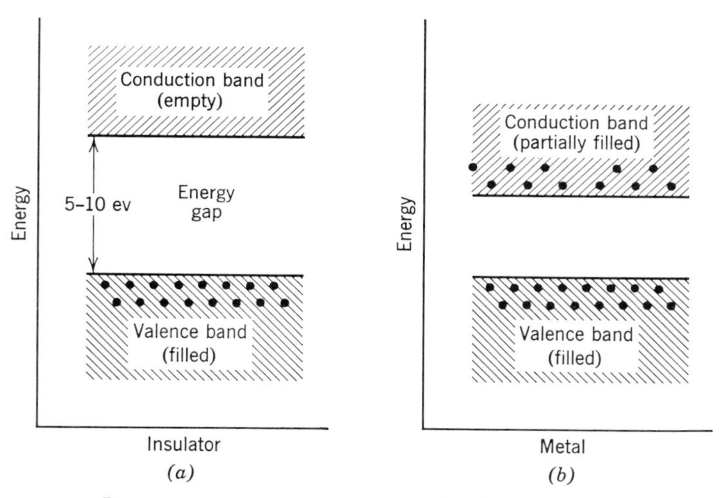

Figure 5.16 Band characteristics of insulators and metals.

material is a poor conductor. Conversely, the electrons in a metal are very easily excited into the levels of slightly higher energies within the same band, and they are quite mobile. The material is therefore an excellent conductor.

A semiconductor has basically the same characteristics as an insulator, except that the energy gap is considerably smaller, on the order of 1 ev as compared with 5–10 ev for the insulator. Consider the intrinsic semiconductor germanium, which has an atomic number of 32 and consequently a valence of 4, i.e., four valence electrons in the outer ring beyond the closed shells of 2, 8, and 18 electrons (see Section 5.9). Therefore, in a crystal structure, each germanium atom shares these four valence electrons with one from each of four neighboring atoms to form the stable

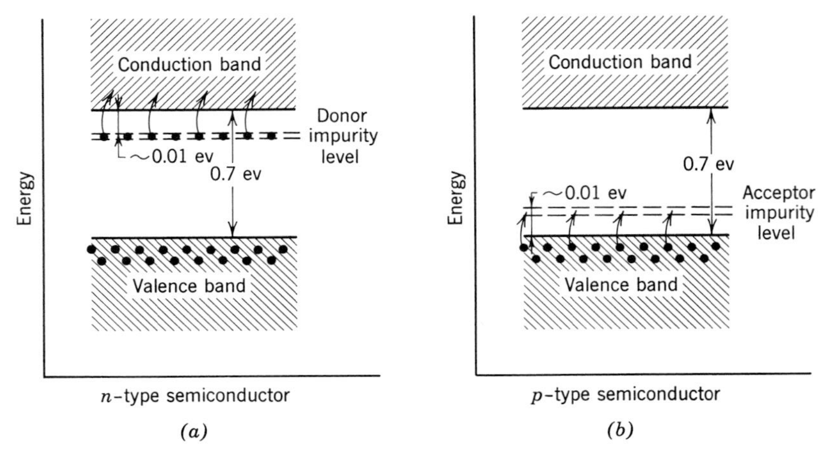

Figure 5.17 Band characteristics of doped semiconductors.

structure of an effective eight-electron outer ring. The energy gap for this pure material is approximately 0.7 ev.

A different situation results when an intrinsic semiconductor is "doped" with controlled small amounts of certain impurities. For example, consider the addition of a small amount of arsenic (atomic number 33, and a valence of 5) to germanium. At the lattice sites now occupied by arsenic atoms, sharing of electrons with neighboring germanium atoms occurs as in pure germanium, except that now there is one extra electron at each of these sites and it is relatively mobile. In terms of the energy bands, the addition of the arsenic atoms therefore corresponds to the addition of an "impurity level" containing electrons. The amount of arsenic can be controlled to locate this impurity level just below the conduction band (~0.01 ev) as shown in Figure 5.17*a*. Thus only small amounts of energy are required to excite electrons from the impurity, or donor, level into the conduction band. Semiconductors of this type are called *n*-type semiconductors because of their negative (electron) charge carriers.

Finally, consider the addition of a small amount of gallium (atomic number 31, and a valence of 3) to germanium. At the lattice sites now occupied by the gallium atoms, the sharing of valence electrons with four neighboring germanium atoms results in an electron deficiency of one, or a positive "hole." Consequently, an electron from another germanium atom jumps into the hole to complete the bond, leaving a hole in that atom. The process continues, resulting in the creation of mobile holes, or positive charge carriers in the material. In terms of band theory, the addition of gallium creates an impurity level of holes, which can be located just above the valence band of pure germanium, as in Figure 5.17*b*. Only small amounts of energy are required to excite electrons from the valence band into the holes in this impurity, or acceptor, level, thereby leaving behind mobile holes in the valence band. Semiconductors of this type are called *p*-type semiconductors, because of their positive (hole) charge carriers.

PROBLEMS

5.1 Consider the Bohr theory of the hydrogen atom. The proportionality constant C in the Coulomb potential has the value 8.98×10^{18} dyne-cm^2/coulomb^2.

(*a*) Show that the radii of the Bohr orbits are proportional to $\mathbf{n}^2$. Calculate the radius of the ground level orbit.

(*b*) Develop an expression for the change in energy corresponding to any change $\mathbf{n}$ to $\mathbf{n}'$.

5.2 (*a*) In 1885, Balmer correlated hydrogen spectral data from the visible and near ultraviolet region according to the expression

$$\nu = c\mathscr{R}\left(\frac{1}{2^2} - \frac{1}{\mathbf{n}^2}\right); \quad \mathbf{n} = 3, 4, 5, \ldots$$

where $\mathscr{R}$ (called the Rydberg constant) has the value 109,678 cm^{-1}. Using the results of Problem 5.1 and Planck's relation, calculate the theoretical value of $\mathscr{R}$ according to Bohr's theory.

(*b*) The Lyman series for H in the far ultraviolet region is expressed by the relation

$$\nu = c\mathscr{R}\left(\frac{1}{1^2} - \frac{1}{\mathbf{n}^2}\right); \quad \mathbf{n} = 2, 3, 4, \ldots$$

Calculate the energy levels for H for both the Lyman and the Balmer series, and compare the results with Figure 5.10. What is the ionization energy for atomic hydrogen?

5.3 A certain spectral series for singly ionized helium (He^+) is correlated by the formula

$$\nu = 4c\mathscr{R}\left(\frac{1}{(\mathbf{n}')^2} - \frac{1}{\mathbf{n}^2}\right)$$

where $\mathscr{R}$ has the value 109,722 cm^{-1}, nearly the same as the Rydberg constant of Problem 5.2. Discuss this expression in terms of the Bohr theory.

5.4 Explanation of the photoelectric effect was one of the early successes of quantum mechanics. For a given metal, it is found that light of less than a certain threshold frequency will not stimulate electron emission from the surface. Above this threshold frequency, the maximum kinetic energy of emitted electrons is directly proportional to frequency and independent of the intensity of the incident light. As explained by the quantum theory, the maximum kinetic energy of an emitted electron is the energy received from a photon of radiation minus an amount φ expended in escaping from the metal. For cesium, φ, called the work function, has the value 1.9 ev. Show that light in the visible region of the spectra (4000–7000 Å) stimulates photoelectron emission in cesium.

5.5 Consider an α particle (the nucleus of a He^4 atom) traveling with a kinetic energy of 6 Mev. What is the deBroglie wavelength associated with this particle?

5.6 Suppose that the position of an electron moving in the x-direction is known to within 0.01 mm. What is the order of magnitude of uncertainty in the velocity of the electron? If the particle is an α particle instead of an electron, how is the result changed?

5.7 Demonstrate that Eq. 5.45 is a solution of the Schrödinger wave equation, Eq. 5.33, for translational energy.

5.8 (*a*) Calculate the spacing of translational energy levels for diatomic hydrogen at the levels where $\mathbf{k}_x$, $\mathbf{k}_y$, $\mathbf{k}_z = 10$ and 1000. Use $V = 1$ cm^3.
(*b*) Repeat part (*a*) for diatomic chlorine.

5.9 Assuming that the contributions of translational and internal modes are independent (Eqs. 5.51 and 5.52), show that the Schrödinger wave equation applied to the molecule can be separated into the two parts, Eqs. 5.53 and 5.54.

5.10 Show that Eq. 5.59 is the representation of the combined rotational kinetic energy for the two atoms comprising a rigid diatomic molecule.

5.11 Demonstrate that Eq. 5.83 is a solution of the differential equation 5.82, and calculate the associated Legendre functions as defined by Eq. 5.83 for the values $\mathbf{j} = 0, 1, 2$.

5.12 (*a*) Calculate the spacing of rotational energy levels for diatomic hydrogen for the first five values of $\mathbf{j}$. Compare the results with the spacings of Problem 5.8. (Use molecular constants from Table A.5.)
(*b*) Repeat the calculation for diatomic chlorine.

5.13 Show that the Schrödinger wave equation for a harmonic oscillator, Eq. 5.103, reduces to the form given by Eq. 5.108 through the changes of variables, Eqs. 5.104 and 5.107.

5.14 Demonstrate that Eq. 5.115, with z defined by Eq. 5.116, is a valid solution of the wave equation for a harmonic oscillator, Eq. 5.103.

5.15 (*a*) Calculate the spacing of vibrational levels for diatomic hydrogen. Compare the result with Problems 5.8 and 5.12.
(*b*) Repeat the calculation for diatomic chlorine.

5.16 Consider the Morse potential, defined by Eq. 5.126. Show that the constant b in this potential is given by Eq. 5.127. Express the potential as a power series in $(r - r_\varepsilon)$ and find the first two correction terms to the harmonic oscillator approximation.

5.17 Plot the Morse potential and the harmonic oscillator potential versus $(r - r_\varepsilon)$ for diatomic nitrogen, using data from Table A.5.

5.18 Calculate the correction constants for hydrogen from Eqs. 5.128 through 5.130. Including the correction terms, compare the spacings of rotational and vibrational levels with the results of Problems 5.12 and 5.15.

5.19 Discuss the first 18 elements of the periodic table (hydrogen through argon) in terms of their respective ground level electron configurations.

5.20 Using values from Table 5.3, compare the energy of the first excited electronic levels for H_2 and Cl_2 with their respective rotational and vibrational energy spacings.

5.21 In this chapter we have been concerned with the stationary states of particles, and consequently we have considered the time-independent Schrödinger wave equation. The general time-dependent Schrödinger equation is

$$\left(\frac{h^2}{8\pi^2 m}\right)\nabla^2\Psi - \Phi\Psi = \left(\frac{h}{2\pi i}\right)\frac{\partial\Psi}{\partial t}$$

Consider this equation applied to the x-direction only, for convenience. Using the procedures followed in Section 5.3, show that, if the potential energy Φ is not a function of time, the general wave function Ψ is separable to

$$\Psi(x, t) = \Psi(x)\varphi(t)$$

with two resulting expressions, one the relation

$$\varphi(t) = e^{-(2\pi i/h)\epsilon t}$$

and the other the time-independent Schrödinger equation, Eq. 5.25.

6 Monatomic Gases

In Chapter 4, certain thermodynamic properties of interest were expressed as functions of the partition function, which is defined in terms of the energy levels and their degeneracies. This partition function development is valid for systems obeying corrected Boltzmann statistics, the approximation to both Bose-Einstein and Fermi-Dirac systems. Then, in Chapter 5, we evaluated the energies and degeneracies for the various energy modes in terms of the appropriate sets of quantum numbers. We now proceed to use these results to evaluate the partition function and, therefore, the thermodynamic properties. In this chapter, we limit our considerations to the properties of monatomic gases, after first discussing general separation of the partition function into translational and internal mode components. A special problem, the representation of electromagnetic radiation by a photon gas, is also introduced. After we consider monatomic gases in this chapter, we shall proceed to more complicated systems of independent particles and deal in turn with solids, diatomic and polyatomic gases, and chemical equilibrium.

6.1. Contributions to the Partition Function and to Properties

For corrected Boltzmann statistics, the partition function has been defined by Eq. 3.55:

$$Z = \sum_j g_j e^{-\epsilon_j/kT}$$

Since the summation is to be made over all energy levels accessible to the molecule, it must include all the various modes of energy. It would prove convenient to separate this into summations over the individual energy modes. The degree to which such a separation can be carried depends upon several factors (the substance and the range of temperature, for example), but in all cases we can separate the energy into translational and internal contributions in accordance with Eq. 5.51,

$$\epsilon = \epsilon_t + \epsilon_{\text{int}}$$

Since the corresponding wave functions Ψ (representing quantum states) are in such a case multiplicative, as given by Eq. 5.52,

$$\Psi = \Psi_t \Psi_{\text{int}}$$

the component energy level degeneracies must also be multiplicative, or

$$g = g_t g_{\text{int}} \tag{6.1}$$

Substitution of Eqs. 5.51 and 6.1 into the partition function Eq. 3.55, results in the separation

$$\begin{aligned} Z &= \sum_{\text{all } j} g_{j_t} g_{j_{\text{int}}} \exp\left[-\frac{\epsilon_{j_t} + \epsilon_{j_{\text{int}}}}{kT}\right] \\ &= \left[\sum_{j_t} g_{j_t} \exp\left(-\frac{\epsilon_{j_t}}{kT}\right)\right]\left[\sum_{j_{\text{int}}} g_{j_{\text{int}}} \exp\left(-\frac{\epsilon_{j_{\text{int}}}}{kT}\right)\right] \\ &= Z_t Z_{\text{int}} \end{aligned} \tag{6.2}$$

where, by definition,

$$Z_t = \sum_{j_t} g_{j_t} \exp\left(-\frac{\epsilon_{j_t}}{kT}\right) \tag{6.3}$$

$$Z_{\text{int}} = \sum_{j_{\text{int}}} g_{j_{\text{int}}} \exp\left(-\frac{\epsilon_{j_{\text{int}}}}{kT}\right) \tag{6.4}$$

Thus the partition function can be written as the product of the translational partition function, Eq. 6.3, and the internal partition function, Eq. 6.4. The first of these can readily be evaluated by using the expression for translational energy found from quantum mechanics in the preceding chapter. The evaluation of the second usually proves more difficult, depending upon the number of modes contributing internally (in general, rotational, vibrational, electronic) and the degree to which they can be separated from one another.

Having found that the partition function can be separated into a product involving translational and internal modes of energy, let us now substitute Eq. 6.4 into our equations for the thermodynamic properties. From Eq. 4.8,

$$\begin{aligned} u &= RT^2\left[\frac{\partial \ln (Z_t Z_{\text{int}})}{\partial T}\right]_V \\ &= RT^2\left(\frac{\partial \ln Z_t}{\partial T}\right)_V + RT^2\left(\frac{\partial \ln Z_{\text{int}}}{\partial T}\right)_V \\ &= u_t + u_{\text{int}} \end{aligned} \tag{6.5}$$

We find that the internal energy can be calculated as a sum of contributions, that due to translation

$$u_t = RT^2\left(\frac{\partial \ln Z_t}{\partial T}\right)_V \tag{6.6}$$

and that resulting from the internal modes

$$u_{\text{int}} = RT^2\left(\frac{\partial \ln Z_{\text{int}}}{\partial T}\right)_V \tag{6.7}$$

The enthalpy of an ideal gas can now be found, from its definition, as

$$h = u + Pv = u + RT = u_t + u_{\text{int}} + RT$$

This can also be expressed as

$$h = h_t + h_{\text{int}} \tag{6.8}$$

if the translational enthalpy is defined as

$$h_t = u_t + RT \tag{6.9}$$

It is then noted that the internal contribution to enthalpy and internal energy are the same, namely,

$$h_{\text{int}} = u_{\text{int}} \tag{6.10}$$

The RT term has been associated with the translation in Eq. 6.9. This can be justified from experimental observation for substances having no internal contributions. We might alternatively justify this by using the expression developed for pressure in Chapter 4, Eq. 4.49,

$$P = NkT\left(\frac{\partial \ln Z}{\partial V}\right)_T$$

We recall from the results of quantum mechanics that the internal modes are not functions of volume and consequently do not contribute anything to the pressure (or to Pv). From the definition of enthalpy, we then arrive at Eq. 6.10.

Expressions for specific heats can also be developed, but these quantities are most readily evaluated from their definitions, Eqs. 4.9 and 4.54, using Eqs. 6.5 to 6.10.

From Eqs. 4.55, 6.2, and 6.5, we find that the entropy can also be expressed as a sum of the contributions of translational and internal modes,

$$s = R\left[\ln\left(\frac{Z_t Z_{\text{int}}}{N}\right) + 1\right] + \frac{u_t + u_{\text{int}}}{T}$$

This can also be written as

$$s = s_t + s_{\text{int}} \tag{6.11}$$

if the translational entropy is defined as

$$s_t = R\left[\ln\left(\frac{Z_t}{N}\right) + 1\right] + \frac{u_t}{T} \tag{6.12}$$

and the internal contribution to this property is defined as

$$s_{\text{int}} = R \ln Z_{\text{int}} + \frac{u_{\text{int}}}{T} \tag{6.13}$$

The factors N and unity are associated with the translational contribution, which is in agreement with observed data for substances lacking internal contributions.

In a similar manner, the Helmholtz function is given by

$$a = a_t + a_{\text{int}} \tag{6.14}$$

with

$$a_t = -RT\left[\ln\left(\frac{Z_t}{N}\right) + 1\right] \tag{6.15}$$

$$a_{\text{int}} = -RT \ln Z_{\text{int}} \tag{6.16}$$

The Gibbs function is given by

$$g = g_t + g_{\text{int}} \tag{6.17}$$

where

$$g_t = -RT \ln\left(\frac{Z_t}{N}\right) \tag{6.18}$$

$$g_{\text{int}} = a_{\text{int}} = -RT \ln Z_{\text{int}} \tag{6.19}$$

6.2. Translation

In the preceding section, we performed a general separation of the partition function into translational and internal mode components. For the remainder of this chapter, we shall discuss the thermodynamic properties for the most elementary class of substances, namely, monatomic gases, for which the sole internal contribution is that of electronic energy states. Before proceeding to this problem, however, in this section we determine the translational contribution to thermodynamic properties.

In order to evaluate the translational partition function, we first recall from Eq. 5.50 that the translational energy,

$$\epsilon_{\mathbf{k}_x,\mathbf{k}_y,\mathbf{k}_z} = \frac{h^2}{8mV^{2/3}}(\mathbf{k}_x^{\,2} + \mathbf{k}_y^{\,2} + \mathbf{k}_z^{\,2})$$

is expressed as the energy of a particular quantum state and is not that for a degenerate level of energy. Thus the translational partition function can be calculated most directly by summing over all quantum or energy states, according to Eq. 3.56, instead of over all levels of energy as implied in Eq. 6.3. That is,

$$Z_t = \sum_{j_t \text{ levels}} g_{j_t} \exp\left[-\frac{\epsilon_{j_t}}{kT}\right]$$
$$= \sum_{\mathbf{k}_x=1}^{\infty} \sum_{\mathbf{k}_y=1}^{\infty} \sum_{\mathbf{k}_z=1}^{\infty} \exp\left[-\frac{\epsilon_{\mathbf{k}_x,\mathbf{k}_y,\mathbf{k}_z}}{kT}\right] \tag{6.20}$$

Now, substituting Eq. 5.50, which gives the energy of a state, into Eq. 6.20, we have

$$Z_t = \sum_{\mathbf{k}_x=1}^{\infty} \sum_{\mathbf{k}_y=1}^{\infty} \sum_{\mathbf{k}_z=1}^{\infty} \exp\left[-\frac{h^2}{8mV^{2/3}kT}(\mathbf{k}_x^{\ 2} + \mathbf{k}_y^{\ 2} + \mathbf{k}_z^{\ 2})\right]$$

Since the three functions and summations in $\mathbf{k}_x$, $\mathbf{k}_y$, $\mathbf{k}_z$ are identical, an arbitrary subscript i can be used for each, so that

$$Z_t = \left\{\sum_{\mathbf{k}_i=1}^{\infty} \exp\left[-\frac{h^2}{8mV^{2/3}kT}\mathbf{k}_i^{\ 2}\right]\right\}^3 \tag{6.21}$$

The summation inside the brackets in this equation is of the form

$$\sum_{\mathbf{k}=1}^{\infty} e^{-r\mathbf{k}^2} = \sum_{\mathbf{k}=0}^{\infty} e^{-r\mathbf{k}^2} - 1$$

with r a constant. The summation on the right side here is evaluated in terms of an integral using the Euler-Maclaurin summation theorem, Eq. 2.78,

$$\sum_{\mathbf{k}=0}^{\infty} e^{-r\mathbf{k}^2} = \int_0^{\infty} e^{-r\mathbf{k}^2}\, d\mathbf{k} + \tfrac{1}{2}(1) - \tfrac{1}{12}(0) + \tfrac{1}{720}(0) + \cdots$$

Therefore only the first two terms are different from zero, and, on combining this expression with the previous one, we have

$$\sum_{\mathbf{k}=1}^{\infty} e^{-r\mathbf{k}^2} = \int_0^{\infty} e^{-r\mathbf{k}^2}\, d\mathbf{k} - \tfrac{1}{2}$$

To apply this result to Eq. 6.21, we anticipate the result that Z_t will be a large number. The difference of $-\frac{1}{2}$ between the summation and integral will then be negligible. In other words, we are making the assumption that the translational energy levels are so closely spaced that the summation of Eq. 6.21 can be directly replaced by an integral. This will be found to be of general validity for translational energy in our applications, but the Euler-Maclaurin theorem gives us a means for justifying the approximation. Therefore

$$Z_t = \left\{\int_0^{\infty} \exp\left[-\frac{h^2}{8mV^{2/3}kT}\mathbf{k}_i^{\ 2}\right] d\mathbf{k}_i\right\}^3$$

On performing this integration, we find

$$Z_t = \left[\tfrac{1}{2}\sqrt{\pi}\sqrt{\frac{8mV^{2/3}kT}{h^2}}\right]^3 = V\left(\frac{2\pi mkT}{h^2}\right)^{3/2} \tag{6.22}$$

This result is identical with that found for a monatomic ideal gas assuming classical mechanics in Section 3.5. We note that the assumption discussed above (neglecting the factor of $\frac{1}{2}$) which led to Eq. 6.22 is equivalent to the assumptions made in the analysis of Section 3.5.

For the translational contributions to thermodynamic properties, it is convenient to write Eq. 6.22 in logarithmic form:

$$\ln Z_t = \ln V + \tfrac{3}{2}\ln\left(\frac{2\pi mk}{h^2}\right) + \tfrac{3}{2}\ln T \tag{6.23}$$

Now, using Eq. 6.23 to evaluate Eq. 4.49,

$$P = NkT\left(\frac{\partial \ln Z}{\partial V}\right)_T$$

we obtain

$$P = NkT\left(\frac{1}{V}\right) = \frac{nRT}{V} = \frac{RT}{v} \tag{6.24}$$

as found previously in Eq. 4.51. This is, of course, the entire contribution to pressure, since the internal modes are not functions of volume. From Eq. 6.6,

$$U_t = RT^2\left(\frac{\partial \ln Z_t}{\partial T}\right)_V$$

the internal energy is

$$u_t = RT^2\left(\frac{3}{2T}\right) = \tfrac{3}{2}RT \tag{6.25}$$

and the enthalpy is found from Eq. 6.9 to be

$$h_t = \tfrac{3}{2}RT + RT = \tfrac{5}{2}RT \tag{6.26}$$

The translational contribution to the constant-volume specific heat is, from Eq. 6.25,

$$C_{v_t} = \left(\frac{\partial u_t}{\partial T}\right)_V = \tfrac{3}{2}R \tag{6.27}$$

while that to the constant-pressure specific heat is, from Eq. 6.26,

$$C_{p_t} = \left(\frac{\partial h_t}{\partial T}\right)_P = \tfrac{5}{2}R \tag{6.28}$$

The translational entropy is found from Eq. 6.12 to be

$$s_t = R\left[\ln\left(\frac{Z_t}{N}\right) + 1\right] + \tfrac{3}{2}R = R\left[\ln\left(\frac{Z_t}{N}\right) + \tfrac{5}{2}\right] \tag{6.29}$$

From Eqs. 6.22 and 6.24,

$$\frac{Z_t}{N} = \frac{V}{N}\left(\frac{2\pi mkT}{h^2}\right)^{3/2} = \left(\frac{2\pi m}{h^2}\right)^{3/2}\frac{(kT)^{5/2}}{P} \tag{6.30}$$

Similarly, the translational contributions to the Helmholtz and Gibbs functions are found from Eqs. 6.15 and 6.18, respectively, using Eq. 6.30 for the quantity (Z_t/N).

Example 6.1

Calculate the translational contributions to enthalpy, entropy, and Gibbs function for helium at 25 C, 1 atm pressure.

For He:

$$m = 1.66 \times 10^{-24}(4.003) = 6.645 \times 10^{-24}\ \text{gm}$$

From Eq. 6.26,

$$h_t = \tfrac{5}{2}RT = \tfrac{5}{2}(1.987)(298.15) = 1480\ \text{cal/mole}$$

From Eq. 6.30,

$$\begin{aligned}\frac{Z_t}{N} &= \left(\frac{2\pi m}{h^2}\right)^{3/2}\frac{(kT)^{5/2}}{P}\\ &= \left(\frac{2\pi \times 6.645 \times 10^{-24}}{(6.625 \times 10^{-27})^2}\right)^{3/2}\frac{(1.3804 \times 10^{-16} \times 298.15)^{5/2}}{1.01325 \times 10^6}\\ &= 3.10 \times 10^5\end{aligned}$$

From Eq. 6.29,

$$s_t = R\left[\ln\left(\frac{Z_t}{N}\right) + \tfrac{5}{2}\right] = 1.987[12.66 + 2.5] = 30.1\ \text{cal/mole-K}$$

Therefore, from Eq. 6.18,

$$\begin{aligned}g_t &= -RT\ln\left(\frac{Z_t}{N}\right)\\ &= -1.987(298.15)(12.66)\\ &= -7500\ \text{cal/mole}\end{aligned}$$

Alternatively, using the definition of the Gibbs function,

$$g_t = h_t - Ts_t = 1480 - 298.15(30.1) = -7500\ \text{cal/mole}$$

6.3. An Alternative Evaluation of the Translational Partition Function

It was noted in Section 3.5 that the partition function defined by Eq. 3.55 in terms of a summation over degenerate levels of energy can alternatively be expressed as a sum over all individual quantum states, according to Eq. 3.56.

$$Z = \sum_{j \text{ levels}} g_j e^{-\epsilon_j/kT} = \sum_{i \text{ states}} e^{-\epsilon_i/kT}$$

The latter approach was used in evaluating the partition function for translation in the previous section, since the translational energy as given by Eq. 5.50 is the energy of a quantum state, and not that of a degenerate level of energy. It is also desirable, however, to analyze this problem from the stand-point of energy levels, and to do so we must develop an expression for energy level degeneracy. That is, the number of individual quantum states (each specified by a set of integers $\mathbf{k}_x$, $\mathbf{k}_y$, $\mathbf{k}_z$) having energy in an interval ϵ to $\epsilon + d\epsilon$ must be found.

Let us consider a three-dimensional space of coordinates $\mathbf{k}_x$, $\mathbf{k}_y$, $\mathbf{k}_z$, a two-dimensional cross section of which is shown in Figure 6.1. Each of the three variables is restricted to positive integral values, and we note that a point in this $\mathbf{k}_x$, $\mathbf{k}_y$, $\mathbf{k}_z$ space represents a translational quantum state.

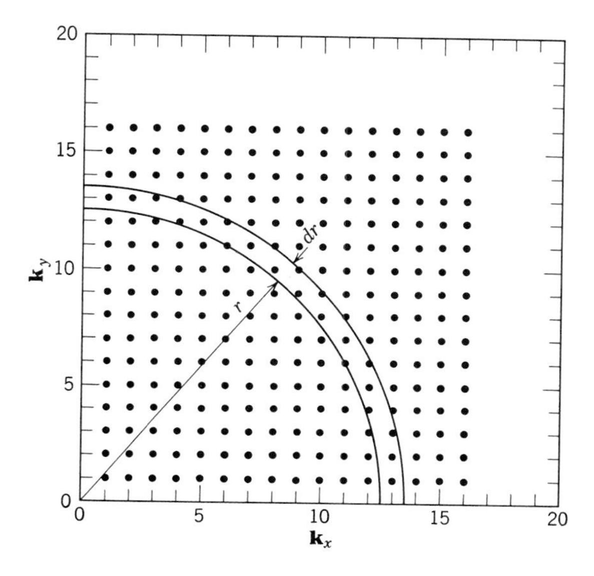

Figure 6.1 Cross section of the three-dimensional $\mathbf{k}_x$, $\mathbf{k}_y$, $\mathbf{k}_z$ space.

The distance r of any point from the origin is given by

$$r = (\mathbf{k}_x^2 + \mathbf{k}_y^2 + \mathbf{k}_z^2)^{1/2} \tag{6.31}$$

We now restrict the discussion to those states (sets of $\mathbf{k}_x$, $\mathbf{k}_y$, $\mathbf{k}_z$) for which r is large compared to unity, so that r can be treated as essentially continuous. (Translational states for which r is small constitute only a negligible fraction of the total number and can therefore be neglected here.) Thus the number of states g_r in the spherical shell from r to $r + dr$ is

$$g_r = \tfrac{1}{8}(4\pi r^2\, dr) \tag{6.32}$$

where the factor $\frac{1}{8}$ accounts for the restriction of $\mathbf{k}_x$, $\mathbf{k}_y$, $\mathbf{k}_z$ to all positive values only.

The quantity of interest to us, however, is the number of states in an interval of energy ϵ to $\epsilon + d\epsilon$. From Eqs. 5.50 and 6.31,

$$\epsilon = \frac{h^2}{8mV^{2/3}}\, r^2 \tag{6.33}$$

and

$$d\epsilon = \frac{h^2}{8mV^{2/3}}\, 2r\, dr \tag{6.34}$$

Using Eqs. 6.33 and 6.34 in Eq. 6.32, we find that the number of states in the interval ϵ to $\epsilon + d\epsilon$ is

$$g_\epsilon = \frac{4\pi mV}{h^3}(2m\epsilon)^{1/2}\, d\epsilon \tag{6.35}$$

If we use Eq. 6.35 (for the degeneracy) in Eq. 6.3 (for the translational partition function) and integrate from $\epsilon = 0$ to $\epsilon = \infty$, the result is once again that given by Eq. 6.22. The details of the evaluation are left as an exercise.

6.4. Electronic Levels

For a monatomic ideal gas, the only internal contribution to the partition function is that of the electronic levels of the atom. By convention, we take the ground electronic level energy as the reference state, that is,

$$\epsilon_{e_0} = 0 \tag{6.36}$$

The internal partition function then becomes

$$Z_{\text{int}} = Z_e = g_{e_0} + g_{e_1} \exp\left[-\frac{\epsilon_{e1}}{kT}\right] + g_{e_2} \exp\left[-\frac{\epsilon_{e2}}{kT}\right] + \cdots \tag{6.37}$$

This can be evaluated by direct summation over the electronic terms as necessary. Using the definitions

$$Z_e' = T\left(\frac{dZ_e}{dT}\right) = \sum_j g_{e_j}\left(\frac{\epsilon_{e_j}}{kT}\right) \exp\left[-\frac{\epsilon_{e_j}}{kT}\right] \tag{6.38}$$

$$Z_e'' = T\left(\frac{dZ_e'}{dT}\right) = \sum_j g_{e_j}\left(\frac{\epsilon_{e_j}}{kT}\right)^2 \exp\left[-\frac{\epsilon_{e_j}}{kT}\right] - Z_e' \tag{6.39}$$

it is found from Eqs. 6.7, 6.10, 6.13, and 6.19 that the electronic contributions to properties are given by the following relations.

$$h_e = u_e = RT\frac{Z_e'}{Z_e} \tag{6.40}$$

$$s_e = R\left[\ln Z_e + \frac{Z_e'}{Z_e}\right] \tag{6.41}$$

$$g_e = a_e = -RT \ln Z_e \tag{6.42}$$

Differentiating Eq. 6.40, we find

$$C_{p_e} = C_{v_e} = R\left[\frac{Z_e'' + Z_e'}{Z_e} - \left(\frac{Z_e'}{Z_e}\right)^2\right] \tag{6.43}$$

Example 6.2

(*a*) Using values from Table 5.2, determine the fraction of atoms in each of the first three electronic levels for monatomic fluorine at 1000 K.

(*b*) Calculate the internal energy, constant-volume specific heat, and entropy of monatomic fluorine at 1000 K, 1 atm pressure.

(*a*) From the data for fluorine in Table 5.2, we have the following:

Term	ϵ/hc, cm^{-1}	$g = 2J + 1$
$^2P_{3/2}$	0	4
$^2P_{1/2}$	404.0	2
$^4P_{5/2}$	102,406.5	6

Therefore, at 1000 K,

$$\frac{\epsilon_{e1}}{kT} = \left(\frac{\epsilon_{e1}}{hc}\right)\left(\frac{hc}{k}\right)\frac{1}{T} = 404.0(1.4388)(\tfrac{1}{1000}) = 0.5813$$

Similarly,

$$\frac{\epsilon_{e2}}{kT} = 102406.5(1.4388)(\tfrac{1}{1000}) = 147.2$$

so that the electronic partition function is, from Eq. 6.37,

$$Z_e = 4 + 2 \times e^{-0.5813} + 6 \times e^{-147.2} = 5.118$$

From the distribution equation, Eq. 3.58,

$$\frac{N_{e_j}}{N} = \frac{g_{e_j} \exp(-\epsilon_{e_j}/kT)}{Z_e}$$

the fraction of particles in the electronic ground level is

$$\frac{N_{e0}}{N} = \frac{g_{e_0}(1)}{Z_e} = \frac{4}{5.118} = 0.782$$

Similarly, the fraction of particles in the first excited level is

$$\frac{N_{e1}}{N} = \frac{g_{e1} \exp(-\epsilon_{e_1}/kT)}{Z_e} = \frac{2 \times e^{-0.5813}}{5.118} = 0.218$$

For the second excited level, we find

$$\frac{N_{e2}}{N} = \frac{g_{e_2} \exp(-\epsilon_{e_2}/kT)}{Z_e} = \frac{6 \times e^{-147.2}}{5.118} \approx 0$$

(*b*) A negligible fraction of the fluorine atoms are in the second and higher excited levels, so we need consider only the ground and first excited levels as contributing to properties. Let us first evaluate the translational contributions. From Eq. 6.25,

$$u_t = \tfrac{3}{2}RT = \tfrac{3}{2}(1.987)(1000) = 2981 \text{ cal/mole}$$

From Eq. 6.27,

$$C_{v_t} = \tfrac{3}{2}R = \tfrac{3}{2}(1.987) = 2.981 \text{ cal/mole}$$

For monatomic fluorine, the atomic mass is

$$m = 1.66 \times 10^{-24}(19) = 31.54 \times 10^{-24} \text{ gm}$$

so that, using Eq. 6.30,

$$\ln \frac{Z_t}{N} = \ln \left[\left(\frac{2\pi m}{h^2} \right)^{3/2} \frac{(kT)^{5/2}}{P} \right]$$

$$= \ln \left[\left(\frac{2\pi \times 31.54 \times 10^{-24}}{(6.625 \times 10^{-27})^2} \right)^{3/2} \frac{(1.3804 \times 10^{-16} \times 1000)^{5/2}}{1.01325 \times 10^6} \right]$$

$$= 18.019$$

Therefore, from Eq. 6.29,

$$s_t = R\left[\ln \left(\frac{Z_t}{N} \right) + \tfrac{5}{2} \right] = 1.987[18.019 + 2.5] = 40.7 \text{ cal/mole-K}$$

For the electronic contributions, it is convenient to construct a table like the accompanying one, using

$$y_j = \frac{\epsilon_{ej}}{kT}$$

ϵ_{e_j}, cm^{-1}	g_{e_j}	y_j	e^{-y_j}	$g_{e_j}e^{-y_j}$	$g_{e_j}y_je^{-y_j}$	$g_{e_j}y_j^2e^{-y_j}$
0	4	0	1	4.0	0	0
404.0	2	0.5813	0.559	1.118	0.650	0.378
102406.5	6	147.2	~0	~0	~0	~0
				$Z_e = 5.118$	$Z_e' = 0.650$	$Z_e'' + Z_e' = 0.378$

The summations for Z_e, Z_e', $Z_e'' + Z_e'$ follow from Eqs. 6.37 to 6.39. From Eq. 6.40,

$$u_e = RT\frac{Z_e'}{Z_e} = 1.987(1000)\left(\frac{0.650}{5.118}\right) = 252 \text{ cal/mole}$$

From Eq. 6.43,

$$C_{v_e} = R\left[\frac{Z_e'' + Z_e'}{Z_e} - \left(\frac{Z_e'}{Z_e}\right)^2\right]$$

$$= 1.987\left[\frac{0.378}{5.118} - \left(\frac{0.650}{5.118}\right)^2\right] = 0.115 \text{ cal/mole-K}$$

From Eq. 6.41,

$$s_e = R\left[\ln Z_e + \frac{Z_e'}{Z_e}\right]$$

$$= 1.987\left[\ln 5.118 + \frac{0.650}{5.118}\right] = 3.50 \text{ cal/mole-K}$$

Therefore the values of the properties to be determined are

$$u = u_t + u_e = 2981 + 252 = 3233 \text{ cal/mole}$$

$$C_v = C_{v_t} + C_{v_e} = 2.981 + 0.115 = 3.096 \text{ cal/mole-K}$$

$$s = s_t + s_e = 40.70 + 3.50 = 44.20 \text{ cal/mole-K}$$

It is of interest to note that, for the very common case in which only the ground electronic level is of significance, the partition function Z_e reduces to the degeneracy of the ground level, and both Z_e' and Z_e'' are zero. Consequently, the electronic level contributions h_e, u_e, C_{p_e}, C_{v_e} are all zero, while s_e, g_e, a_e, are zero only if the ground level degeneracy is unity.

6.5. The Photon Gas

In this section, we consider electromagnetic radiation as represented by a gas of photons, an application of statistical thermodynamics in which corrected Boltzmann statistics is not a reasonable model. It was found in Section 3.3 that, if the multiplier e^{α} is very large, then both the quantum statistical models, Bose-Einstein and Fermi-Dirac, reduce to a common approximate form, which we call the corrected Boltzmann model. Whenever e^{α} is not large compared with unity, however, the appropriate quantum statistical model must be employed in its general form.

The atoms comprising a system change from one electronic state to a higher or lower energy state as a result of absorbing or emitting photons of radiation of the appropriate frequency, as discussed in Chapter 5. Since these photons are strictly independent of one another, we can reasonably treat a collection of photons constituting an amount of radiant energy as an ideal gas. In doing so, we realize that photons behave as elementary particles of zero spin. Therefore the Pauli exclusion principle is not applicable, and the Bose-Einstein model is the appropriate quantum statistical model for such a system.

The behavior of a body emitting or absorbing radiation at some steady state temperature T can be analyzed in terms of the radiation field or photon gas in equilibrium with the body. Let us consider, then, a simple model consisting of a box of sides $x = y = z = l$, having perfectly reflecting walls and containing a quantity of radiant energy U. We shall proceed to determine the equilibrium distribution of photons comprising this radiation among the various available energy levels (actually frequencies). This problem is in many respects identical with those considered previously, namely, determination of the equilibrium distribution of particles in a system by maximizing the entropy subject to the system constraints.

Let us assume for the moment a set of discrete energy levels ϵ_j, with corresponding degeneracies g_j. A particular distribution is specified by the number of photons, N_j, in the various levels of energy, such that the system energy is given by

$$U = \sum_j N_j \epsilon_j \tag{6.44}$$

The problem at hand is different from previous ones in one very important aspect. In earlier developments, the total number of particles in the system was always fixed, providing a second constraint on the system variables. Photons, however, can be emitted or absorbed at the walls of the system in such a manner that the number of photons in the system is not necessarily conserved. For example, at some instant of time, two photons of energy ϵ_j could be absorbed at the wall and another of energy $2\epsilon_j$ emitted, with the system energy U remaining constant as required.

Thus the equilibrium distribution in this system is determined through the maximization of the thermodynamic probability for Bose-Einstein statistics, which from Eq. 3.21 is

$$d \ln w = \sum_j \ln \left(\frac{g_j + N_j}{N_j} \right) dN_j = 0 \tag{6.45}$$

subject to the single constraint

$$dU = \sum_j \epsilon_j \, dN_j = 0 \tag{6.46}$$

Multiplying Eq. 6.46 by β and adding to the negative of Eq. 6.45, we obtain

$$\sum_j \left[-\ln \left(\frac{g_j + N_j}{N_j} \right) + \beta \epsilon_j \right] dN_j = 0 \tag{6.47}$$

By the method of undetermined multipliers, we conclude that the coefficients of all the dN_j must be zero. Therefore the equilibrium distribution of photons is specified by

$$N_j = \frac{g_j}{e^{\beta \epsilon_j} - 1} \tag{6.48}$$

As in earlier developments,

$$\beta = \frac{1}{kT}$$

so that

$$N_j = \frac{g_j}{e^{\epsilon_j/kT} - 1} \tag{6.49}$$

This result is the Bose-Einstein equilibrium distribution equation for ordinary particles, Eq. 4.64, with the multiplier α taken as zero.

Having developed the distribution equation, Eq. 6.49, we now turn to the problem of evaluating the energy levels ϵ_j and degeneracies g_j. From Planck's equation, Eq. 5.1,

$$\epsilon = h\nu$$

we recognize that such a level ϵ_j corresponds to a small range of frequencies ν_j to $\nu_j + \Delta\nu$, and also that the degeneracy g_j represents the number of quantum states in this interval $\Delta\nu$.

The differential wave equation describing the photons is identical in form with that for the translation of a particle in a box analyzed in Section 5.4. As in the solution to that equation, the equilibrium condition is specified by standing waves having nodes at the boundaries of the container. The solution is that given by Eq. 5.48 in terms of the set of positive integers $\mathbf{k}_x$, $\mathbf{k}_y$, $\mathbf{k}_z$,

$$\Psi_{\mathbf{k}_x, \mathbf{k}_y, \mathbf{k}_z}(x, y, z) = C \sin \left(\mathbf{k}_x \pi \frac{x}{l} \right) \sin \left(\mathbf{k}_y \pi \frac{y}{l} \right) \sin \left(\mathbf{k}_z \pi \frac{z}{l} \right)$$

In order that the wave nodes coincide with the wall of the container (the wave function equals zero at the walls), the wave number (reciprocal of wavelength λ) components are restricted to the values

$$\omega_x = \frac{\mathbf{k}_x}{2l}, \quad \omega_y = \frac{\mathbf{k}_y}{2l}, \quad \omega_z = \frac{\mathbf{k}_z}{2l} \tag{6.50}$$

Therefore the allowed wave numbers ω can be represented as

$$\omega = \frac{1}{\lambda} = (\omega_x^2 + \omega_y^2 + \omega_z^2)^{1/2} = \frac{r}{2l} \tag{6.51}$$

where

$$r = (\mathbf{k}_x^2 + \mathbf{k}_y^2 + \mathbf{k}_z^2)^{1/2} \tag{6.52}$$

which is the same as the vector r defined in Section 6.3 for translational states. Proceeding in the same manner as followed in that section, we find that the number of sets of all positive $\mathbf{k}_x$, $\mathbf{k}_y$, $\mathbf{k}_z$ in a spherical shell between r and $r + dr$ is

$$\tfrac{1}{8}(4\pi r^2\, dr)$$

This can alternatively be expressed in terms of the wavelength λ using Eq. 6.51, or in terms of the frequency ν, where

$$\nu = \frac{c}{\lambda} = \frac{cr}{2l}$$

Thus the number of sets of $\mathbf{k}_x$, $\mathbf{k}_y$, $\mathbf{k}_z$ can be written as

$$\tfrac{1}{8}(4\pi)\left(\frac{2l\nu}{c}\right)^2\left(\frac{2l}{c}\right) d\nu = \frac{4\pi V}{c^3}\nu^2\, d\nu$$

For any direction of wave propagation, the radiation has two planes of polarization normal to the direction of propagation. Consequently, this result must be multiplied by a factor of 2 in order to give the total number of quantum states, g_ν, for photons in the frequency band ν to $\nu + d\nu$. Therefore we have

$$g_\nu = \frac{8\pi V}{c^3}\nu^2\, d\nu \tag{6.53}$$

corresponding to an energy ϵ_ν given by Eq. 5.1 or by

$$\epsilon_\nu = h\nu = \frac{hcr}{2l} = \frac{hc}{2V^{1/3}}(\mathbf{k}_x^2 + \mathbf{k}_y^2 + \mathbf{k}_z^2)^{1/2} \tag{6.54}$$

If we now substitute Eqs. 6.53 and 5.1 into Eq. 6.49, the equilibrium distribution can be expressed in terms of the number of photons, dN_ν, in the frequency band ν to $\nu + d\nu$ as

$$dN_\nu = \frac{8\pi V}{c^3}\frac{\nu^2}{e^{h\nu/kT} - 1}\, d\nu \tag{6.55}$$

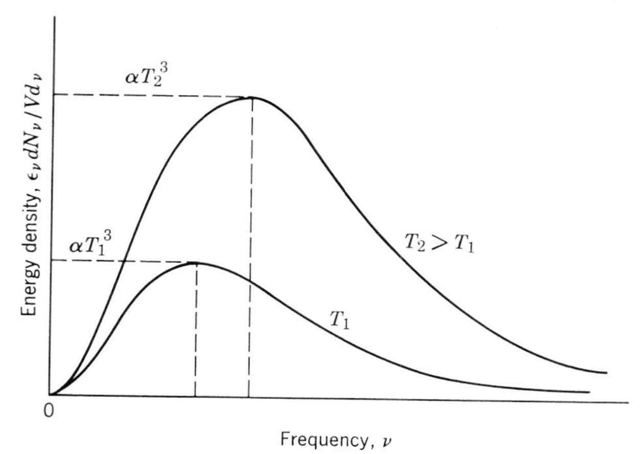

Figure 6.2 Planck's distribution law for black body radiation.

Using this result, we see that the radiant energy per unit volume, or the energy density, for the frequency band ν to $\nu + d\nu$ is

$$\frac{\epsilon_\nu \, dN_\nu}{V} = \frac{8\pi h}{c^3} \frac{\nu^3}{e^{h\nu/kT} - 1} d\nu \tag{6.56}$$

which is called Planck's distribution equation for a black body (ideal) radiator. If the energy density according to Eq. 6.56 is plotted versus frequency at a given temperature, the distribution is as indicated in Figure 6.2. We note from this figure that the maximum in energy density for the distribution curve shifts to higher frequency as the temperature is increased, and it can be shown that the height at this maximum point is proportional to T^3.

It is of interest to examine two special cases of Eq. 6.56, the very low and the very high frequency ranges, respectively. If, at a given temperature, we consider only the very low frequency end, so that

$$\frac{h\nu}{kT} \ll 1$$

then

$$e^{h\nu/kT} - 1 \approx \frac{h\nu}{kT}$$

With this approximation, the radiant energy distribution, Eq. 6.56, reduces to

$$\frac{\epsilon_\nu \, dN_\nu}{V} = \frac{8\pi\nu^2 kT}{c^3} d\nu \tag{6.57}$$

This is the Rayleigh-Jeans distribution equation for low frequency radiation, derived originally from classical mechanics. This expression predicts a continually increasing energy density with frequency, which obviously cannot be reasonable at high frequencies.

The other special case to be analyzed is that of very high frequency, such that

$$\frac{h\nu}{kT} \gg 1$$

for which

$$e^{h\nu/kT} - 1 \approx e^{h\nu/kT}$$

We note that neglecting unity compared with the exponential term is the approximation made in corrected Boltzmann statistics. With this approximation, the energy density (Eq. 6.56) reduces to

$$\frac{\epsilon_\nu \, dN_\nu}{V} = \frac{8\pi h}{c^3} \nu^3 e^{-h\nu/kT} \, d\nu \tag{6.58}$$

This is called Wien's formula, the form of which was proposed originally as an empirical correlation of high frequency distributions.

Our primary interest in considering these two special cases, the Rayleigh-Jeans and Wien formulas, is that historically both preceded the correct expression, Eq. 6.56. Planck originally proposed this relation as a means for interpolating between the low frequency (Rayleigh-Jeans) and high frequency (Wien) equations. Of course, his development was not based on Bose-Einstein statistics, but it did incorporate the concept of photons and proved to be the foundation of the science of quantum mechanics.

Our final analysis in treating the photon gas as a representative model of radiation is the determination of the radiant energy and entropy, from which other thermodynamic properties of interest can then be found. The total energy per unit volume, or energy density, at any temperature is found as the area beneath the distribution curve of Figure 6.2. Using Eq. 6.56, this is found to be

$$\frac{U}{V} = \int_0^\infty \left(\frac{\epsilon_\nu \, dN_\nu}{V \, d\nu} \right) d\nu = \frac{8\pi h}{c^3} \int_0^\infty \frac{\nu^3}{e^{h\nu/kT} - 1} \, d\nu \tag{6.59}$$

Substituting $x = h\nu/kT$, and integrating Eq. 6.59 (Table A.3),

$$\frac{U}{V} = \frac{8\pi(kT)^4}{(hc)^3} \int_0^\infty \frac{x^3}{e^x - 1} \, dx = \frac{8\pi^5}{15} \frac{(kT)^4}{(hc)^3} \tag{6.60}$$

The energy density at a given temperature is therefore found to be proportional to T^4. If we consider a small rate of flow of radiant energy (through a small pinhole in the wall), the rate of flow is found to be proportional to the energy density inside the box and consequently also proportional to the fourth power of the temperature. This relation is the well-known Stefan-Boltzmann radiation law.

The entropy of a photon gas is that for a Bose-Einstein system with α taken to be zero. From Eq. 4.67, this is found to be

$$S = (k \ln w_{\text{mp}})_{\text{B-E}} = -k \sum_j g_j \ln (1 - e^{-\epsilon_j/kT}) + \frac{U}{T} \tag{6.61}$$

Using Eqs. 5.1, 6.53, and 6.60, the photon gas entropy density is found to be

$$\begin{aligned}\frac{S}{V} &= -k\,\frac{8\pi}{c^3}\int_0^\infty \nu^2 \ln (1 - e^{-h\nu/kT})\, d\nu + \frac{U/V}{T} \\ &= -8\pi k^4\left(\frac{T}{hc}\right)^3\int_0^\infty x^2 \ln (1 - e^{-x})\, dx + \frac{U/V}{T} \\ &= \frac{32\pi^5}{45}\, k^4\left(\frac{T}{hc}\right)^3 \end{aligned} \tag{6.62}$$

As mentioned above, other thermodynamic properties of interest can be found from their definitions and the classical relations.

PROBLEMS

6.1 Calculate the translational partition function for a particle contained in a box of dimensions X, Y, Z, all of which are different.

6.2 Consider the evaluation of the translational partition function, Eq. 6.22, from the quantum mechanical energy expression, Eq. 6.21. For argon at -100 C, 1 atm pressure, estimate the error introduced by neglecting the 1/2 term in this development.

6.3 The equation for translational entropy of an ideal gas can be written in the form

$$s = R[C_1 \ln T + C_2 \ln M - \ln P + C_3]$$

which is known as the Sackur-Tetrode equation. If T is in degrees K, M is the molecular weight, and P is in atmospheres, determine the values of the constants C_1, C_2, C_3.

6.4 A two-dimensional ideal gas has been considered in Problem 3.5. Consider such a gas confined to a surface of dimensions $X = Y$ and possessing only translational energy.

(*a*) Show that the energy of a quantum state is given by

$$\epsilon_{\mathbf{k}_x,\mathbf{k}_y} = \frac{h^2}{8m\mathscr{A}}(\mathbf{k}_x^2 + \mathbf{k}_y^2)$$

where $\mathscr{A}$ is the surface area.

(*b*) Find expressions for the internal energy and entropy of this gas.

(*c*) Using the thermodynamic relation

$$P = -\left(\frac{\partial A}{\partial \mathscr{A}}\right)_T$$

show that P is given by the result found in Problem 3.5.

6.5 Neon gas at 1700 F, 5 atm, undergoes a steady flow, adiabatic expansion through a turbine and exhausts at 1 atm. If the entropy increases by 5% during this process, determine:

(*a*) The exit temperature and work output.

(*b*) The isentropic efficiency, ratio of actual to isentropic work.

6.6 Starting with Eq. 6.35, show that the translational partition function is given by Eq. 6.22, the result found by using the approach of degenerate levels of energy.

6.7 Consider a 1 cm^3 sample of O_2 at 100 C, 1 atm pressure. Using Eq. 6.35, calculate the number of translational quantum states in the interval of energy kT to $1.001kT$.

6.8 Considering translational energy only, use Eq. 6.35 and the result of Problem 3.9 to find an expression for the fractional occupancy dN_ϵ/g_ϵ at any level of energy ϵ to $\epsilon + d\epsilon$. Is the result a surprising one? Plot this expression versus ϵ/kT.

6.9 Calculate the entropy per mole of monatomic hydrogen at 300 K, 1 atm pressure.

6.10 Repeat Example 6.2 for monatomic fluorine at 500 K and at 2000 K, both for 1 atm pressure. Compare the results with that example.

6.11 Calculate the values of h, C_p, and s for monatomic oxygen at 100 K and 1000 K, 1 atm pressure.

6.12 Plot the electronic contribution to specific heat versus temperature for monatomic oxygen.

6.13 Calculate the Gibbs function at 1 atm pressure for monatomic nitrogen at 1000° intervals from 4000 K to 10,000 K. These values are necessary to evaluate chemical equilibrium constants, which will be discussed in Chapter 9.

6.14 Consider a monatomic gas having two non-degenerate electronic states $\epsilon_{e_0} = 0$ and ϵ_{e_1}.

(*a*) Plot the electronic contribution to specific heat versus "reduced" temperature, kT/ϵ_{e_1}.

(*b*) Determine analytically the value of reduced temperature at which the specific heat contribution is a maximum.

6.15 The electronic partition function is evaluated by direct summation over the levels that make a significant contribution. At high temperature, this number of terms increases greatly, and the series tends to diverge. An approach to this problem is to consider a cut-off point for the summation, a level above which the electron is no longer considered bound to the atom. A number of quantum mechanical theories give such a cut-off point. An alternative approach utilizes an empirical representation of two or three terms to approximate the true energy levels and degeneracies over a certain range of temperature. Such an approximation for argon is

$$Z_e \approx 1 + 60e^{-162,500/T}$$

Calculate the Gibbs function for argon at 1 atm pressure, 10,000 K and 20,000 K. Does the electronic contribution seem to be significant? (In Chapter 9, one large Gibbs function will be subtracted from another.)

6.16 Calculate the entropy per mole for a mixture of 50% helium, 50% neon, at 500 C, 10 atm.

6.17 Consider the black body radiation distribution equation, Eq. 6.56. At any given temperature, find the frequency at which the energy density is a maximum. Show that this maximum energy density is proportional to T^3.

6.18 At 500 K, what fraction of the total radiant energy (assuming black body distribution) has a frequency within 1% of that corresponding to the maximum energy density?

6.19 Plot the black body distribution versus frequency for a temperature of 1000 K. Compare the result with the Rayleigh-Jeans and Wien equations.

6.20 Consider a photon gas as representative of thermal radiation in equilibrium with its container. Using the thermodynamic relation

$$P = -\left(\frac{\partial A}{\partial V}\right)_T$$

calculate the radiation pressure at 1000 K.

6.21 In Problem 3.13 an expression for the flux of particles escaping a tank was developed. An analogous result can be found for the energy flux of radiant energy.

(*a*) Show that the radiant energy flux through a small hole (such that the equilibrium distribution at temperature T inside the container is not disturbed) is given by the relation

$$\frac{\text{Energy}}{\text{cm}^2\text{-sec}} = \frac{1}{4}\left(\frac{U}{V}\right)c$$

in terms of the equilibrium energy density and speed of photons inside the container.

(*b*) The radiant energy flux found in part (*a*) is also equal to σT^4, where σ is the Stefan-Boltzmann constant. Determine the numerical value of σ.

6.22 The intensity of solar radiation is a maximum at approximately 5000 Å.

(*a*) Assuming the sun to be an ideal radiator, calculate its surface temperature.

(*b*) Use the results of part (*a*) and Problem 6.21 to find the flux of solar radiant energy.

7 Monatomic Solids

Up to this point in our development and application of statistical thermodynamics, we have been concerned with the analysis of systems of independent particles. In these systems, the forces between particles are negligibly small, and we may therefore correctly speak of single-particle states and the energy per particle. The characteristic structure of a solid, however, is that of atoms which are held tightly together in a crystal lattice by strong interatomic forces. The individual atoms possess energy as a result of vibrations about their equilibrium positions in the lattice. These vibrations become stronger with respect to one another as energy is added to the crystal. Certainly, the behavior of each atom in the lattice is strongly influenced by its neighboring atoms in the lattice structure. Nevertheless, it is found to be reasonable to treat the crystal as a collection of independent or semi-independent harmonic oscillators, and to calculate the energy and other properties of the solid phase on the basis of this assumption. In this chapter, we discuss first the properties of the individual harmonic oscillator and, on the basis of that discussion, expressions for various thermodynamic properties of solids are derived. Finally, the electronic contributions to thermodynamic properties are evaluated on the basis of the development of the Fermi-Dirac statistical model.

7.1. The Harmonic Oscillator

In Chapter 5 it was found that the energy of a harmonic oscillator is given in terms of the fundamental frequency ν_ε and the vibrational quantum number $\mathbf{v}$ by Eq. 5.112 as

$$\epsilon_v = h\nu_\varepsilon(\mathbf{v} + \tfrac{1}{2})$$

for $\mathbf{v} = 0, 1, 2, 3, \ldots$. Consequently, in the ground state ($\mathbf{v} = 0$) there is a corresponding ground state energy

$$\epsilon_{v_0} = \tfrac{1}{2}h\nu_\varepsilon \tag{7.1}$$

such that the vibrational energy above that of the ground state is

$$\epsilon_v - \epsilon_{v_0} = h\nu_\varepsilon \mathbf{v} \tag{7.2}$$

For convenience, the vibrational contribution to the partition function will be assumed to consist of the vibrational energy above the ground state, as given by Eq. 7.2. The contribution resulting from the ground state energy, Eq. 7.1, can then be incorporated into the arbitrary selection of an overall zero-reference level for energy. It is apparent that the value of vibrational energy is dependent upon such a zero-reference selection, but values of entropy and specific heat are not.

It was found in Chapter 5 that a harmonic energy level is non-degenerate, since each vibrational wave function as given by Eq. 5.115 has a different value of energy. Therefore the vibrational partition function is, using Eq. 7.2,

$$Z_v = \sum_{\mathbf{v}=0}^{\infty} e^{-(\epsilon_v - \epsilon_{v_0})/kT} = \sum_{\mathbf{v}=0}^{\infty} e^{-(h\nu_\varepsilon/kT)\mathbf{v}} \tag{7.3}$$

For most oscillators the frequency of vibration, ν_ε, is large, and the vibrational energy levels are therefore widely spaced. As a result, it is not possible for us to replace the summation of Eq. 7.3 by a corresponding integral as we did in Chapter 6 for translational energy. We note, however, that the particular summation of Eq. 7.3 is given by the closed-form expression

$$Z_v = \sum_{\mathbf{v}=0}^{\infty} e^{-(h\nu_\varepsilon/kT)\mathbf{v}} = \frac{1}{1 - e^{-\theta_v/T}} \tag{7.4}$$

in which the quantity θ_v has been defined as

$$\theta_v = \frac{h\nu_e}{k} = \frac{hc\omega_e}{k} \tag{7.5}$$

The parameter θ_v is called the characteristic vibrational temperature, and is a function only of the fundamental frequency of the oscillator.

To evaluate the vibrational contribution to thermodynamic properties, we express Eq. 7.4 in logarithmic form

$$\ln Z_v = -\ln\left(1 - e^{-\theta_v/T}\right) \tag{7.6}$$

It is convenient to evaluate the thermodynamic properties on a molal basis, in terms of the internal-mode expressions of Chapter 6 (Eqs. 6.7, 6.10, 6.13, and 6.19). These expressions therefore give the contribution of N_0 oscillators of frequency ν_ε. If we consider the ground state molal energy u_{v_0}, which is equal to $N_0\epsilon_{v_0}$, then the contribution of a mole of harmonic oscillators to the internal energy above u_{v_0} is, from Eq. 6.7,

$$(u_v - u_{v_0}) = RT^2\left(\frac{\partial \ln Z_v}{\partial T}\right)_V = RT^2\left[-\frac{e^{-\theta_v/T}}{1-e^{-\theta_v/T}}\left(-\frac{\theta_v}{T^2}\right)\right]$$
$$= \frac{R\theta_v}{e^{\theta_v/T}-1} \tag{7.7}$$

Therefore, since u_{v_0} is a constant, the vibrational constant-volume specific heat is

$$C_{v_v} = \left(\frac{\partial u_v}{\partial T}\right)_V = \frac{d(u_v - u_{v_0})}{dT} = \frac{R(\theta_v/T)^2 e^{\theta_v/T}}{(e^{\theta_v/T}-1)^2} \tag{7.8}$$

The contribution to other thermodynamic properties for a harmonic oscillator may be found in a similar manner. From Eq. 6.10, for a mole of oscillators,

$$(h_v - u_{v_0}) = (u_v - u_{v_0}) = \frac{R\theta_v}{e^{\theta_v/T}-1} \tag{7.9}$$

so that

$$C_{p_v} = \left(\frac{\partial h_v}{\partial T}\right)_P = \frac{d(h_v - u_{v_0})}{dT} = C_{v_v} = \frac{R(\theta_v/T)^2 e^{\theta_v/T}}{(e^{\theta_v/T}-1)^2} \tag{7.10}$$

We note that C_{v_v} and C_{p_v} are equal for the system consisting of harmonic oscillators. From Eq. 6.13,

$$s = R \ln Z_{\text{int}} + \frac{u_{\text{int}}}{T}$$

the entropy is

$$s_v = R\left[-\ln(1 - e^{-\theta_v/T}) + \frac{\theta_v/T}{e^{\theta_v/T}-1}\right] \tag{7.11}$$

From Eqs. 6.16 and 6.19, the values of a_v and g_v are found to be

$$(a_v - u_{v_0}) = (g_v - u_{v_0}) = RT\ln(1 - e^{-\theta_v/T}) \tag{7.12}$$

The mathematical functions required to evaluate Eqs. 7.6 to 7.12 are given in Table A.4 of the Appendix.

7.2. The Einstein Solid

In Einstein's simple model of the solid phase, it is assumed that each of the N atoms in a crystal vibrates harmonically and independently in three directions about its equilibrium position in the crystal lattice. The system is therefore assumed to consist of $3N$ independent harmonic oscillators, all of the same fundamental frequency ν_ε, since the vibrational modes are independent of one another. The properties of such a system are then $3N$

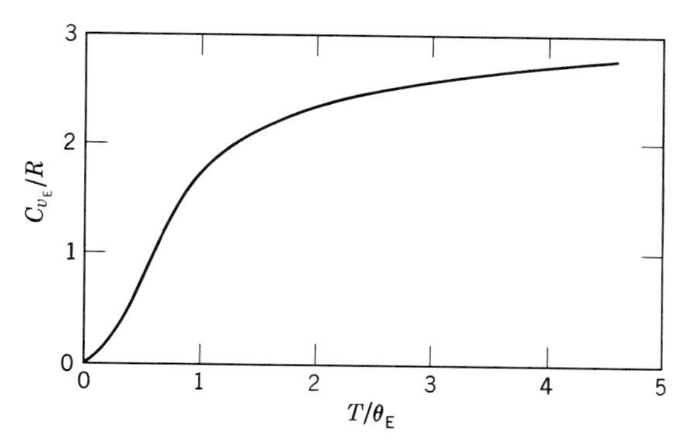

Figure 7.1 The specific heat of an Einstein solid.

times those for a single oscillator. In Section 7.1, the thermodynamic properties were expressed on a molal basis (N_0 oscillators), so that for an Einstein solid the properties per mole are simply three times those expressed by Eqs. 7.7 to 7.12.

Of particular interest is the specific heat, which can be compared with experimentally determined values. It follows from Eq. 7.8 that, for the Einstein solid, the specific heat C_{v_E} is

$$C_{v_E} = \frac{3R(\theta_E/T)^2 e^{\theta_E/T}}{(e^{\theta_E/T} - 1)^2} \tag{7.13}$$

where θ_E, which is called the Einstein temperature, has been used instead of θ_v. Figure 7.1 shows a plot of the Einstein specific heat as given by Eq. 7.13 versus the dimensionless temperature T/θ_E.

At high temperature, C_{v_E} approaches a limit of $3R$, which is in agreement with most observed data. This high temperature limiting value of $3R$ was observed experimentally by Dulong and Petit in the early nineteenth century; this is the analytical result that one obtains from classical mechanics for all temperatures. It should be pointed out that for some solids the specific heat does not approach this value, because of anharmonicity or additional contributions from electronic states.

At low temperature, the Einstein specific heat expression, Eq. 7.13, reduces to approximately

$$C_{v_E} \cong 3R\left(\frac{\theta_E}{T}\right)^2 e^{-\theta_E/T} \tag{7.14}$$

and approaches zero as the temperature approaches absolute zero. This zero limit is also in agreement with experimental observation, but the quantitative behavior of the Einstein model at low temperature is not particularly good. It is found experimentally that the specific heat of a crystal at low temperature is very closely proportional to T^3. If the value for θ_E is so selected that reasonable agreement with experimental observations is obtained over a wide range of temperature, then the calculated specific heats at low temperature are considerably smaller than the experimental values. One concludes that the Einstein model, which assumes $3N$ harmonic oscillators of the same frequency, is certainly a vast oversimplification of the problem. Nevertheless, it demonstrates a qualitatively correct behavior, and was a tremendous advance beyond classical mechanics, which predicts a constant value of $3R$ for the specific heat.

7.3. The Debye Solid

The failure of the Einstein model of the solid phase to provide accurate correlation with experimental specific heat data, especially at low temperature, led Debye, in 1912, to improve the mathematical model. Einstein assumed that the vibrational modes in the crystal are independent of one another, and consequently that all the modes are of the same frequency. It is only reasonable to expect, however, that the atoms in a crystal vibrate with many different frequencies, and also that low frequency vibrations are of particular importance in analyzing low temperature behavior of the crystal.

Debye therefore assumed that there are a large number of very closely spaced frequencies of vibration, such that a continuous distribution function $f(\nu)$ can be used in representing the number of normal vibrations in any interval ν to $\nu + d\nu$. From the point of view of a distribution function, the Einstein model having $3N$ oscillators of frequency ν_ε is as shown in Figure 7.2a, and the Debye distribution of frequencies is shown in Figure

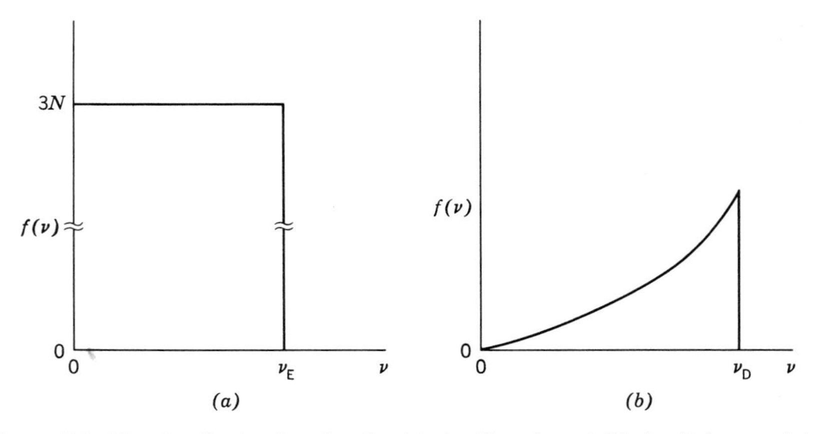

Figure 7.2 The distribution function for (*a*) the Einstein and (*b*) the Debye models.

7.2b. Since the atoms are held in a lattice having certain atomic distances and orientations, there is some limiting high frequency value corresponding to a wavelength of the order of magnitude of the interatomic distances. This maximum value ν_{D} is called the Debye frequency. It follows that the area under the Debye distribution curve of Figure 7.2b must equal the total number of vibrational modes, $3N$, or in terms of the distribution function,

$$\int_0^{\nu_{\text{D}}} f(\nu)\, d\nu = 3N \tag{7.15}$$

To determine the nature of the distribution $f(\nu)$, Debye assumed that the vibrations in the solid can be treated as elastic waves in a continuous medium. The normal vibrational modes are represented by the various standing waves, with nodes coinciding with the crystal boundary. Since this is the same mathematical problem as that for the translation of a particle in a box (and also for the photon gas representing radiation), the solution is given by Eq. 5.48. For a crystal of side l, the wave function is, in terms of the three positive integers $\mathbf{k}_x$, $\mathbf{k}_y$, $\mathbf{k}_z$,

$$\Psi_{\mathbf{k}_x,\mathbf{k}_y,\mathbf{k}_z}(x, y, z) = C \sin\left(\mathbf{k}_x\pi\,\frac{x}{l}\right) \sin\left(\mathbf{k}_y\pi\,\frac{y}{l}\right) \sin\left(\mathbf{k}_z\pi\,\frac{z}{l}\right) \tag{7.16}$$

There is, consequently, a wave function, Eq. 7.16, for each set of integers $\mathbf{k}_x$, $\mathbf{k}_y$, $\mathbf{k}_z$. We now define

$$r = (\mathbf{k}_x^{\,2} + \mathbf{k}_y^{\,2} + \mathbf{k}_z^{\,2})^{1/2}$$

where r is the radius of a sphere in a space of cartesian coordinates $\mathbf{k}_x$, $\mathbf{k}_y$, $\mathbf{k}_z$. The number of sets of these positive integers, $N_{\mathbf{k}_x,\mathbf{k}_y,\mathbf{k}_z}$, in the interval r to $r + dr$ is then given by

$$N_{\mathbf{k}_x,\mathbf{k}_y,\mathbf{k}_z} = \tfrac{1}{8}(4\pi r^2\, dr) \tag{7.17}$$

The factor $\frac{1}{8}$ is introduced into Eq. 7.17 because only $\frac{1}{8}$ of the surface of the sphere of radius r corresponds to all positive $\mathbf{k}_x$, $\mathbf{k}_y$, $\mathbf{k}_z$. This analysis is the same as that discussed for translational states in Section 6.3.

For wave nodes coincidental with the crystal boundaries, the allowed wavelength λ must be, as in Eq. 6.51 for the photon gas,

$$\lambda = 2\,\frac{l}{r} \tag{7.18}$$

and the corresponding frequency ν for a given wave velocity V is

$$\nu = \frac{\mathsf{V}}{\lambda} = \frac{\mathsf{V}r}{2l} \tag{7.19}$$

Therefore, from Eqs. 7.17 and 7.19, the number of frequencies of oscillation, N_ν (sets of $\mathbf{k}_x, \mathbf{k}_y, \mathbf{k}_z$), in the interval ν to $\nu + d\nu$ for a given wave velocity V is

$$N_\nu = \frac{4\pi}{8}\left(\frac{2l\nu}{\mathsf{V}}\right)^2\left(\frac{2l}{\mathsf{V}}\right) d\nu = 4\pi \frac{l^3}{\mathsf{V}^3} \nu^2 \, d\nu = 4\pi \frac{V}{\mathsf{V}^3} \nu^2 \, d\nu \tag{7.20}$$

In considering wave propagation in some direction of an elastic medium, we must include both the longitudinal waves of velocity V_l and the simultaneous transverse waves of velocity V_t, the latter having two normal directions of polarization. Therefore the total number of vibrational modes in the interval ν to $\nu + d\nu$ is, using Eq. 7.20,

$$f(\nu)\, d\nu = \left(\frac{1}{\mathsf{V}_l^{\,3}} + \frac{2}{\mathsf{V}_t^{\,3}}\right) 4\pi V \nu^2 \, d\nu = \frac{12\pi V}{\overline{\mathsf{V}}^3} \nu^2 \, d\nu \tag{7.21}$$

where the average wave velocity $\overline{\mathsf{V}}$ has been defined by the expression

$$\frac{3}{\overline{\mathsf{V}}^3} = \frac{1}{\mathsf{V}_l^{\,3}} + \frac{2}{\mathsf{V}_t^{\,3}} \tag{7.22}$$

Having determined the distribution function $f(\nu)$ in terms of Eq. 7.21, we can now integrate Eq. 7.15. The result is

$$3N = \int_0^{\nu_{\mathrm{D}}} \frac{12\pi V}{\overline{\mathsf{V}}^3} \nu^2 \, d\nu = \frac{4\pi V}{\overline{\mathsf{V}}^3} \nu_{\mathrm{D}}^{\;3}$$

or

$$\nu_{\mathrm{D}}^{\;3} = \frac{3N\overline{\mathsf{V}}^3}{4\pi V} \tag{7.23}$$

The distribution function $f(\nu)$ can alternatively be expressed in terms of the Debye frequency ν_{D} as

$$f(\nu)\, d\nu = \frac{9N}{\nu_{\mathrm{D}}^{\;3}} \nu^2 \, d\nu \tag{7.24}$$

for all values of ν between zero and ν_{D}.

The internal energy with respect to the ground state for a mole of oscillators of frequency ν is given by Eq. 7.7. Thus, for a Debye solid, using Eq. 7.24 we have, on a mole basis,

$$(u - u_{v0}) = \int_0^{\nu_{\mathrm{D}}} \left[\frac{k(h\nu/k)}{e^{h\nu/kT} - 1}\right] f(\nu)\, d\nu = \frac{9N_0kT}{\nu_{\mathrm{D}}^{\;3}} \int_0^{\nu_{\mathrm{D}}} \left[\frac{h\nu/kT}{e^{h\nu/kT} - 1}\right] \nu^2 \, d\nu \tag{7.25}$$

If we let

$$x = \frac{h\nu}{kT} = \frac{\theta}{T} \tag{7.26}$$

this relation can be rewritten in the form

$$(u - u_{v0}) = 3RT\left[\frac{3}{x_D{}^3}\int_0^{x_D}\frac{x^3}{e^x - 1}\,dx\right] = 3RT[D(x_D)] \tag{7.27}$$

where

$$x_D = \frac{h\nu_D}{kT} = \frac{\theta_D}{T} \tag{7.28}$$

The quantity θ_D is called the Debye temperature, and is defined in terms of ν_D by Eq. 7.28.

TABLE 7.1

$x_D = \theta_D/T$	$D(x_D)$	$x_D = \theta_D/T$	$D(x_D)$
0	1.0000	3	0.2836
0.1	0.9630	4	0.1817
0.2	0.9270	5	0.1176
0.5	0.8250	10	0.0193
1	0.6744	20	0.0024
2	0.4411	∞	0

The function $D(x_D)$ defined by Eq. 7.27 is called the Debye function; this must be evaluated numerically as a function of x_D or θ_D/T. A few values of the Debye function are given in Table 7.1.

The specific heat for the Debye solid is found by differentiating Eq. 7.27:

$$C_v = \left(\frac{\partial u}{\partial T}\right)_V = d\,\frac{(u - u_{v0})}{dT} = 3R\left[4 \times D(x_D) - \frac{3x_D}{e^{x_D} - 1}\right] \tag{7.29}$$

At high temperature, the Debye function approaches unity and the specific heat C_v approaches a limiting value of $3R$, which is identical with the result obtained for the Einstein model. As mentioned previously, this is in excellent agreement with observed specific heat data for most monatomic solids. At very low temperature, the Debye function reduces to the form

$$D(x_D) = \frac{\pi^4}{5}(x_D)^{-3} \tag{7.30}$$

so that, at low temperature,

$$C_v = \frac{12\pi^4}{5}R\left(\frac{T}{\theta_D}\right)^3 \tag{7.31}$$

From this expression, we note the proportionality of C_v to T^3 at temperatures near absolute zero, which is in good agreement with experimental values. We also note that C_v goes to zero at absolute zero temperature.

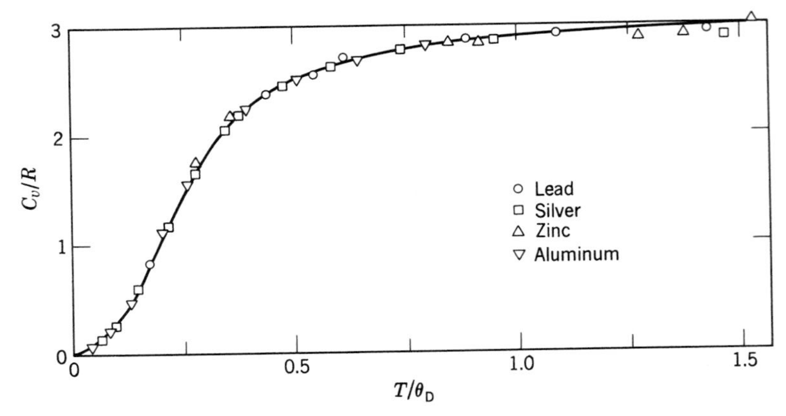

Figure 7.3 The specific heat of a Debye solid.

Over the entire range of temperature, the shape of the specific heat from the Debye theory is qualitatively the same as that for the simpler Einstein model shown in Figure 7.1. The Debye specific heat function is shown in Figure 7.3 and is compared with a few experimental points for lead, silver, zinc, and aluminium.

Quantitatively, the Debye model is excellent and is superior to the Einstein model, particularly at low temperatures. If the theory were exact, it would be possible to select a Debye temperature θ_D that would correlate the experimental data over a very wide temperature range. For many solids this can be done, but for some materials the Debye temperature varies somewhat with the range of temperature in which the data are to be correlated. Average values for the parameter θ_D for several substances are listed in Table 7.2.

The entropy for a Debye solid is, of course, independent of the arbitrary

TABLE 7.2

Element	θ_D, °K	Element	θ_D, °K
Pb	86	Zn	240
Hg	90	Cu	315
K	99	Al	398
Bi	111	Cr	405
Na	160	Co	445
Sn	165	Fe	453
Ag	215	Ni	456
Pt	225	Be	980
Ca	230	C (diamond)	1860

zero reference for energy. Equation 7.11 gives the entropy for N_0 vibrational modes of a given frequency ν. Therefore, using the Debye frequency distribution function, Eq. 7.24, the molal entropy is found to be

$$\begin{aligned} s &= k\int_0^{\nu_D}\left[\frac{x}{e^x - 1} - \ln(1 - e^{-x})\right] f(\nu)\, d\nu \\ &= \frac{9R}{x_D{}^3}\int_0^{x_D}\left[\frac{x}{e^x - 1} - \ln(1 - e^{-x})\right] x^2\, dx \\ &= R[4 \times D(x_D) - 3\ln(1 - e^{-x_D})] \end{aligned} \tag{7.32}$$

At low temperature, Eq. 7.32 reduces to the form

$$s = \frac{4\pi^4}{5} R\left(\frac{T}{\theta_D}\right)^3 \tag{7.33}$$

and is proportional to T^3. The entropy then approaches zero as temperature approaches absolute zero; this, as noted earlier, is in agreement with the reference state for entropy. On a physical basis, this is seen to be a result of each of the $3N$ vibrational modes dropping into its non-degenerate ground state as the temperature becomes close to absolute zero.

Example 7.1

Calculate the specific heat and entropy of aluminum at 100 K. From Table 7.2, for aluminum,

$$\theta_D = 398 \text{ K}$$

Therefore, at 100 K,

$$x_D = \frac{\theta_D}{T} = \frac{398}{100} = 3.98$$

Interpolating in Table 7.1, the Debye function at 100 K is found to be

$$D(x_D) = 0.1837$$

From Table A.4 for $x_D = 3.98$,

$$\frac{x_D}{e^{x_D} - 1} = 0.07580, \qquad -\ln(1 - e^{-x_D}) = 0.01887$$

Therefore, the specific heat is, from Eq. 7.29,

$$C_v = 3(1.987)[4(0.1837) - 3(0.07580)] = 3.02 \text{ cal/mole-K}$$

From Eq. 7.32, the entropy is

$$s = 1.987[4(0.1837) + 3(0.01887)] = 1.570 \text{ cal/mole-K}$$

The Debye model of the solid phase has been found to give good quantitative agreement with experimental observations for many substances.

More complex models of the solid have been developed to explain a number of problems and features of solid behavior for which the Debye model is inadequate, but these models will not be discussed here.

7.4. The Electron Gas in a Metal

In previous sections of this chapter, we have discussed the behavior of solids and have noted that the Debye model, which considers vibration of the atoms in the lattice sites, gives quite accurate representation of the thermodynamic properties of crystals, even for many metals. Such a model has, of course, not included the contributions to properties of the conduction, or valence, electrons in metals. As discussed in Section 5.11, the electrons in the conduction band of a metal are quite mobile. Because of the large zero-point energy of a Fermi-Dirac system, the conduction electrons have sufficient energy to move quite freely through the lattice of positive ions. Since forces between the electrons are relatively small, we assume that it is reasonable to treat the collection of free electrons in a metal as a system of independent particles, which we term the electron gas.

The Fermi-Dirac model must be used to represent this system, for which the equilibrium distribution of electrons among energy levels is (Eq. 4.76),

$$N_j = \frac{g_j}{e^{(\epsilon_j - \mu)/kT} + 1}$$

in which the chemical potential μ has been substituted for G/N. It was found in Section 4.7 that, at absolute zero temperature, all energy levels below the Fermi level μ_0 are filled, while all levels above μ_0 are empty, resulting in the distribution shown in Figure 4.5.

If we wish to evaluate the Fermi energy μ_0, it will therefore be necessary to determine the number of quantum states available at all levels of energy less than μ_0. The number of translational states in an interval of energy ϵ to $\epsilon + d\epsilon$ has been given previously by Eq. 6.35,

$$g_{\epsilon\ \text{trans}} = \frac{4\pi m V}{h^3}(2m\epsilon)^{1/2}\, d\epsilon$$

Since this expression considers translation only, this result must be multiplied by a factor of 2 for the electron gas to account for electron spin degeneracy. That is, an electron in any translational state can have a spin of either $+\frac{1}{2}$ or $-\frac{1}{2}$. Thus the total number of quantum states available to the electron gas in an interval of energy ϵ to $\epsilon + d\epsilon$ is

$$g_\epsilon = \frac{8\pi m V}{h^3}(2m\epsilon)^{1/2}\, d\epsilon \tag{7.34}$$

At absolute zero temperature, Eq. 7.34 also represents the number of particles, dN_ϵ, in the energy range ϵ to $\epsilon + d\epsilon$ for all energies less than the Fermi level μ_0, since all quantum states below μ_0 are filled at this condition. Therefore the total number of electrons, N, in the system is found to be

$$N = \int_{N_{\epsilon=0}}^{N_{\epsilon=\mu_0}} dN_\epsilon = \frac{8\pi m V}{h^3}(2m)^{1/2}\int_0^{\mu_0} \epsilon^{1/2}\, d\epsilon = \frac{8\pi V}{3h^3}(2m\mu_0)^{3/2} \tag{7.35}$$

The Fermi energy μ_0, which is also the chemical potential G/N of the system at zero temperature, is found from Eq. 7.35 to be

$$\mu_0 = \frac{h^2}{8m}\left(\frac{3N}{\pi V}\right)^{2/3} \tag{7.36}$$

The total energy U_0 of the system at absolute zero temperature is

$$\begin{aligned} U_0 &= \int_{N_{\epsilon=0}}^{N_{\epsilon=\mu_0}} \epsilon\, dN_\epsilon = \frac{8\pi m V}{h^3}(2m)^{1/2}\int_0^{\mu_0} \epsilon^{3/2}\, d\epsilon \\ &= \tfrac{3}{5}N\mu_0 \end{aligned} \tag{7.37}$$

We note that the average energy per electron, U_0/N, at absolute zero is equal to three-fifths of the maximum energy μ_0.

It was determined in Section 4.7 that the entropy of a Fermi-Dirac system at absolute zero is zero, or

$$S_0 = 0 \tag{7.38}$$

The zero-point Helmholtz function A_0 is seen to be

$$A_0 = U_0 - T_0 S_0 = U_0 \tag{7.39}$$

and the Gibbs function is

$$G_0 = A_0 + P_0 V = U_0 + P_0 V \tag{7.40}$$

However, from Eq. 7.37,

$$G_0 = N\mu_0 = \tfrac{5}{3}U_0 \tag{7.41}$$

so that, for the electron gas,

$$P_0 V = \tfrac{2}{3}U_0 \tag{7.42}$$

which is of the same form as the relation found for the ordinary classical Boltzmann gas, Eq. 3.51, at normal temperatures.

A representative calculation of the Fermi energy for the electron gas in a metal by using Eq. 7.36 indicates that μ_0 is an extremely large number, of the order of magnitude of $10^5 \times k$. Consequently, it is reasonable to treat the behavior of the electron gas at ordinary temperatures, room temperature for example, as merely small perturbations from the zero temperature behavior. This procedure involves a number of series expansions and approximations and is quite complex mathematically. It is found that the chemical potential for the electron gas in a metal is, in terms of the zero-point value,

$$\mu = \mu_0\left[1 - \frac{\pi^2}{12}\left(\frac{kT}{\mu_0}\right)^2 + \cdots\right] \tag{7.43}$$

Similarly, the energy is

$$U = U_0\left[1 + \frac{5\pi^2}{12}\left(\frac{kT}{\mu_0}\right)^2 - \cdots\right] \tag{7.44}$$

The specific heat contribution is therefore

$$C_v = \left[\frac{\partial(U/n)}{\partial T}\right]_V = \frac{N_0\pi^2 k^2 T}{2\mu_0} - \cdots \tag{7.45}$$

The entropy of the electron gas is readily found from the classical expression, as the integral of $(C_v/T)\,dT$ from 0 to T.

Because of the large value of μ_0 mentioned above, it is seen that the specific heat contribution of the electron gas is quite small at ordinary temperature, of the order of magnitude of $10^{-2} \times R$. This is in basic agreement with the conclusions reached in studying the results of the Debye model of solids, which includes only the lattice contribution to specific heat. Of course, if the electron gas had been treated strictly as an ordinary Boltzmann gas, the contribution to specific heat would be $\frac{3}{2}R$, which is not at all in agreement with experimental data. As a final observation on the electron contribution to specific heat, we note that, at low temperature, only the first term of Eq. 7.45 is significant, and the specific heat contribution is proportional to T. The lattice contribution has been found previously to be proportional to T^3 at low temperature. Thus it should be possible to distinguish between these contributions at temperatures close to absolute zero; this behavior has been confirmed experimentally.

PROBLEMS

7.1 The specific heat of graphitic carbon at 300 K is 2.06 cal/mole-K. Calculate the characteristic Einstein temperature from this value, and use it to compute the specific heat at 50 K and at 100 K, assuming an Einstein solid. Compare the results with the experimental values 0.12 and 0.40 cal/mole-K.

7.2 Repeat Problem 7.1, assuming the Debye model of the solid phase.

7.3 Calculate the specific heat and the Helmholtz function for lead at the boiling point temperature of hydrogen, 20.4 K.

7.4 Experimental values for the constant-volume specific heat of solid argon are given in the accompanying table.

T, °K	C_v, cal/mole-K
10	0.80
15	1.88
20	2.94
30	4.25
40	4.61
60	5.12
80	5.74

Calculate a characteristic Debye temperature from each of these data points, and plot the results versus temperature. If the Debye theory were exact, θ_D would be a constant, the same value from each experimental point. Is this approximately true for argon?

7.5 Recently measured values of the specific heat of copper are listed below.

T, °K	C, joules/gm-K
40	0.060
60	0.137
100	0.254
180	0.346
300	0.386

Find values for the characteristic Debye and Einstein temperatures from each of these values. Which theory results in a better correlation of the data?

7.6 Helium gas is contained in a 1-liter aluminum tank at 300 K, 10 atm pressure. This system (tank and gas) is then cooled using liquid nitrogen to 77.3 K. If the mass of the aluminum tank is 1500 gm, how much heat is rejected to the liquid nitrogen during this cooling process?

7.7 Consider a monatomic solid in equilibrium with its vapor at very low temperature. Using the requirement for equilibrium between two phases of a pure substance (equal chemical potentials), show that the sublimation pressure for the solid is expressed as

$$\ln P_{\text{sub}} = \frac{C_1}{T} + C_2 \ln T + C_3 T^3 + C_4$$

Neglect the specific volume of the solid and any electronic contribution.

7.8 (*a*) Using the Clapeyron equation,

$$\frac{dP_{\text{sub}}}{dT} = \frac{\Delta h_{\text{sub}}}{T\,\Delta v_{\text{sub}}}$$

and the result of Problem 7.7, find an expression for the enthalpy of sublimation for a solid at low temperature.

(*b*) Calculate the sublimation pressure and Δh_{sub} for solid argon at 4.2 K, using $\theta_D = 83.5$ K.

7.9 The coefficient of thermal expansion, α, is defined as

$$\alpha = \frac{1}{V}\left(\frac{\partial V}{\partial T}\right)_P$$

Using the Maxwell relation

$$\left(\frac{\partial V}{\partial T}\right)_P = -\left(\frac{\partial S}{\partial P}\right)_T$$

show that α for a Debye solid can be expressed as

$$\alpha = \frac{nC_v}{V}\left(\frac{\partial \ln \theta_D}{\partial P}\right)_T$$

7.10 Calculate the Fermi energy μ_0 and the pressure exerted by the electron gas in a metal at 0 K. Assume one conduction electron per atom of metal and a metal density of 0.1 mole/cm^3.

7.11 Consider the electron gas in a metal.
(*a*) Using the thermodynamic relation

$$P = -\left(\frac{\partial A}{\partial V}\right)_T$$

show that, at room temperature,

$$PV = \tfrac{2}{3}U$$

(*b*) Using the data of Problem 7.10, calculate μ and the electron gas pressure at 300 K.

7.12 (*a*) Consider the electron gas in tungsten. Assume a metal density of 19.3 gm/cm^3 and two conduction electrons per atom. Find the number density N/V of the electron gas, the Fermi energy μ_0, and a characteristic temperature μ_0/k.
(*b*) Now consider the electrons in a vacuum tube as a Fermi-Dirac gas with a number density N/V of 10^{11} per cm^3. Calculate the Fermi energy μ_0 and μ_0/k for this gas, and compare the values with those of part (*a*). What is the significance of this comparison?

7.13 Using a metal density of 10.5 gm/cm^3 and assuming one conduction electron per atom, calculate the lattice and electronic contributions to the specific heat of silver at 30 K and 300 K.

7.14 Recent experimental specific heat data for copper are correlated by the expression

$$C_p = 30.6\left(\frac{T}{344.5}\right)^3 + 10.8 \times 10^{-6}T$$

at temperatures below 10 K. Neglecting the difference between C_p and C_v, discuss this expression and the numerical constants in terms of the Debye theory—electron gas model.

7.15 (*a*) Using Eqs. 4.76 and 7.34 for an electron gas, show that the number of electrons, dN_ϵ, in an interval of energy ϵ to $\epsilon + d\epsilon$ is

$$dN_\epsilon = \frac{4\pi V}{h^3}\frac{(2m)^{3/2}\epsilon^{1/2}}{e^{(\epsilon-\mu)/kT} + 1}\,d\epsilon$$

(*b*) Plot the function $dN_\epsilon/d\epsilon$ vs. ϵ for the electron gas of a 1 cm^3 sample of copper at 0 K and 300 K. Use a density of 8.96 gm/cm^3 and one conduction electron per atom for copper.

7.16 (*a*) Use the result of Problem 7.15*a* to find the speed distribution in an electron gas. Show that the number of electrons, dN_V in an interval V to V + dV is

$$dN_V = \frac{8\pi m^3 V}{h^3} \frac{V^2}{e^{(mV^2/2-\mu)/kT} + 1} dV$$

(*b*) Find an expression for the average speed of an electron gas at 0 K. What is the order of magnitude of this value for a typical metal?

7.17 Substitute the relation $dV_x\, dV_y\, dV_z = 4\pi V^2\, dV$ into the result of Problem 7.16*a* to obtain a relation for the distribution of component velocities in an electron gas. Using this result, show that the number of electrons having an x-component in the interval V_x to $V_x + dV_x$ and any V_y, V_z is

$$dN_{V_x} = \frac{4\pi V m^2 kT}{h^3} \ln\,[e^{(\mu-\epsilon_x)/kT} + 1]\, dV_x$$

(Transform V_y, V_z, to polar coordinates to perform the necessary integration.)

7.18 The problem of particles meeting a potential barrier at a surface was analyzed for Boltzmann statistics in Problem 3.20. To analyze the phenomenon of thermionic emission, the emission of electrons from an incandescent cathode, however, it is necessary to utilize Fermi-Dirac statistics for the electron gas in the cathode metal.

(*a*) Show that the rate at which electrons of a given V_x strike a unit surface (from the interior) is, as was found in Chapter 3,

$$\frac{\text{No. of } V_x}{\text{cm}^2\text{-sec}} = \frac{V_x}{V} dN_{V_x}$$

where dN_{V_x} is the number of electrons in the metal in the interval V_x to $V_x + dV_x$, and x is taken normal to the surface.

(*b*) If ϵ_m is the minimum kinetic energy normal to the surface with which an electron can escape the surface, show that the total rate of electron flow per unit area of cathode is given by

$$\frac{\text{No.}}{\text{cm}^2\text{-sec}} = CT^2 e^{-\varphi/kT}$$

where $\varphi = (\epsilon_m - \mu)$ is called the work function of the cathode. Compare this result with that of Problem 3.20.

(*c*) Each electron escaping from the cathode surface carries the charge e. Therefore, the electrical current density J_{sat}/A is given by

$$\frac{J_{sat}}{A} = CeT^2 e^{-\varphi/kT}$$

which is termed the Richardson-Dushman equation for thermionic emission. Using the value $\varphi = 4.5$ ev and the theoretical value for C, calculate J_{sat}/A for tungsten at 3000 K. What assumption is implied in using the theoretical value of C in such a calculation?

8 *Diatomic and Polyatomic Gases*

In this chapter we extend the developments of Chapter 6 to evaluate the thermodynamic properties for an ideal gas consisting of diatomic or polyatomic molecules. We consider the simple internal model, which was discussed in Chapter 5, and for the diatomic molecule we also discuss the more general model that is necessary at high temperature. Some special problems related to homonuclear molecules at low temperature are also considered.

8.1. The Diatomic Gas—Simple Internal Model

The evaluation of thermodynamic properties of diatomic gases is approached in the same manner as that utilized in Section 6.1 for monatomic gases. The method involves separating the translational and the internal energy modes. The translational contributions to the thermodynamic properties can then be determined by the equations developed for translation in Section 6.2. Evaluation of the internal contributions for diatomic gases is much more difficult than for monatomic gases, since a diatomic molecule has rotational and vibrational energies in addition to the electronic states.

In this section, we consider the simple internal model, consisting of rigid rotator, harmonic oscillator, ground electronic level, with the energy modes separated as outlined in Chapter 5. With this assumption, the component energies are additive, as indicated in Eq. 5.55:

$$\epsilon_{\text{int}} = \epsilon_r + \epsilon_v + \epsilon_{e0}$$

The component wave functions (and, therefore, energy level degeneracies) for this model are multiplicative, as in Eq. 5.56. By convention, the energy of the ground electronic level, ϵ_{e_0}, is taken to be zero, but we include the term here as a reminder that the degeneracy of this level must be included in our calculations.

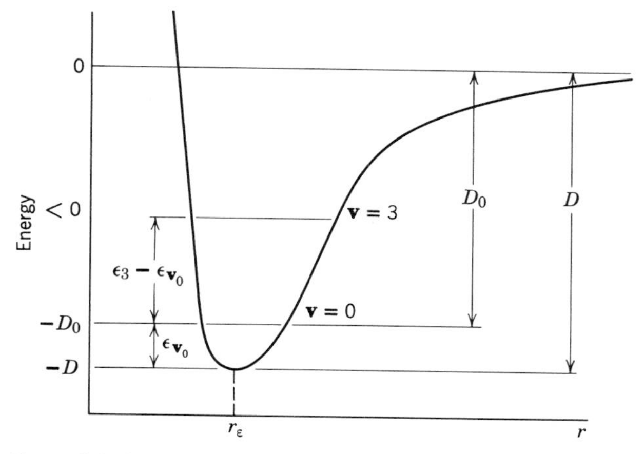

Figure 8.1 The zero-energy reference for a diatomic molecule.

Before we proceed, we must consider the reference state or "zero-of-energy." We have already noted that the vibrational energy is written with respect to the minimum point on the potential curve, which is shown in Figure 8.1 (in the vicinity of the minimum point, the assumption of harmonic oscillator is a good one). Thus, when $\mathbf{v} = 0$, the ground vibrational state energy ϵ_{v_0}, is given by the equation

$$\epsilon_{v_0} = \tfrac{1}{2}h\nu_\varepsilon \tag{8.1}$$

(If we wish to include corrections for anharmonicity, ϵ_{v_0} is given more accurately by Eq. 5.121.) Since it is more convenient to calculate vibrational contributions with respect to the ground vibrational state, we write Eq. 5.55 in the form

$$\epsilon_{\text{int}} = \epsilon_r + (\epsilon_v - \epsilon_{v0}) + \epsilon_{v0} + \epsilon_{e0} \tag{8.2}$$

One other point to be considered in selecting the energy reference state is that the reference must be made consistent with that for monatomic gases. This is particularly important in the analysis of chemical reactions, as will be found in Chapter 9. In order that the two be consistent, we should select for the zero-of-energy of the diatomic molecule that energy at which the molecule is dissociated to the monatomic form as was discussed previously and shown in Figure 5.9. Therefore, as is evident from Figure 8.1, ϵ_{int} relative to this base is given by the expression

$$\epsilon_{\text{int}} = \epsilon_r + (\epsilon_v - \epsilon_{v0}) + \epsilon_{v0} + \epsilon_{e0} - D = \epsilon_r + (\epsilon_v - \epsilon_{v0}) + \epsilon_{e0} - D_0 \tag{8.3}$$

in terms of the observed dissociation energy of the molecule, D_0.

Equation 6.4, which is

$$Z_{\text{int}} = \sum_{j_{\text{int}}} g_{j_{\text{int}}} \exp\left[\frac{-\epsilon_{j_{\text{int}}}}{kT}\right]$$

gives an expression for the internal partition function in terms of the various internal modes of energy. Substituting Eq. 8.3 into Eq. 6.4, we have

$$\begin{aligned} Z_{\text{int}} &= \sum_{\text{all } j} g_{r_j} g_{v_j} g_{e0} \exp\left[\frac{-(\epsilon_{r_j} + (\epsilon_{v_j} - \epsilon_{v0}) + \epsilon_{e0} - D_0)}{kT}\right] \\ &= Z_r Z_v Z_{e_0} Z_{\text{chem}} \end{aligned} \tag{8.4}$$

The rotational partition function Z_r in Eq. 8.4 is

$$Z_r = \sum_{rj} g_{r_j} \exp\left[\frac{-\epsilon_{r_j}}{kT}\right] \tag{8.5}$$

It was found in Chapter 5 that the vibrational energy levels are non-degenerate. Therefore g_v is unity for all levels, and the vibrational partition function Z_v with respect to the ground vibrational state energy ϵ_{v_0} is

$$Z_v = \sum_{vj} \exp\left[\frac{-(\epsilon_{v_j} - \epsilon_{v0})}{kT}\right] \tag{8.6}$$

This expression is the same as that discussed in Chapter 7 for the harmonic oscillations in a solid. Since the electronic ground level energy ϵ_{e_0} is by convention taken as zero, the partition function Z_{e_0} for the electronic ground state is

$$Z_{e_0} = g_{e0} \exp\left[\frac{-\epsilon_{e0}}{kT}\right] = g_{e0} \tag{8.7}$$

Finally, we define the chemical partition function Z_{chem} in terms of the dissociation energy D_0 by the relation

$$Z_{\text{chem}} = \exp\left[\frac{D_0}{kT}\right] \tag{8.8}$$

If Eq. 8.4 is written in logarithmic form, we have

$$\ln Z_{\text{int}} = \ln Z_r + \ln Z_v + \ln Z_{e_0} + \ln Z_{\text{chem}} \tag{8.9}$$

From this equation, it is apparent that each of the four terms has a contribution to the thermodynamic properties as given by Eqs. 6.7, 6.10, 6.13, 6.16, and 6.19 for u, h, s, a, g, respectively.

8.2. Rotation

Using the results of quantum mechanics to evaluate these individual contributions, let us consider first the rotational energy. Substituting the expressions for energy (Eq. 5.87) and rotational level degeneracy (Eq. 5.90) into Eq. 8.5, we have

$$Z_r = \sum_{j=0}^{\infty}(2j + 1) \exp\left[-\frac{h^2}{8\pi^2 I_\varepsilon kT} j(j + 1)\right] \tag{8.10}$$

Let us for the moment assume that the rotational energy levels are spaced sufficiently close together that the summation in this expression can be replaced directly by an integral. If we also change variable according to the relation

$$y^2 = j(j + 1)$$

Eq. 8.10 becomes

$$Z_r = \int_0^{\infty} 2y \exp\left[-\frac{h^2}{8\pi^2 I_\varepsilon kT} y^2\right] dy = \frac{8\pi^2 I_\varepsilon kT}{h^2} = \frac{T}{\theta_r} \tag{8.11}$$

where the constant θ_r is defined by

$$\theta_r = \frac{h^2}{8\pi^2 I_\varepsilon k} \tag{8.12}$$

and, having the units of temperature, is termed the characteristic rotational temperature.

There is a one additional factor that must be considered in the evaluation of the rotational partition function for homonuclear molecules (for example, N_2, O_2, H_2, etc.). Since the two atoms of which the molecule is comprised are identical, the molecule can be rotated through 180° and the new configuration will be indistinguishable from the original one. To correct for this double counting, we introduce a symmetry number σ into the result, so that the correct expression for the rotational partition function is

$$Z_r = \frac{T}{\sigma\theta_r} \tag{8.13}$$

in which $\sigma = 1$ for heteronuclear molecules, $\sigma = 2$ for homonuclear molecules. The argument used here in introducing the symmetry number is admittedly quite artificial and incomplete. The complete explanation for the justification of this number is based on the symmetry characteristics of the molecular wave function and is a rather complete subject in itself. Discussion of this matter will be delayed until Section 8.6.

For the rotational contributions to thermodynamic properties, we rewrite Eq. 8.13 in logarithmic form:

$$\ln Z_r = \ln T - \ln \sigma\theta_r \tag{8.14}$$

Then, on applying Eq. 6.7,

$$u_{\text{int}} = RT^2\left(\frac{\partial \ln Z_{\text{int}}}{\partial T}\right)_V$$

to the rotation, we find that

$$u_r = RT^2\left(\frac{1}{T}\right) = RT \tag{8.15}$$

Similarly, it follows from Eq. 6.10 that

$$h_r = u_r = RT \tag{8.16}$$

When Eqs. 8.15 and 8.16 are differentiated with respect to temperature, it follows from the definitions of the specific heats that

$$C_{v_r} = C_{p_r} = R \tag{8.17}$$

The rotational contribution to entropy can be found by applying Eq. 6.13,

$$s_{\text{int}} = R \ln Z_{\text{int}} + \frac{u_{\text{int}}}{T}$$

to the rotation. The result is

$$s_r = R\left[\ln\left(\frac{T}{\sigma\theta_r}\right) + 1\right] \tag{8.18}$$

The rotational contributions to Helmholtz and Gibbs functions are, from Eqs. 6.16 and 6.19,

$$a_r = g_r = -RT \ln\left(\frac{T}{\sigma\theta_r}\right) \tag{8.19}$$

Example 8.1

Calculate the rotational partition function and the rotational contributions to internal energy and entropy for diatomic nitrogen at 500 K.

From Table A.5, for N_2,

$$r_\varepsilon = 1.0976 \text{ Å}$$

The mass of a single nitrogen atom is

$$m_{\text{N}} = 1.66 \times 10^{-24}(14.008) = 23.25 \times 10^{-24} \text{ gm}$$

or half the mass of the diatomic molecule. The reduced mass is found from Eq. 5.64:

$$m_r = \frac{m_{\text{N}} m_{\text{N}}}{m_{\text{N}} + m_{\text{N}}} = \frac{m_{\text{N}}}{2} = 11.625 \times 10^{-24} \text{ gm}$$

Then, from Eq. 5.63,

$$I_\varepsilon = m_r r_\varepsilon^2 = 11.625 \times 10^{-24}(1.0976 \times 10^{-8})^2 = 14.0 \times 10^{-40} \text{ gm cm}^2$$

The characteristic rotational temperature θ_r is found from Eq. 8.12,

$$\theta_r = \frac{h^2}{8\pi^2 I_\varepsilon k} = \frac{(6.625 \times 10^{-27})^2}{8\pi^2(14.0 \times 10^{-40})(1.3804 \times 10^{-16})} = 2.875 \text{ K}$$

Since diatomic nitrogen is a homonuclear molecule (two N atoms), the symmetry number σ is 2. The rotational partition function can be found from Eq. 8.13,

$$Z_r = \frac{T}{\sigma\theta_r} = \frac{500}{2(2.875)} = 87.0$$

The internal energy of rotation is, from Eq. 8.15,

$$u_r = RT = 1.987(500) = 994 \text{ cal/mole}$$

The rotational entropy is calculated from Eq. 8.18:

$$s_r = R[\ln Z_r + 1] = 1.987[\ln 87.0 + 1] = 10.85 \text{ cal/mole-K}$$

The expression for the rotational partition function for a rigid rotator as given by Eq. 8.13 was obtained under the assumption that the rotational energy levels are very closely spaced. To determine the error resulting from this approximation, we must evaluate the summation of Eq. 8.10, using the Euler-Maclaurin summation theorem, as we did for translational energy in Chapter 6. The procedure here is straightforward but involves the rather tedious job of taking the derivatives of the function in that equation; this will be left as an exercise. The result, after symmetry number σ has been included can be written in the form†

$$Z_r = \frac{T}{\sigma\theta_r}\left[1 + \frac{1}{3}\left(\frac{\theta_r}{T}\right) + \frac{1}{15}\left(\frac{\theta_r}{T}\right)^2 + \frac{4}{315}\left(\frac{\theta_r}{T}\right)^3 + \cdots\right] \tag{8.20}$$

Note that in this equation Z_r is given in terms of the simplified result plus a series of correction terms. We find from Eq. 8.20 that the simplification made in Eq. 8.11 is reasonable only for $T \gg \theta_r$ for, in this case, the correction terms are negligibly small. When correction terms are of importance, it is necessary to evaluate the rotational contributions to properties directly from Eq. 8.20, instead of determining properties from Eqs. 8.15 to 8.19.

If all the terms of Eq. 8.20 are used in expressing the rotational partition function, the following relations for various thermodynamic properties result.

$$h_r = u_r = RT\left(\frac{Z_r'}{Z_r}\right) \tag{8.21}$$

$$s_r = R\left[\ln Z_r + \left(\frac{Z_r'}{Z_r}\right)\right] \tag{8.22}$$

$$g_r = a_r = -RT\ln Z_r \tag{8.23}$$

$$C_{p_r} = C_{v_r} = R\left[\frac{Z_r'' + Z_r'}{Z_r} - \left(\frac{Z_r'}{Z_r}\right)^2\right] \tag{8.24}$$

† For $T/\theta_r > 1$; otherwise, the rotational partition function can be evaluated by direct expansion of Eq. 8.10.

We note that this set of expressions is identical in form with Eqs. 6.40 to 6.43, developed for electronic energy contributions. In this case, the quantities Z_r' and Z_r'' are

$$Z_r' = T\left(\frac{dZ_r}{dT}\right) = \frac{T}{\sigma\theta_r}\left[1 - \frac{1}{15}\left(\frac{\theta_r}{T}\right)^2 - \frac{8}{315}\left(\frac{\theta_r}{T}\right)^3\right] \tag{8.25}$$

$$Z_r'' = T\left(\frac{dZ_r'}{dT}\right) = \frac{T}{\sigma\theta_r}\left[1 + \frac{1}{15}\left(\frac{\theta_r}{T}\right)^2 + \frac{16}{315}\left(\frac{\theta_r}{T}\right)^3\right] \tag{8.26}$$

Example 8.2

Calculate the rotational contributions to internal energy and specific heat C_v for hydrogen fluoride, HF, at 100 K.

From Table A.5 for HF,

$$r_\varepsilon = 0.9168 \text{ Å}$$

The atomic masses are

$$m_{\text{H}} = 1.66 \times 10^{-24}(1.008) = 1.673 \times 10^{-24} \text{ gm}$$

$$m_{\text{F}} = 1.66 \times 10^{-24}(19) = 31.54 \times 10^{-24} \text{ gm}$$

From Eqs. 5.64 and 5.63,

$$m_r = \frac{m_{\text{H}} m_{\text{F}}}{m_{\text{H}} + m_{\text{F}}} = \frac{1.673(31.54)}{33.21} \times 10^{-24} = 1.59 \times 10^{-24} \text{ gm}$$

$$I_\varepsilon = m_r r_\varepsilon^2 = 1.59 \times 10^{-24}(0.9168 \times 10^{-8})^2 = 1.34 \times 10^{-40} \text{ gm cm}^2$$

Thus, from Eq. 8.12,

$$\theta_r = \frac{h^2}{8\pi^2 I_\varepsilon k} = \frac{(6.625 \times 10^{-27})^2}{8\pi^2(1.34 \times 10^{-40})(1.3804 \times 10^{-16})} = 30.3 \text{ K}$$

We note that, at 100 K, the ratio

$$\frac{T}{\theta_r} = \frac{100}{30.3} = 3.30$$

is not a large number. Therefore the rotational partition function should be evaluated from the more general expression of Eq. 8.20, as the use of Eq. 8.13 would introduce a significant error. From Eq. 8.20, with $\sigma = 1$ for the HF molecule, we have

$$\begin{aligned} Z_r &= \frac{T}{\sigma\theta_r}\left[1 + \frac{1}{3}\left(\frac{\theta_r}{T}\right) + \frac{1}{15}\left(\frac{\theta_r}{T}\right)^2 + \frac{4}{315}\left(\frac{\theta_r}{T}\right)^3 + \cdots\right] \\ &= 3.30[1 + \tfrac{1}{3}(0.303) + \tfrac{1}{15}(0.303)^2 + \tfrac{4}{315}(0.303)^3] \\ &= 3.30[1 + 0.101 + 0.0061 + 0.0004] \\ &= 3.30[1.1075] = 3.655 \end{aligned}$$

From Eq. 8.25,

$$Z_r' = \frac{T}{\sigma\theta_r}\left[1 - \frac{1}{15}\left(\frac{\theta_r}{T}\right)^2 - \frac{8}{315}\left(\frac{\theta_r}{T}\right)^3\right]$$
$$= 3.30[1 - \tfrac{1}{15}(0.303)^2 - \tfrac{8}{315}(0.303)^3] = 3.277$$

Therefore, from Eq. 8.21,

$$u_r = RT\left(\frac{Z_r'}{Z_r}\right) = 1.987(100)\left(\frac{3.277}{3.655}\right) = 178.0 \text{ cal/mole}$$

Had the simplified expression, Eq. 8.15, been used, the result would have been

$$u_r = RT = 198.7 \text{ cal/mole}$$

which is in error by nearly 12%. From Eq. 8.26,

$$Z_r'' = \frac{T}{\sigma\theta_r}\left[1 + \frac{1}{15}\left(\frac{\theta_r}{T}\right)^2 + \frac{16}{315}\left(\frac{\theta_r}{T}\right)^3\right]$$
$$= 3.30[1 + \tfrac{1}{15}(0.303)^2 + \tfrac{16}{315}(0.303)^3] = 3.325$$

Therefore, using Eq. 8.24, we find

$$C_{v_r} = R\left[\frac{Z_r'' + Z_r'}{Z_r} - \left(\frac{Z_r'}{Z_r}\right)^2\right]$$
$$= 1.987\left[\left(\frac{3.325 + 3.277}{3.655}\right) - \left(\frac{3.277}{3.655}\right)^2\right]$$
$$= 1.999 \text{ cal/mole-K}$$

as compared with the value 1.987 cal/mole-K from the simplified expression, Eq. 8.17.

8.3. Vibration

The partition function and the resulting contributions to thermodynamic properties for a harmonic oscillator have already been discussed in detail in Section 7.1 in connection with the properties of a monatomic solid. However, because vibrational energy is an important aspect of the behavior of diatomic molecules, that development will be outlined here for the purpose of continuity.

For the vibrational partition function in our simple internal model, we consider the molecule to behave as a harmonic oscillator, the energy being given by Eq. 5.112. In the vibrational partition function (Eq. 8.6), the energy considered is that with respect to the ground state $\mathbf{v} = 0$, which is

$$\epsilon_v - \epsilon_{v_0} = h\nu_\varepsilon \mathbf{v} \tag{8.27}$$

where $\mathbf{v}$ is the vibrational quantum number. Substituting Eq. 8.27 into the vibrational partition function (Eq. 8.6), we obtain

$$Z_v = \sum_{\mathbf{v}=0}^{\infty} e^{-(h\nu_\varepsilon/kT)\mathbf{v}} \tag{8.28}$$

As pointed out in Chapter 7, vibrational energy levels are so widely spaced that it is not possible to replace the summation in Eq. 8.28 by an integral, even if the Euler-Maclaurin summation theorem is used. It was noted, however, that this expression is given by

$$Z_v = \sum_{\mathbf{v}=0}^{\infty} e^{-(h\nu_\varepsilon/kT)\mathbf{v}} = \frac{1}{1 - e^{-\theta_v/T}} \tag{8.29}$$

where the characteristic vibrational temperature θ_v is defined as

$$\theta_v = \frac{h\nu_\varepsilon}{k} = \frac{hc\omega_\varepsilon}{k} \tag{8.30}$$

In the evaluation of the vibrational contributions to thermodynamic properties, it is convenient to express Eq. 8.29 in the logarithmic form

$$\ln Z_v = -\ln (1 - e^{-\theta_v/T}) \tag{8.31}$$

The internal energy contribution is found from Eq. 6.7,

$$u_{\text{int}} = RT^2\left(\frac{\partial \ln Z_{\text{int}}}{\partial T}\right)_V$$

However, since the partition function Z_v to be used in this expression is evaluated with respect to the ground vibrational state energy ϵ_{v_0}, it follows that the molal energy calculated from the equation will be that with respect to the molal ground state energy u_{v_0}, which equals $N_0\epsilon_{v_0}$. Therefore, from Eqs. 6.7 and 8.31, we find

$$(u_v - u_{v0}) = RT^2\left[-\frac{e^{-\theta_v/T}}{1 - e^{-\theta_v/T}}\left(-\frac{\theta_v}{T^2}\right)\right] = \frac{R\theta_v}{e^{\theta_v/T} - 1} \tag{8.32}$$

Similarly, using Eq. 6.10, we have

$$(h_v - u_{v0}) = (u_v - u_{v0}) = \frac{R\theta_v}{e^{\theta_v/T} - 1} \tag{8.33}$$

Differentiating Eqs. 8.32 and 8.33, we obtain

$$C_{v_v} = C_{p_v} = \frac{R(\theta_v/T)^2 e^{\theta_v/T}}{(e^{\theta_v/T} - 1)^2} \tag{8.34}$$

The entropy contribution is found directly from Eq. 6.13,

$$s_{\text{int}} = R \ln Z_{\text{int}} + \frac{u_{\text{int}}}{T}$$

It is easily shown that this expression is independent of the zero-point energy u_{v_0}. Therefore, substituting Eqs. 8.31 and 8.32 into Eq. 6.13, we have

$$s_v = R\left[-\ln(1 - e^{-\theta_v/T}) + \frac{\theta_v/T}{e^{\theta_v/T} - 1}\right] \tag{8.35}$$

It follows also from Eqs. 6.16 and 6.19 that

$$(a_v - u_{v0}) = (g_v - u_{v0}) = RT\ln(1 - e^{-\theta_v/T}) \tag{8.36}$$

Example 8.3

Calculate the specific heat C_p for diatomic iodine at 0 C.

The translational, rotational, and vibrational contributions to the specific heat must all be considered. The translational contribution is found from Eq. 6.28,

$$C_{p_t} = \tfrac{5}{2}R = \tfrac{5}{2}(1.987) = 4.968 \text{ cal/mole-K}$$

For rotation, it is found from the values in Table A.5 that, for diatomic iodine,

$$\theta_r = 0.054 \text{ K}$$

Therefore, at $T = 273.15$ K, we find that $T \gg \theta_r$; hence the rotational partition function is accurately given by Eq. 8.13 and, correspondingly, the specific heat contribution is given by Eq. 8.17,

$$C_{p_r} = R = 1.987 \text{ cal/mole-K}$$

To determine the vibrational contribution, we use the value of ω_ε from Table A.5 to calculate the characteristic vibrational temperature from Eq. 8.30:

$$\theta_v = \left(\frac{hc}{k}\right)\omega_\varepsilon = 1.4388(214.5) = 308.8 \text{ K}$$

Therefore

$$\frac{\theta_v}{T} = \frac{308.8}{273.15} = 1.131$$

The value for C_{p_v} can now be found from Eq. 8.34 and Table A.4:

$$C_{p_v} = \frac{R(\theta_v/T)^2 e^{\theta_v/T}}{(e^{\theta_v/T} - 1)^2} = 1.987(0.8999) = 1.788 \text{ cal/mole-K}$$

Adding the translational, rotational, and vibrational contributions, we find, for the specific heat C_p,

$$C_p = C_{p_t} + C_{p_r} + C_{p_v} = 4.968 + 1.987 + 1.788 = 8.743 \text{ cal/mole-K}$$

8.4. Electronic Ground Level and Chemical Energy

Since the electronic energy in the ground electronic level is by convention assigned the value zero, only the ground level degeneracy is significant for electronic contributions in the simple internal model. In logarithmic form, the electronic partition function (Eq. 8.7) is

$$\ln Z_e = \ln g_{e_0} = \text{constant} \tag{8.37}$$

It therefore follows from Eqs. 6.7, 6.10, and the definitions of C_v and C_p that

$$u_e = h_e = C_{v_e} = C_{p_e} = 0 \tag{8.38}$$

However, there is an electronic contribution to entropy as given by Eq. 6.13, from which we find

$$s_e = R \ln g_{e_0} \tag{8.39}$$

For the Helmholtz and Gibbs functions we have, from Eq. 6.19,

$$a_e = g_e = -RT \ln g_{e_0} \tag{8.40}$$

Note that s_e, a_e, and g_e are zero only if the ground electronic level degeneracy is unity.

Finally, we must determine the contribution of the chemical energy to thermodynamic properties. As noted in Section 8.1, we have selected the zero-of-energy for the molecule as the dissociated state, so that the reference state will be consistent with that chosen previously for a monatomic substance. This selection introduces a constant value of $-D_0$ into the internal mode energy of the molecule, with a resulting contribution to the thermodynamic properties. Therefore, in accordance with Eq. 8.8,

$$\ln Z_{\text{chem}} = \frac{D_0}{kT} = \frac{N_0 D_0}{RT} = -\frac{h^0}{RT} \tag{8.41}$$

where h^0 is the enthalpy of formation of the molecule at absolute zero temperature and is by definition

$$h^0 = -N_0 D_0 \tag{8.42}$$

With the molecule in the ground electronic level, the enthalpy of formation is a negative number, since the energy of the dissociated molecule has been selected as zero, as discussed with reference to Figure 8.1.

From Eqs. 6.7 and 6.10,

$$h_{\text{chem}} = u_{\text{chem}} = RT^2\left(+\frac{h^0}{RT^2}\right) = h^0 \tag{8.43}$$

which is a constant. Therefore

$$C_{v_{\text{chem}}} = C_{p_{\text{chem}}} = 0 \tag{8.44}$$

It also follows from Eq. 6.13 that

$$s_{\text{chem}} = R\left(-\frac{h^0}{RT}\right) + \frac{u_{\text{chem}}}{T} = 0 \tag{8.45}$$

Since s_{chem} is zero, we find from Eqs. 6.19 and 8.43 that

$$a_{\text{chem}} = g_{\text{chem}} = h^0 \tag{8.46}$$

8.5. Diatomic Gases—General Internal Model

From our discussions of Chapter 5 we realize that the error associated with the assumption of the simple internal model increases significantly as the temperature is increased. At high temperature, the rotational and vibrational energies may both be quite large, and corrections for centrifugal and vibrational stretching and for vibrational anharmonicity should be introduced into the calculations of internal energy modes.

We have found previously that, if essentially all the molecules are in the electronic ground level, the energy for coupled rotation-vibration can be expressed in the form (Eq. 5.125)

$$\begin{aligned}\frac{\epsilon_{rv}}{hc} &= \omega_\varepsilon(\mathbf{v} + \tfrac{1}{2}) - x_\varepsilon\omega_\varepsilon(\mathbf{v} + \tfrac{1}{2})^2 \\ &\quad + B_\varepsilon\mathbf{j}(\mathbf{j} + 1) - D_\varepsilon\mathbf{j}^2(\mathbf{j} + 1)^2 \\ &\quad - \alpha(\mathbf{v} + \tfrac{1}{2})\mathbf{j}(\mathbf{j} + 1) + \cdots\end{aligned}$$

In this equation, B_ε is given in terms of the equilibrium point moment of inertia I_ε by Eq. 5.118,

$$B_\varepsilon = \frac{h}{8\pi^2 I_\varepsilon c}$$

The anharmonicity constant x_ε, the centrifugal stretching constant D_ε, and the vibrational stretching constant α are determined from observed spectral data for the molecule; in the absence of such information they are calculated from Eqs. 5.128 to 5.130.

As in the development of the simple internal model, we again express the energy relative to the vibrational ground state. However, as noted in Eq. 5.121, for this general model the zero-point energy is, in units of reciprocal centimeters,

$$\frac{\epsilon_{v0}}{hc} = \tfrac{1}{2}\omega_\varepsilon - \tfrac{1}{4}x_\varepsilon\omega_\varepsilon$$

instead of merely the first term as given by Eq. 8.1. Therefore, substituting Eq. 5.121 into Eq. 5.125 and rearranging, we obtain

$$\begin{aligned}\frac{\epsilon_{rv} - \epsilon_{v0}}{hc} &= \omega_\varepsilon \mathbf{v} - x_\varepsilon \omega_\varepsilon \mathbf{v}(\mathbf{v} + 1) + (B_\varepsilon - \tfrac{1}{2}\alpha)\mathbf{j}(\mathbf{j} + 1) \\ &\quad - D_\varepsilon \mathbf{j}^2(\mathbf{j} + 1)^2 - \alpha \mathbf{v}\mathbf{j}(\mathbf{j} + 1) \\ &= \omega^* \mathbf{v} - x^* \omega^* \mathbf{v}(\mathbf{v} - 1) + B^* \mathbf{j}(\mathbf{j} + 1) \\ &\quad - D_\varepsilon \mathbf{j}^2(\mathbf{j} + 1)^2 - \alpha \mathbf{v}\mathbf{j}(\mathbf{j} + 1)\end{aligned} \tag{8.47}$$

where the variables ω^*, x^*, and B^* are defined as follows:

$$\omega^* = \omega_\varepsilon(1 - 2x_\varepsilon) \tag{8.48}$$

$$x^*\omega^* = x_\varepsilon \omega_\varepsilon \tag{8.49}$$

$$B^* = B_\varepsilon - \tfrac{1}{2}\alpha \tag{8.50}$$

Equation 8.47 then represents the total internal contribution to the energy of a molecule with respect to the ground state, the electronic ground level being assumed and the chemical energy h^0 of the molecule not being included. Before substituting this expression into the internal partition function, Eq. 6.4, let us rewrite Eq. 8.47 in the form

$$\frac{\epsilon_{\text{int}} - \epsilon_{v0}}{kT} = t\mathbf{v} - x^* t\mathbf{v}(\mathbf{v} - 1) + y[1 - \gamma \mathbf{j}(\mathbf{j} + 1) - \delta \mathbf{v}]\mathbf{j}(\mathbf{j} + 1) \tag{8.51}$$

where

$$t = \frac{hc\omega^*}{kT} = \frac{\omega^*}{\omega_\varepsilon}\left(\frac{\theta_v}{T}\right) \tag{8.52}$$

$$y = \frac{hcB^*}{kT} = \frac{B^*}{B_\varepsilon}\left(\frac{\theta_r}{T}\right) \tag{8.53}$$

$$\gamma = \frac{D_\varepsilon}{B^*} \tag{8.54}$$

$$\delta = \frac{\alpha}{B^*} \tag{8.55}$$

Now, including the electronic ground level degeneracy g_{e_0} and the rotational symmetry number σ (but not h^0), the internal partition function is, from Eqs. 6.4 and 8.51,

$$\begin{aligned} Z_{\text{int}} &= \frac{g_{e0}}{\sigma} \sum_{\mathbf{v}=0}^{\infty} \sum_{\mathbf{j}=0}^{\infty} (2\mathbf{j} + 1) \\ &\quad \times \exp\{-y[1 - \gamma \mathbf{j}(\mathbf{j} + 1) - \delta \mathbf{v}]\mathbf{j}(\mathbf{j} + 1)\} \exp\{-t[\mathbf{v} - x^* \mathbf{v}(\mathbf{v} - 1)]\} \end{aligned} \tag{8.56}$$

Given the molecular constants, Eq. 8.56 can be evaluated at a given temperature by a double summation over the necessary number of terms.

This is, of course, a very tedious process. We can, however, obtain an approximate result, suitable for all but the most precise calculations. To accomplish this, we first evaluate the summation over **j**, using the Euler-Maclaurin summation theorem. Approximating the resulting integral, we find that, if the higher order terms are neglected, Eq. 8.56 reduces to

$$Z_{\text{int}} = \frac{g_{e0}}{\sigma}\sum_{\mathbf{v}=0}^{\infty}\frac{1}{y}\left(1 + \frac{y}{3} + \frac{2\gamma}{y} + \delta\mathbf{v}\right)\exp(-t\mathbf{v})\exp[tx^*\mathbf{v}(\mathbf{v}-1)] \quad (8.57)$$

The summation over **v** is now approximated by expanding the second exponential of Eq. 8.57 in a series, again neglecting higher order terms, and recognizing the resulting summations as closed-form expressions (in a manner similar to that in Eq. 8.29). The resulting expression is

$$Z_{\text{int}} = \frac{g_{e0}}{\sigma y}\left(\frac{1}{1 - e^{-t}}\right)\left(1 + \frac{y}{3} + \frac{2\gamma}{y} + \frac{\delta}{e^t - 1} + \frac{2x^*t}{(e^t - 1)^2}\right) \quad (8.58)$$

For convenience, we write this result in the form

$$Z_{\text{int}} = Z_{\text{ideal}}Z_{\text{corr}} \quad (8.59)$$

where

$$Z_{\text{ideal}} = \frac{g_{e0}}{\sigma y}\left(\frac{1}{1 - e^{-t}}\right) \quad (8.60)$$

This is termed the ideal internal partition function. On comparison of this result with the product of Eqs. 8.7, 8.13, and 8.29 (the simple internal model expressions), which is

$$Z_{\text{int (simple)}} = (g_{e0})\left(\frac{T}{\sigma\theta_r}\right)\left(\frac{1}{1 - e^{-\theta_v/T}}\right)$$

we find that, while the two are of the same form, they are not identical because of the difference between y and (θ_r/T) (which is given by Eq. 8.53) and that between t and (θ_v/T) (which is given in Eq. 8.52). These differences, however, are relatively small. Since Eq. 8.60 is of the same form as the result for the simple internal model, the contributions from Eq. 8.60 to thermodynamic properties are found by using the equations of the preceding section, but replacing (θ_r/T) by y and (θ_v/T) by t.

The second part of Eq. 8.59,

$$Z_{\text{corr}} = 1 + \frac{y}{3} + \frac{2\gamma}{y} + \frac{\delta}{e^t - 1} + \frac{2x^*t}{(e^t - 1)^2} \quad (8.61)$$

is a correction to the ideal internal partition function. Since the contributions to properties for the ideal partition function have already been determined, we need now evaluate only property corrections in terms of Eq. 8.61. This is accomplished in the same manner as for electronic contributions (Eqs. 6.40 to 6.43) and rotational contributions (Eqs. 8.21 to 8.24).

The result is

$$h_{\text{corr}} = u_{\text{corr}} = RT^2\left(\frac{d \ln Z_{\text{corr}}}{dT}\right) = RT\left(\frac{Z'_{\text{corr}}}{Z_{\text{corr}}}\right) \tag{8.62}$$

where, from Eq. 8.61,

$$Z'_{\text{corr}} = T\left(\frac{dZ_{\text{corr}}}{dT}\right) = -\frac{y}{3} + \frac{2\gamma}{y} + \frac{t(\delta e^t - 2x^*)}{(e^t - 1)^2} + \frac{4x^* t^2 e^t}{(e^t - 1)^3} \tag{8.63}$$

Similarly,

$$s_{\text{corr}} = R\left[\ln Z_{\text{corr}} + \frac{Z'_{\text{corr}}}{Z_{\text{corr}}}\right] \tag{8.64}$$

and

$$g_{\text{corr}} = a_{\text{corr}} = -RT \ln Z_{\text{corr}} \tag{8.65}$$

The effort involved in evaluating a specific heat correction using an expression analogous to Eq. 8.24, which was developed for the rotational partition function (Eq. 8.20) is not warranted. Approximations have already been made in the derivation of Eq. 8.61, and, in addition, the parameters y, t, γ, δ are small. Thus the approximate equation obtained by differentiating Eq. 8.62 and neglecting higher order terms,

$$C_{p_{\text{corr}}} = C_{v_{\text{corr}}} = R\left\{\frac{4\delta}{y} + \frac{t^2 e^t}{(e^t - 1)^3}\left[\delta(e^t + 1) + 8x^*(t - 1) + \frac{12x^* t}{e^t - 1}\right]\right\} \tag{8.66}$$

is nearly as accurate and requires less effort.

Example 8.4

Calculate the internal contribution to the Gibbs function for diatomic nitrogen at 5000 K.

From Table A.5 for N_2,

$$r_\varepsilon = 1.0976 \text{ Å}$$
$$\omega_\varepsilon = 2357.6 \text{ cm}^{-1}$$
$$g_{e_0} = 1$$
$$D_0 = 9.757 \text{ ev} = 1.565 \times 10^{-11} \text{ erg}$$

Given only the above data, the constants x_ε, D_ε, and α must be calculated from the expressions in Chapter 5. (It would be more precise to use values of these constants found by empirically fitting the energy equation, Eq. 5.125.) The value of θ_r was calculated in Example 8.1 as 2.875 K; thus, from Eq. 5.118,

$$B_\varepsilon = \frac{h}{8\pi^2 I_\varepsilon c} = \frac{\theta_r}{(hc/k)} = \frac{2.875}{1.4388} = 2.00 \text{ cm}^{-1}$$

From Eqs. 5.118 and 5.128,

$$\frac{\epsilon_{v0}}{hc} = \frac{D - D_0}{hc} = \tfrac{1}{2}\omega_\varepsilon - \tfrac{1}{4}x_\varepsilon\omega_\varepsilon = \tfrac{1}{2}\omega_\varepsilon - \frac{hc\omega_\varepsilon^{\,2}}{16D}$$

$$\frac{hc\omega_\varepsilon}{2} = \frac{(6.625 \times 10^{-27})(2.998 \times 10^{10})(2357.6)}{2} = 0.0234 \times 10^{-11} \text{ erg}$$

Therefore,

$$D = 1.565 \times 10^{-11} + 0.0234 \times 10^{-11} - \frac{1}{4D}(0.0234 \times 10^{-11})^2$$

Solving for D,

$$D = 1.588 \times 10^{-11} \text{ erg}$$

From Eq. 5.128,

$$x_\varepsilon = \frac{hc\omega_\varepsilon}{4D} = \frac{0.0234 \times 10^{-11}}{2(1.588 \times 10^{-11})} = 0.0737$$

and, from Eq. 5.129,

$$D_\varepsilon = \frac{4B_\varepsilon^{\,3}}{\omega_\varepsilon^{\,2}} = \frac{4(2.00)^3}{(2357.6)^2} = 5.76 \times 10^{-6} \text{ cm}^{-1}$$

Using Eq. 5.130 gives

$$\alpha = \frac{6B_\varepsilon^{\,2}}{\omega_\varepsilon}\left[\left(\frac{x_\varepsilon\omega_\varepsilon}{B_\varepsilon}\right)^{1/2} - 1\right]$$

$$= \frac{6(2.00)^2}{2357.6}\left[\left(\frac{0.00737 \times 2357.6}{2.00}\right)^{1/2} - 1\right] = 0.0198 \text{ cm}^{-1}$$

The values of the variables used in Eq. 8.47 are determined from the expressions (Eqs. 8.48 to 8.55)

$$\omega^* = \omega_\varepsilon(1 - 2x_\varepsilon) = 2357.6(1 - 2 \times 0.00737) = 2322.8 \text{ cm}^{-1}$$

$$x^* = x_\varepsilon\frac{\omega_\varepsilon}{\omega^*} = 0.00737\left(\frac{2357.6}{2322.8}\right) = 0.00748$$

$$B^* = B_\varepsilon - \tfrac{1}{2}\alpha = 2.00 - \tfrac{1}{2}(0.0198) = 1.99 \text{ cm}^{-1}$$

The variables and constants used in the final expressions are found from Eqs. 8.52 to 8.55. For a temperature of 5000 K,

$$t = \frac{hc\omega^*}{kT} = \frac{1.4388(2322.8)}{5000} = 0.668$$

$$y = \frac{hcB^*}{kT} = \frac{1.4388(1.99)}{5000} = 5.726 \times 10^{-4}$$

$$\gamma = \frac{D_\varepsilon}{B^*} = \frac{5.76 \times 10^{-6}}{1.99} = 2.90 \times 10^{-6}$$

$$\delta = \frac{\alpha}{B^*} = \frac{0.0198}{1.99} = 0.010$$

When Eq. 8.60 is written in logarithmic form,

$$\ln Z_{\text{ideal}} = \ln g_{eo} + \ln \left(\frac{1}{\sigma y}\right) - \ln (1 - e^{-t})$$

the ideal internal partition function is expressed as a sum of electronic, rotational, and vibrational contributions, and the various ideal contributions to the Gibbs function can be found. These ideal contributions are

$$
\begin{aligned}
g_{eo} &= -RT \ln g_{eo} = -1.987(5000) \ln 1 = 0 \\
g_r &= -RT \ln \left(\frac{1}{\sigma y}\right) = +9935 \ln (2 \times 5.726 \times 10^{-4}) \\
&= -67{,}200 \text{ cal/mole} \\
(g_v - u_{vo}) &= +RT \ln (1 - e^{-t}) = 9935(-0.7189) \\
&= -7140 \text{ cal/mole}
\end{aligned}
$$

Therefore, adding these contributions yields

$$
\begin{aligned}
(g_{\text{int}} - u_{v_0})_{\text{ideal}} &= (g_{e_0} + g_r + g_v - u_{v_0})_{\text{ideal}} \\
&= 0 - 67{,}200 - 7140 = -74{,}340 \text{ cal/mole}
\end{aligned}
$$

We must also determine g_{corr} from Z_{corr} and add this correction to the ideal value that we have just calculated. Using Eq. 8.61,

$$
\begin{aligned}
Z_{\text{corr}} &= 1 + \frac{y}{3} + \frac{2\gamma}{y} + \frac{\delta}{e^t - 1} + \frac{2x^* t}{(e^t - 1)^2} \\
&= 1 + \frac{0.0005726}{3} + \frac{2(2.90 \times 10^{-6})}{5.726 \times 10^{-4}} + \frac{0.01}{0.950} + \frac{2(0.00748)(0.668)}{(0.950)^2} \\
&= 1.03188
\end{aligned}
$$

Therefore, from Eq. 8.65,

$$g_{\text{corr}} = -RT \ln Z_{\text{corr}} = -9935 \ln (1.03188) = -312 \text{ cal/mole}$$

so that

$$(g_{\text{int}} - u_{v_0}) = (g_{\text{int}} - u_{v_0})_{\text{ideal}} + g_{\text{corr}} = -74{,}652 \text{ cal/mole}$$

We note that the correction term for the Gibbs function constitutes a very small percentage of the total value for nitrogen, even at 5000 K. The correction would be even less significant if the translational contribution (−224,300 cal/mole) had been included. One should not conclude, however, that the correction is totally negligible, because chemical equilibrium constants are evaluated in terms of differences in the Gibbs function for the substances taking part in the reaction. Since the difference between two large numbers may be quite small, corrections such as those found here may in turn become relatively important. This point will be considered further in Chapter 9 when chemical reactions are discussed. We should also note that the corrections to the other properties of nitrogen may be considerably larger (in per cent) than that for the Gibbs function determined in the example above.

Throughout the present section we have assumed electronic ground level only, this being reasonable for the majority of diatomic molecules. In special substances such as oxygen, for which the low-lying electronic terms were pictured in Figure 5.14, the methods used in this section can be generalized quite easily. The completely general internal partition function is, relative to the electronic ground level energy,

$$Z_{\text{int}} = \sum_{\text{all } e_i} g_{e_i} e^{-\epsilon_{e_i}/kT} \left[\frac{1}{\sigma y} \left(\frac{1}{1 - e^{-t}} \right) \right]_{e_i} (Z_{\text{corr}})_{e_i} \tag{8.67}$$

It is necessary to write the general expression in this form because the rotational and vibrational constants and variables are not necessarily the same for the different electronic levels of the molecule. If these constants are known for each electronic term, the equations of this section can then be applied to evaluate each of the electronic terms of Eq. 8.67 individually.

8.6. Homonuclear Diatomic Molecules at Low Temperature

Throughout our development of quantum mechanics and the subsequent application of its results to the determination of thermodynamic properties, we have always neglected any consideration of nuclear states. It has been mentioned at several points that the nuclear energy levels are so very widely separated that for our purposes the nuclei are always in the ground level. We have also neglected to include the nuclear ground level degeneracy in our calculations, this being the standard convention. This omission normally introduces no problems, even in chemical reaction analyses, since such constants, unlike the electronic ground level degeneracies, cancel out of the calculations. There are, however, a few special cases in which it is necessary to consider the nuclear ground level degeneracy (and wave functions). For example, in the analysis of molecular symmetry characteristics, which is the topic developed in this section, as well as in the consideration of the isotopes of a pure substance, the nuclear ground level degeneracy must be considered.

In Section 5.10, we discussed the symmetry characteristics of wave functions with respect to coordinate interchange for a pair of identical particles. We found there that particles of integral spin (including nuclei of even mass number) have only symmetric states and system wave functions, while those particles of half-integral spin (including nuclei of odd mass number) have only antisymmetric states and wave functions.

In this section, we wish to apply these results to diatomic molecules. Since these remarks apply only to pairs of identical particles, we consider only homonuclear diatomic molecules (as, for example, H_2, N_2, O_2). Assuming the simple internal model for which the various energy modes can be separated, the system function with which we must concern ourselves can be written as the product of the separate energy mode wave functions, or

$$\Psi = \Psi_t \Psi_r \Psi_v \Psi_e \Psi_n \tag{8.68}$$

The electronic wave function Ψ_e and the nuclear wave function Ψ_n refer to the ground levels only.

Now we wish to investigate the symmetry characteristics of the system wave function given by Eq. 8.68 with respect to interchange of the two identical nuclei. From Chapter 5, if the nuclei are of odd mass number, the system function, Eq. 8.68, must be antisymmetric, whereas, if the nuclei are of even mass number, that function must be symmetric for nuclear exchange. The translational function Ψ_t depends only upon the coordinates of the molecule's center of mass, and the vibrational function Ψ_v depends only upon the distance of separation of the nuclei. Therefore neither influences the symmetry nature of Eq. 8.68. The electronic ground level function Ψ_e is commonly symmetric with respect to nuclear exchange, and it will be assumed as such in the present development. Thus we find that the system wave function Ψ has the same symmetry characteristics as the product $\Psi_r \Psi_n$. We conclude that, for a molecule comprised of identical nuclei of odd mass number, $\Psi_r \Psi_n$ is antisymmetric (since Ψ is antisymmetric, according to the Pauli exclusion principle) and, conversely, for a molecule comprised of identical nuclei of even mass number, $\Psi_r \Psi_n$ is symmetric (since Ψ is symmetric).

The ground energy level of a single nucleus may be degenerate because of nuclear spin. This degeneracy $\mathbf{g}_n$ for each nucleus is given by

$$\mathbf{g}_n = 2\mathbf{m}_n + 1 \tag{8.69}$$

where $\mathbf{m}_n$ is the nuclear spin quantum number. Thus, for each nucleus, there must be $\mathbf{g}_n$ nuclear ground level wave functions, which can be represented as $\Psi_{n_1}, \Psi_{n_2}, \ldots, \Psi_{ng_n}$. To determine the symmetry characteristics of the wave function describing the pair of identical nuclei, which we refer to as N_1, N_2, we proceed as in Section 5.10. One symmetric function is

$$\Psi_{S_n} = \Psi_{n_1}(N_1)\Psi_{n_2}(N_2) + \Psi_{n_1}(N_2)\Psi_{n_2}(N_1) \tag{8.70}$$

Since there are two nuclei and g_n single-nucleus functions, it follows that there must be

$$\frac{(g + N - 1)!}{(g - 1)!N!} = \frac{(g_n + 2 - 1)!}{(g_n - 1)!2!} = \frac{g_n(g_n + 1)}{2} \tag{8.71}$$

symmetric functions of the type given by Eq. 8.70. We conclude that Eq. 8.71 must then also represent the number of symmetric nuclear states for the molecule. Similarly, we may construct antisymmetric functions of the type

$$\Psi_{A_n} = \Psi_{n_1}(N_1)\Psi_{n_2}(N_2) - \Psi_{n_1}(N_2)\Psi_{n_2}(N_1) \tag{8.72}$$

It then follows that there is a total of

$$\frac{g!}{(g - N)!N!} = \frac{g_n!}{(g_n - 2)!2!} = \frac{g_n(g_n - 1)}{2} \tag{8.73}$$

such antisymmetric wave functions, and consequently the same number of antisymmetric nuclear states for the molecule.

From Eqs. 8.71 and 8.73, we find a total of

$$\frac{g_n(g_n + 1)}{2} + \frac{g_n(g_n - 1)}{2} = g_n^{\,2} \tag{8.74}$$

nuclear states for the diatomic molecule corresponding to the nuclear ground level.

The problem remaining to be solved concerns the symmetry characteristics of the molecule's rotational wave function with respect to exchange of the nuclei. Using Figure 5.4, we note that such an interchange corresponds to replacing the angle θ by $(\pi - \theta)$, since θ is restricted to $0 \leq \theta \leq \pi$, and replacing the angle ϕ by $(\phi + \pi)$. From the rotational wave function (Eq. 5.88) and the definition of the associated Legendre functions from Eq. 5.83, we find that

$$\Psi_r[(\pi - \theta), (\phi + \pi)] = (-1)^{\mathbf{j}}\Psi_r(\theta, \phi) \tag{8.75}$$

From this result it is apparent that the rotational wave function is symmetric for rotational states of even **j**, and antisymmetric for rotational states of odd **j**.

We are now able to draw a most interesting and important conclusion. We recall that, for a homonuclear molecule comprised of nuclei of odd mass number, the product $\Psi_r\Psi_n$ must be antisymmetric. Therefore for rotational states of even **j** only antisymmetric nuclear states are accessible, whereas for rotational states of odd **j** only the symmetric nuclear states are accessible. The combined partition function for rotational energy and nuclear ground level is consequently a sum of the contributions of these two pairs, instead of a single summation over all **j** as performed earlier for the simple internal model. This result is

$$Z_{rn(\mathrm{odd})} = \frac{g_n(g_n - 1)}{2} \sum_{j=0,2,4,6,\ldots} (2j + 1) \exp\left[-\frac{\theta_r}{T} j(j + 1)\right] + \frac{g_n(g_n + 1)}{2} \sum_{j=1,3,5,7,\ldots} (2j + 1) \exp\left[-\frac{\theta_r}{T} j(j + 1)\right] \tag{8.76}$$

Conversely, we recall that for a homonuclear molecule comprised of nuclei of even mass number, the product $\Psi_r\Psi_n$ must always be symmetric. Therefore we conclude that, for rotational states of even **j**, only symmetric nuclear states are accessible, while, for rotational states of odd **j**, only the antisymmetric nuclear states are permitted. For this case, the combined partition function for rotation and nuclear ground level is

$$Z_{rn(\mathrm{even})} = \frac{g_n(g_n + 1)}{2} \sum_{j=0,2,4,6,\ldots} (2j + 1) \exp\left[-\frac{\theta_r}{T} j(j + 1)\right] + \frac{g_n(g_n - 1)}{2} \sum_{j=1,3,5,7,\ldots} (2j + 1) \exp\left[-\frac{\theta_r}{T} j(j + 1)\right] \tag{8.77}$$

If $\mathbf{m}_n = 0$ (this occurs only for even mass number), then, from Eq. 8.69, $g_n = 1$ and it is seen that the odd rotational states do not occur at all. This finding is in agreement with observed spectral data.

The expressions given by Eqs. 8.76 and 8.77 are the correct equations for odd and even mass number homonuclear molecules, respectively. However, at relatively high temperature where $T \gg \theta_r$,

$$\sum_{j=0,2,4,\ldots} (2j + 1) \exp\left[-\frac{\theta_r}{T} j(j + 1)\right] \approx \sum_{j=1,3,5,\ldots} (2j + 1) \exp\left[-\frac{\theta_r}{T} j(j + 1)\right] \approx \tfrac{1}{2} \sum_{j=0,1,2,3,\ldots} (2j + 1) \exp\left[-\frac{\theta_r}{T} j(j + 1)\right] \tag{8.78}$$

Therefore, at high temperatures, both Eqs. 8.76 and 8.77 reduce to the form

$$Z_{rn} = \left[\frac{g_n(g_n + 1)}{2} + \frac{g_n(g_n - 1)}{2}\right] \times (\tfrac{1}{2}) \times \sum_{j=0,1,2,3,\ldots} (2j + 1) \exp\left[-\frac{\theta_r}{T} j(j + 1)\right] = g_n^{\,2} \times (\tfrac{1}{2}) \times \sum_{j=0,1,2,3,\ldots} (2j + 1) \exp\left[-\frac{\theta_r}{T} j(j + 1)\right] \tag{8.79}$$

But, for $T \gg \theta_r$, the summation in Eq. 8.79 can be replaced by an integral according to Eq. 8.11. Therefore Eq. 8.79 reduces to the form

$$Z_{rn} = g_n{}^2 \frac{1}{2}\left(\frac{T}{\theta_r}\right) \tag{8.80}$$

The term $g_n{}^2$ is merely the nuclear degeneracy, which, as mentioned before, is normally not included in the calculations (this point will be discussed in Chapter 9). The factor of 2 in the denominator is now recognized as the so-called symmetry number σ for a homonuclear diatomic molecule, this factor having been introduced into Eq. 8.13 through a rather superficial argument. Of course, the expressions derived in this section do not apply to heteronuclear molecules; thus, no such factor of 2 is introduced into the partition functions for those molecules.

Before proceeding, it should perhaps be noted that, for homonuclear molecules for which the electronic wave function Ψ_e is antisymmetric in the nuclei (opposite that assumed at the beginning of this section), it follows that the results are exactly opposite to those found here. In such a case, Eq. 8.76 then applies for even mass number nuclei and Eq. 8.77 for odd mass number nuclei.

Inasmuch as we have discussed several different methods for evaluating the rotational partition function, let us at this point briefly review the applicability and validity of each. Beginning at very low temperature for homonuclear molecules only, either Eq. 8.76 or 8.77 is the correct expression, depending upon the molecule, and must be evaluated by direct summation. As the temperature is increased, the approximation of Eq. 8.79 becomes reasonable for both (usually far below room temperature except for H_2), and both Eqs. 8.76 and 8.77 reduce to the simple rotational model (Eq. 8.80), which is the same as Eq. 8.13. (In some cases, at intermediate temperature, the approximation of Eq. 8.79 may be made, and the Euler-Maclaurin result (Eq. 8.20) used to evaluate the partition function.) The corrections discussed in Section 8.5 (for rotational and vibrational stretching and vibrational anharmonicity) become significant only at high temperature.

For heteronuclear molecules, we use the Euler-Maclaurin result (Eq. 8.20) at low temperature, except at extremely low temperature where it may be necessary to sum Eq. 8.10 directly. As the temperature is increased, Eq. 8.20 reduces to the simple model result (Eq. 8.13). The corrections of Section 8.5 apply only at very high temperature, as was the case for homonuclear molecules.

Let us now apply the results of this section to diatomic hydrogen, for which $\theta_r = 85.4$ K. This is a relatively large number, and therefore the approximation given by Eq. 8.79 is not reasonable at temperatures below room temperature. For the H atom, $\mathbf{m}_n = \frac{1}{2}$, and, from Eq. 8.69, $\mathbf{g}_n = 2$. Also, H has an odd mass number (1) and Ψ'_e is symmetric in the nuclei for H_2, so that Eq. 8.76 applies to diatomic hydrogen. The partition function is

$$Z_{rn} = 1 \sum_{\mathrm{j}=0,2,4,\ldots} (2\mathrm{j}+1) \exp\left[-\frac{\theta_r}{T}\mathrm{j}(\mathrm{j}+1)\right] + 3 \sum_{\mathrm{j}=1,3,5,\ldots} (2\mathrm{j}+1) \exp\left[-\frac{\theta_r}{T}\mathrm{j}(\mathrm{j}+1)\right] \quad (8.81)$$

The first term represents the contribution of molecules in symmetric rotational states and antisymmetric nuclear states. Such molecules are by convention termed para-hydrogen. The second term in Eq. 8.81 then is the contribution of molecules in antisymmetric rotational states and symmetric nuclear states, and these molecules are referred to as ortho-hydrogen. It is apparent from Eq. 8.81 that the two contribute differently to the partition function and to thermodynamic properties. Thus ortho-H_2 and para-H_2 individually have different properties, and any sample of hydrogen consists of some mixture of the two. From the equilibrium distribution equation and the definition of the partition function, it can be shown that the equilibrium ratio of the two types is temperature-dependent and is given by the relation

$$\frac{N_{\mathrm{ortho}}}{N_{\mathrm{para}}} = \frac{3 \sum_{\mathrm{j}=1,3,5,\ldots} (2\mathrm{j}+1) \exp\left[-\frac{\theta_r}{T}\mathrm{j}(\mathrm{j}+1)\right]}{1 \sum_{\mathrm{j}=0,2,4,\ldots} (2\mathrm{j}+1) \exp\left[-\frac{\theta_r}{T}\mathrm{j}(\mathrm{j}+1)\right]} \quad (8.82)$$

At high temperature (around room temperature) the two summations of Eq. 8.82 become approximately equal as in Eq. 8.78, so that the equilibrium ratio approaches the limit

$$\frac{N_{\mathrm{ortho}}}{N_{\mathrm{para}}} \rightarrow 3 \quad \text{(high temperature limit)} \quad (8.83)$$

This high temperature equilibrium mixture, consisting of 75% ortho-molecules and 25% para-molecules, is called "normal" hydrogen and is the equilibrium composition hydrogen at room temperature and above.

If Eq. 8.82 is evaluated at several points, it is found that the percentage of ortho-molecules in equilibrium hydrogen steadily decreases from the value given above as the temperature is reduced below room temperature. As temperature approaches absolute zero, it is seen from Eq. 8.82 that the equilibrium ratio approaches the limit

$$\frac{N_{\text{ortho}}}{N_{\text{para}}} \to 0 \quad \text{(low temperature limit)} \tag{8.84}$$

That is, equilibrium-H_2 at very low temperature consists of nearly all para-molecules.

Example 8.5

Calculate the composition of equilibrium-hydrogen at 20.4 K, the normal boiling point for hydrogen. Use the value $\theta_r = 85.4$ K† for the characteristic rotational temperature.

From Eq. 8.82 at low temperature,

$$\frac{N_{\text{ortho}}}{N_{\text{para}}} = \frac{3[3e^{-2(\theta_r/T)} + 7e^{-12(\theta_r/T)} + 11e^{-30(\theta_r/T)} + \cdots]}{1 + 5e^{-6(\theta_r/T)} + 9e^{-20(\theta_r/T)} + \cdots}$$

At 20.4 K,

$$\frac{\theta_r}{T} = \frac{85.4}{20.4} = 4.186$$

Therefore

$$\frac{N_{\text{ortho}}}{N_{\text{para}}} = \frac{3[3(2.32 \times 10^{-4}) + 7(1.56 \times 10^{-22}) + \cdots]}{1 + 5(1.25 \times 10^{-11}) + \cdots} = 0.00209$$

It follows that the fraction of ortho-molecules in equilibrium-H_2 at 20.4 K is

$$\frac{N_{\text{ortho}}}{N_{\text{ortho}} + N_{\text{para}}} = \frac{0.00209}{1.00209} = 0.00208$$

That is, at 20.4 K, equilibrium-H_2 consists of 0.208% ortho-H_2 and 99.792% para-H_2.

By calculations similar to those used in this example, we can determine the composition of equilibrium-H_2 over the entire range of temperature. A plot of composition versus temperature is shown in Figure 8.2.

Experimental determination of the properties of hydrogen at low temperatures may yield most curious results. For example, at a given low temperature, it may be found that the properties do not agree at all with those predicted for equilibrium-H_2 but are instead in very close agreement with those calculated for normal-H_2 (75% ortho, 25% para) at the same

† The constants for H_2 in Table A.5 give the value $\theta_r = 87.5$ K, which is intended for the calculation of normal-H_2 properties at high temperature. The value of 85.4 K is the accepted value for low temperature calculations.

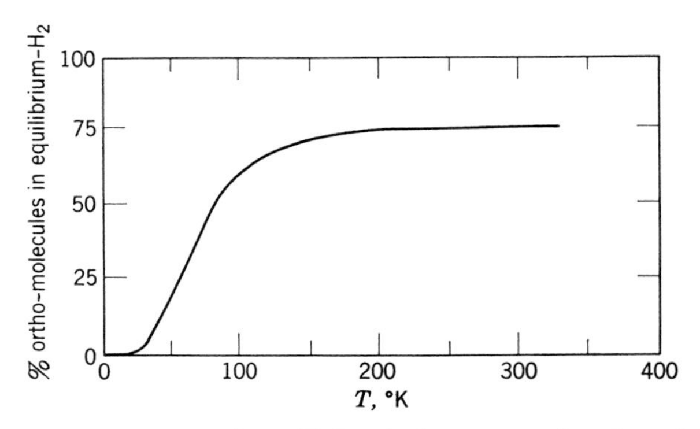

Figure 8.2 Composition of equilibrium-hydrogen as a function of temperature.

temperature. The explanation of this phenomenon is that such experiments are often conducted over relatively short periods of time using hydrogen cooled from room temperature. The rate of conversion from ortho- to para-molecules, that is, the rate at which equilibrium is approached, is extremely slow unless the hydrogen is cooled in the presence of a suitable catalyst. Thus normal-H_2 remains for relatively long periods of time in a state of metastable equilibrium at these reduced temperatures. A typical example of the conversion rate for hydrogen in the absence of a catalyst is shown in Figure 8.3. The curve shows the per cent of ortho-H_2 as a function of time, beginning with normal-hydrogen at 20.4 K. The per cent of ortho-H_2 would eventually fall to the equilibrium value 0.208% calculated for equilibrium-H_2. It has also been observed that, if the hydrogen is cooled in the presence of activated hydrous ferric oxide, which acts as a catalyst for the ortho-para conversion, equilibrium is then established very rapidly and the observed behavior is in agreement with that predicted for equilibrium-hydrogen.

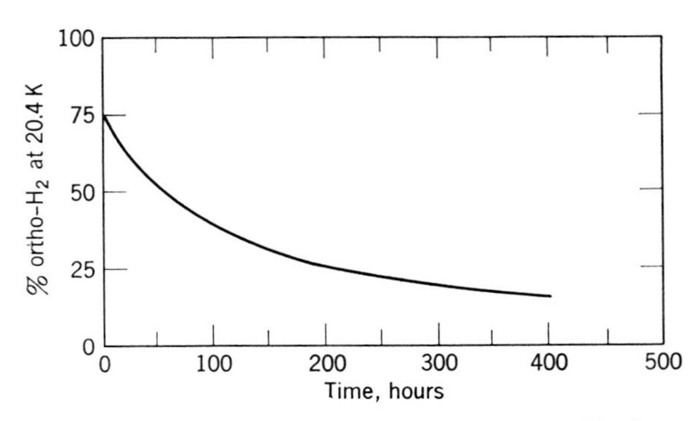

Figure 8.3 An example of the spontaneous conversion of hydrogen at 20.4 K.

8.7. The Polyatomic Gas

In developing the expressions of quantum mechanics in Chapter 5, we consider the energy modes only for atoms and diatomic molecules, and, because of the complexity involved, did not discuss polyatomic molecules. For such substances, we can separate the energy into the same modes—translational, rotational, vibrational, electronic, although the evaluation of certain of these energy modes may become very difficult.

The translational portion presents no problems, because the energy, and consequently the translational contributions to partition function and properties, are the same as for monatomic and diatomic substances. We need only be careful to use the total mass of the molecule in the calculations. For a molecule comprised of a atoms,

$$m = \sum_{i=1}^{a} m_i \tag{8.85}$$

where the m_i are the masses of the component atoms.

It is in evaluating internal modes for polyatomic molecules that we encounter considerable difficulty. Fortunately, nearly all polyatomic molecules are sufficiently heavy (have large moments of inertia) that we can make the assumption of classical rigid rotator (very closely spaced rotational levels). In this section we consider only the simple internal model: rigid rotator, harmonic oscillator, electronic ground level. However, problems encountered in applying this simple model to diatomic gases must also be considered for polyatomic molecules. These include wave function symmetry considerations (accounted for here by the classical inclusion of a symmetry number), low temperature rotational evaluation from the Euler-Maclaurin theorem, high temperature corrections for vibrational anharmonicity and rotation-vibration coupling, and, for polyatomic molecules, interaction between vibrational modes as well.

Even for the simple model, we must consider separately the two classes of molecules, linear and non-linear, which behave differently. Linear polyatomic molecules are similar in behavior to diatomic molecules. The rotational contributions to the partition function and properties for such molecules are of exactly the same form as those for diatomic substances. The partition function is found from Eq. 8.13 with the characteristic rotational temperature given by Eq. 8.12. Rotational contributions to the thermodynamic properties are found from Eqs. 8.15 to 8.19, as before. Evaluation of the equilibrium-point moment of inertia, I_ε, which is used in Eq. 8.12, is somewhat more complicated for the polyatomic gas. For

Figure 8.4 The alignment of two linear polyatomic molecules.

the linear polyatomic molecule, I_ε is given by the relation

$$I_\varepsilon = \sum_{i=1}^{a} m_i r_i^2 \tag{8.86}$$

where r_i is the equilibrium distance of atom i from the center of mass of the molecule. This equilibrium distance r_i is defined by the expression

$$\sum_{i=1}^{a} m_i r_i = 0 \tag{8.87}$$

It is also necessary to know the symmetry number σ for the molecule. This requires a knowledge of the structure of the molecule, two of which are shown in Figure 8.4.

For the vibration of a linear polyatomic molecule, it was noted in Section 1.5 that there are $(3a - 5)$ vibrational degrees of freedom. The contributions of the $(3a - 5)$ vibrational modes can be considered as being independent in the simple model, and the vibration can be assumed harmonic for each mode. Thus we have $(3a - 5)$ vibrational partition function contributions of the form of Eq. 8.29. The $(3a - 5)$ characteristic vibrational temperatures θ_v are given in terms of the normal frequencies or wave numbers, as in Eq. 8.30. Two or more of these modes may have the same frequency, depending upon the molecule. Contributions to properties from these modes are given by Eqs. 8.32 to 8.36 for each of the $(3a - 5)$ vibrational modes of the molecule. We should keep in mind that there is a $\frac{1}{2}h\nu_\varepsilon$ zero-point contribution for each of the vibrational modes, which by convention is included in the chemical energy term. These zero-point terms add a constant value to u, h, a, g, but of course not to s, C_p, C_v for the molecule.

Electronic ground state contributions to properties are found, as for diatomic substances, by using Eqs. 8.38 to 8.40.

Example 8.6

Calculate the entropy of carbon dioxide at 1200 K, 1 atm pressure.

From Table A.6 for CO_2 the C—O distances are 1.162 Å, and the vibrational wave numbers are

$$\omega_1 = 1342.9 \text{ cm}^{-1}$$
$$\omega_2 = 667.3 \text{ (2 modes)}$$
$$\omega_3 = 2349.3$$

For the ground electronic level, $g_{e_0} = 1$. The mass of the CO_2 molecule is

$$m = 1.66 \times 10^{-24}(44.01) = 73.05 \times 10^{-24} \text{ gm}$$

In solving this problem, we consider in turn the translational, rotational, vibrational, and electronic modes, and find the entropy contribution of each. We consider first the translational contribution to entropy. From Eq. 6.30, at 1200 K, 1 atm,

$$\begin{aligned}\frac{Z_t}{N} &= \left(\frac{2\pi m}{h^2}\right)^{3/2} \frac{(kT)^{5/2}}{P} \\ &= \left(\frac{2\pi \times 73.05 \times 10^{-24}}{(6.625 \times 10^{-27})^2}\right)^{3/2} \frac{(1.3804 \times 10^{-16} \times 1200)^{5/2}}{1.01325 \times 10^6} \\ &= 3.70 \times 10^8\end{aligned}$$

The translational entropy is, from Eq. 6.29,

$$s_t = R\left[\ln \frac{Z_t}{N} + \frac{5}{2}\right] = 1.987[19.73 + 2.5] = 44.15 \text{ cal/mole-K}$$

Next we consider the rotational contribution. From Eq. 8.87, the center of mass is at the center of the C atom. Therefore $r_C = 0$, $r_O = \pm 1.162$ Å, and, using Eq. 8.86, we have

$$\begin{aligned}I_\varepsilon &= \sum_i m_i r_i^2 = 2(16.0 \times 1.66 \times 10^{-24})(1.162 \times 10^{-8})^2 \\ &= 71.5 \times 10^{-40} \text{ gm-cm}^2\end{aligned}$$

From Eq. 8.12,

$$\theta_r = \frac{h^2}{8\pi^2 I_\varepsilon k} = \frac{(6.625 \times 10^{-27})^2}{8\pi^2 \times 71.5 \times 10^{-40} \times 1.3804 \times 10^{-16}} = 0.562 \text{ K}$$

The rotational partition function is found from Eq. 8.13:

$$Z_r = \frac{T}{\sigma\theta_r} = \frac{1200}{2(0.562)} = 1067$$

The resulting contribution to entropy is, from Eq. 8.18,

$$s_r = R[\ln Z_r + 1] = 1.987[\ln 1067 + 1] = 15.80 \text{ cal/mole-K}$$

As regards the vibrational contribution, there are four vibrational degrees of freedom, two of which have the same frequency. Thus

$$\begin{aligned}\theta_{v_1} &= \frac{hc}{k}\,\omega_1 = 1.4388(1342.9) = 1932 \text{ K} \\ \theta_{v_2} &= 1.4388(667.3) = 960 \text{ K} \quad \text{(double)} \\ \theta_{v_3} &= 1.4388(2349.3) = 3380 \text{ K}\end{aligned}$$

and

$$\frac{\theta_{v1}}{T} = \frac{1932}{1200} = 1.61$$

$$\frac{\theta_{v2}}{T} = \frac{960}{1200} = 0.80 \quad \text{(double)}$$

$$\frac{\theta_{v3}}{T} = \frac{3380}{1200} = 2.82$$

Using Eq. 8.35 for each of the four vibrational modes and Table A.4, the vibrational entropy is found to be

$$s_v = R \sum_i \left\{ -\ln \left[1 - \exp\left(-\frac{\theta_{v_i}}{T} \right) \right] + \frac{\theta_{v_i}/T}{\exp(\theta_{v_i}/T) - 1} \right\}$$
$$= 1.987[(0.2230 + 0.4022) + 2(0.5966 + 0.6528) + (0.0615 + 0.1787)]$$
$$= 6.70 \text{ cal/mole-K}$$

From Eq. 8.39, the electronic contribution is

$$s_e = R \ln g_{e_0} = 0$$

Therefore the entropy of CO_2 at 1200 K, 1 atm, is found by summing the various contributions. The result is

$$s = s_t + s_r + s_v + s_e = 44.15 + 15.80 + 6.70 + 0$$
$$= 66.65 \text{ cal/mole-K}$$

For non-linear polyatomic molecules, as discussed briefly in Section 1.5, there are three degrees of freedom for rotation, and therefore only $(3a - 6)$ for vibrational modes. We consider again only the simple internal model, and assume the rotational levels to be closely spaced so that the classical approximation is valid. However, the rotational energy for a non-linear molecule is not the same as that for a linear one. It was pointed out in Section 1.5 that, for a linear molecule, the moments of inertia about two of the mutually perpendicular axes are equal while that about the axis joining the nuclei is negligible. For the general non-linear molecule this is not the case, and the development of rotational energy and partition function expressions becomes extremely complex, even for the classical rigid rotator. We shall not derive these expressions here but will simply present the result, which is given in terms of the three principal moments of inertia, I_x, I_y, I_z, and the rotational symmetry number σ. The relation is

$$Z_r = \frac{\pi^{1/2}}{\sigma} \left(\frac{8\pi^2 I_x kT}{h^2} \right)^{1/2} \left(\frac{8\pi^2 I_y kT}{h^2} \right)^{1/2} \left(\frac{8\pi^2 I_z kT}{h^2} \right)^{1/2}$$
$$= \frac{8\pi^2}{\sigma h^3} (I_x I_y I_z)^{1/2} (2\pi kT)^{3/2} \tag{8.88}$$

One should note the similarity between this result and that for the linear molecule, which has only two moments, and these are equal in magnitude.

The three principal moments of inertia for the molecule must be found in the following manner. Let us consider first any set of mutually perpendicular coordinate axes x, y, z, having their origin at the center of mass of the molecule. The center of mass is defined by the expressions

$$\sum_{i=1}^{a} m_i x_i = \sum_{i=1}^{a} m_i y_i = \sum_{i=1}^{a} m_i z_i = 0 \tag{8.89}$$

The moments of inertia corresponding to the selected coordinate directions are given by

$$\begin{aligned} I_{xx} &= \sum_{i=1}^{a} m_i(y_i^2 + z_i^2) \\ I_{yy} &= \sum_{i=1}^{a} m_i(z_i^2 + x_i^2) \\ I_{zz} &= \sum_{i=1}^{a} m_i(x_i^2 + y_i^2) \end{aligned} \tag{8.90}$$

The products of inertia for this coordinate system are

$$\begin{aligned} I_{xy} &= I_{yx} = \sum_{i=1}^{a} m_i x_i y_i \\ I_{yz} &= I_{zy} = \sum_{i=1}^{a} m_i y_i z_i \\ I_{zx} &= I_{xz} = \sum_{i=1}^{a} m_i z_i x_i \end{aligned} \tag{8.91}$$

The six values given by Eqs. 8.90 and 8.91 depend upon the selection of coordinate axes and are changed byrotation of the coordinate system about the origin. The particular set of axes for which the products of inertia (Eq. 8.91) are all zero is called the principal set of axes. The corresponding moments of Eq. 8.90 are called the principal moments of inertia, designated simply by I_x, I_y, I_z, and are the values to be used in calculating the partition function from Eq. 8.88. The principal set of axes can usually be selected quite easily by letting any one of the coordinate directions x, y, z coincide with a line of symmetry in the molecule.

Non-linear molecules are classified according to the relative values of their three principal moments of inertia. A molecule for which all three principal moments are equal is referred to as a spherical top molecule. One for which two of the three principal moments are the same is called a symmetric top molecule, and a molecule with all three moments different is termed an asymmetric top molecule.

The rotational symmetry number for non-linear molecules can be very difficult to determine, especially for complex substances. The methane

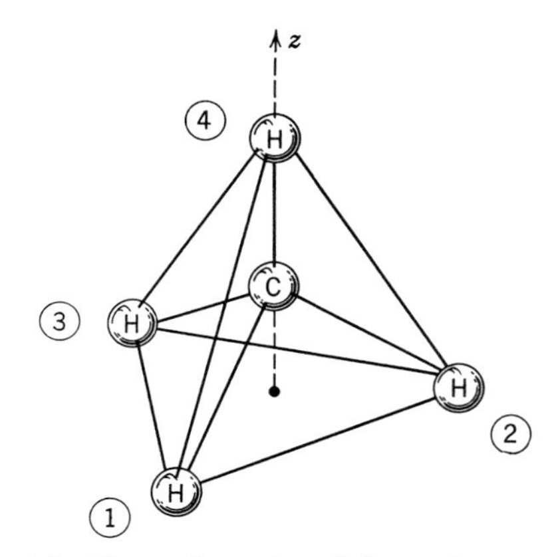

Figure 8.5 The configuration of the methane molecule.

molecule is shown in Figure 8.5. It has the four H atoms arranged in the shape of a tetrahedron with the single C atom at the center. To determine σ, let us label the H atoms as shown in the figure, and refer to the plane formed by atoms 1, 2, and 3 in the diagram as the base plane. Successive rotations of 120° about the z-axis result in three indistinguishable configurations of the molecule. However, the molecule could also be rotated about an axis normal to z, putting 1, 2, 4, for example, in the base plane. Subsequently 120° rotations about the z-axis result in three additional indistinguishable positions. After similar rotations to put either 1, 3, 4 or 2, 3, 4 in the base plane, each with three positions about the z-axis, we conclude that there is a total of twelve indistinguishable configurations for the CH_4 molecule and that $\sigma = 12$. For many molecules, the determination of rotational symmetry number is even more involved.

The rotational contributions to thermodynamic properties are found from the partition function expression (Eq. 8.88). Using Eqs. 6.7 and 6.10 gives

$$h_r = u_r = RT^2\left(\frac{d \ln Z_r}{dT}\right) = RT^2\left(\frac{3}{2T}\right) = \tfrac{3}{2}RT \tag{8.92}$$

Differentiating this result with respect to temperature, we find

$$C_{p_r} = C_{v_r} = \tfrac{3}{2}R \tag{8.93}$$

From Eqs. 6.13 and 8.92,

$$s_r = R \ln Z_r + \frac{u_r}{T} = R[\ln Z_r + \tfrac{3}{2}] \tag{8.94}$$

while, from Eqs. 6.16 and 6.19,

$$g_r = a_r = -RT \ln Z_r \tag{8.95}$$

Example 8.7

Calculate the rotational contribution to Gibbs function for water at 1000 K.

From Table A.6, for H_2O the O—H distance is 0.9584 Å and the H—O—H angle 104.45°, as shown in Figure 8.6. It is first necessary to find the center of mass and the principal moments for the molecule. We select a coordinate system with the z-axis coinciding with the line of symmetry and the x—z plane in the plane of the molecule as shown in the diagram. The y-axis is then normal to the page.

The atomic masses are

$$m_{\mathrm{O}} = 1.66 \times 10^{-24}(16.0) = 26.55 \times 10^{-24}\ \mathrm{gm}$$
$$m_{\mathrm{H}} = 1.66 \times 10^{-24}(1.008) = 1.673 \times 10^{-24}\ \mathrm{gm}$$

For this coordinate system,

$$x_{\mathrm{O}} = 0$$
$$x_{\mathrm{H}} = \pm 0.9584 \sin 52.225^\circ = \pm 0.756\ \text{Å}$$
$$y_{\mathrm{O}} = y_{\mathrm{H}} = 0$$
$$z_{\mathrm{O}} - z_{\mathrm{H}} = 0.9584 \cos 52.225^\circ = 0.5855$$

From Eq. 8.89,

$$\sum_i m_i z_i = 26.55 \times 10^{-24} \times z_{\mathrm{O}} + 2 \times 1.673 \times 10^{-24}(z_{\mathrm{O}} - 0.5855) = 0$$

Therefore

$$z_{\mathrm{O}} = +0.0655\ \text{Å}$$
$$z_{\mathrm{H}} = -0.5200\ \text{Å}$$

Since $y_{\mathrm{O}} = y_{\mathrm{H}} = 0$, it is apparent that

$$I_{xy} = I_{yz} = 0$$

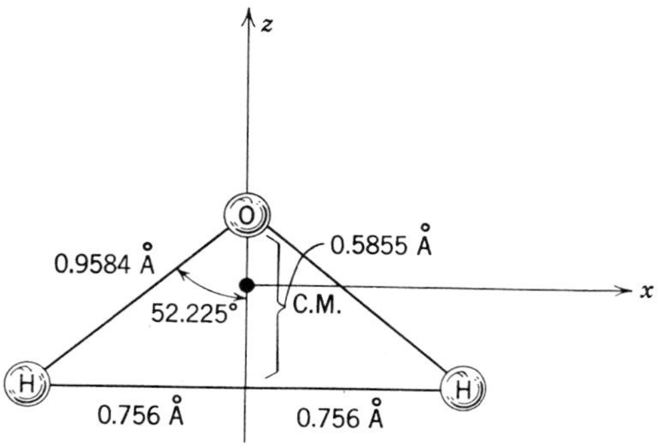

Figure 8.6 Diagram of the water molecule for Example 8.7.

and

$$I_{zx} = \sum_i m_i z_i x_i$$
$$= 26.55 \times 10^{-24}(0.0655 \times 10^{-8})(0)$$
$$+ 1.673 \times 10^{-24}(-0.5200 \times 10^{-8})(0.756 \times 10^{-8})$$
$$+ 1.673 \times 10^{-24}(-0.5200 \times 10^{-8})(-0.756 \times 10^{-8}) = 0$$

so that the axes selected are the principal set, and Eq. 8.90 therefore gives the principal moments. These are found to be

$$I_x = \sum_i m_i(y_i^2 + z_i^2)$$
$$= 26.55 \times 10^{-24}[0 + (0.0655 \times 10^{-8})^2]$$
$$+ 2 \times 1.673 \times 10^{-24}[0 + (-0.5200 \times 10^{-8})^2]$$
$$= 1.019 \times 10^{-40} \text{ gm-cm}^2$$

Similarly,

$$I_y = \sum_i m_i(z_i^2 + x_i^2) = 2.932 \times 10^{-40} \text{ gm-cm}^2$$
$$I_z = \sum_i m_i(x_i^2 + y_i^2) = 1.913 \times 10^{-40} \text{ gm-cm}^2$$

The water molecule is consequently found to be an asymmetric top. The symmetry number of the molecule is 2, since the molecule could be rotated 180° about the z-axis into an indistinguishable position. Therefore, from Eq. 8.88, using $\sigma = 2$, we find

$$Z_r = \frac{8\pi^2}{\sigma h^3}(I_x I_y I_z)^{1/2}(2\pi kT)^{3/2}$$
$$= \frac{8\pi^2}{2(6.625 \times 10^{-27})^3}$$
$$\times (1.019 \times 10^{-40} \times 2.932 \times 10^{-40} \times 1.913 \times 10^{-40})^{1/2}$$
$$\times (2\pi \times 1.3804 \times 10^{-16} \times 1000)^{3/2}$$
$$= 263$$

Then, from Eq. 8.95, the rotational Gibbs function for H_2O at 1000 K is

$$g_r = -RT \ln Z_r = -1.987(1000) \ln 263 = -11{,}070 \text{ cal/mole}$$

The vibrational contributions for the non-linear molecule can be handled in exactly the same manner as for the linear polyatomic molecule. The sole difference is that in this case there are only $(3a - 6)$ vibrational modes and separate contributions to properties, instead of $(3a - 5)$ as for the linear molecule. Each of the modes contributes to properties according to the set of equations 8.32 to 8.36. The electronic ground level contributions are found from Eqs. 8.38 to 8.40 as before.

There is an exception to the use of Eqs. 8.32 to 8.36 for each of the vibrational modes that should be pointed out here. As an example of this

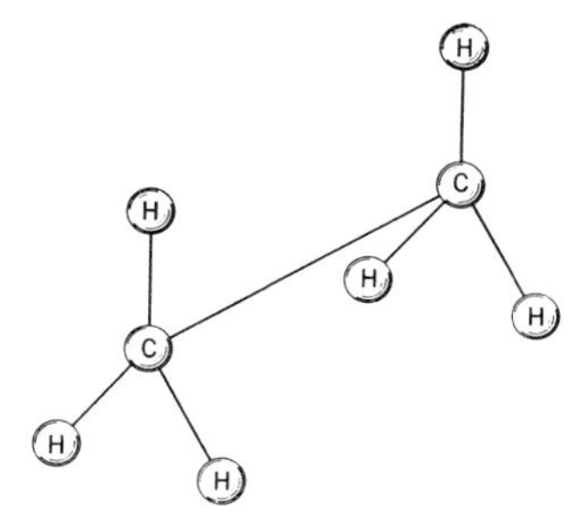

Figure 8.7 Representation of the ethane molecule.

special problem, we consider ethane, the structure of which is represented in Figure 8.7.

Since C_2H_6 has 8 atoms, there must be 18 vibrational degrees of freedom, of which only 17 are observed spectroscopically. The eighteenth degree of freedom represents a torsional mode, or twisting of one CH_3 methyl group with respect to the other about the C—C bond. The contribution to a property, specific heat for example, can be found from the difference between the measured calorimetric value and that calculated for the remaining contributions—translational, rotational, and seventeen vibrational modes. From such a procedure, we find that it is not possible to select (empirically) a vibrational frequency that will correlate the data using the harmonic oscillator model, except at low temperature. On the other hand, it cannot be assumed that this degree of freedom is a free rotation of one methyl group with respect to the other. The measured values lie between these two extremes, and consequently this degree of freedom is termed a restricted internal rotation of the molecule.

We must consider a potential Φ between the two methyl groups that varies periodically with angle as one group is rotated with respect to the other. For the three-fold symmetry of a methyl group, such a potential is of the form depicted in Figure 8.8. With this potential, the Schrödinger

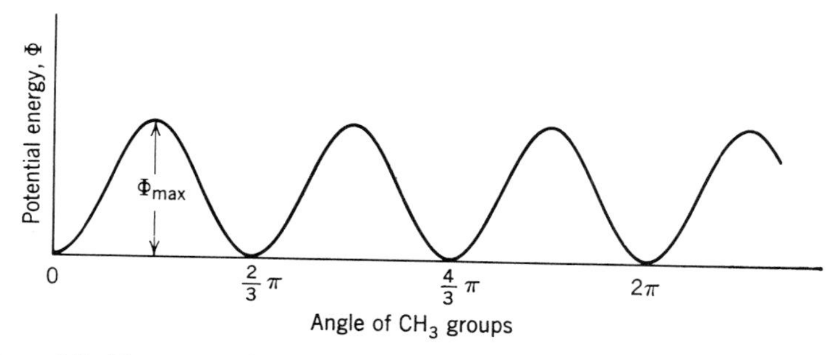

Figure 8.8 The potential energy between methyl groups in the ethane molecule.

wave equation can be evaluated for energy, and the partition function and the property contributions for this mode subsequently can be found. There is no theoretical basis for selection of the maximum potential Φ_{max}, but a value can be found empirically that will accurately represent the contribution over the range of temperature for which calorimetric data are available. Such correlations for this and other molecules indicate that this analysis is valid. It is also seen, using Figure 8.8, that at low temperature, $kT \ll \Phi_{max}$, the behavior is essentially that of a harmonic oscillator (near the bottom of a potential well at $\frac{2}{3}\pi, \frac{4}{3}\pi, 2\pi, \ldots$), while at high temperature, $kT \gg \Phi_{max}$, the potential of Figure 8.8 presents no significant restriction and the behavior is essentially that of free rotation.

PROBLEMS

8.1 Plot the distribution of molecules among the various rotational levels for carbon monoxide at 100 K and at 300 K.

8.2 For HCl, find the temperature at which exactly 50% of the molecules have a rotational quantum number greater than 3.

8.3 (*a*) Using the Euler-Maclaurin summation theorem, show that Eq. 8.20 is valid.

(*b*) Estimate the error in calculating the rotational partition function for carbon monoxide at 50 K, 100 K, and 500 K by the integrated expression, Eq. 8.13.

8.4 When T/θ_r is less than unity, the rotational partition function for a heteronuclear molecule can be evaluated by direct expansion of Eq. 8.10.

(*a*) For such a region, develop an expression for the rotational internal energy.

(*b*) Plot the rotational internal energy vs. T for HF at temperatures below 100 K, using the appropriate method for determining u_r.

8.5 Repeat Problem 8.4 for the rotational specific heat of HF, and discuss the results of the two problems.

8.6 Hydrogen deuteride (HD), the diatomic molecule comprised of a hydrogen atom and a deuteron (1 proton and 1 neutron in the nucleus), has a characteristic rotational temperature $\theta_r = 63$ K. Calculate the rotational entropy of HD at 20, 100, and 500 K, using whatever method is appropriate at each temperature.

8.7 For diatomic nitrogen at 2000 K, calculate the probability that a molecule will have a vibrational quantum number less than 3.

8.8 Plot the distribution of molecules among the various vibrational levels for diatomic bromine at 300 K and 3000 K.

8.9 Prove that the vibrational entropy is independent of the selection for the ground level reference of energy, u_{v_0}.

8.10 For diatomic fluorine at 1 atm pressure, compare the relative magnitudes of the various contributions to C_p and s at 300 K and also at 3000 K.

8.11 One gram-mole of nitric oxide (NO) is contained in a closed tank at 500 R, 1 atm pressure. Calculate the heat transfer and change in entropy for the NO if the gas is heated to 2000 R. Assume that the molecular constants r_ε and ω_ε are the same for both electronic levels.

8.12 Diatomic chlorine at 1000 K, 1 atm, is cooled in a constant-pressure, steady flow process to 300 K. Calculate the heat transfer and entropy change per mole during this process.

8.13 The Gibbs function for monatomic nitrogen at 1 atm was calculated at 1000° intervals from 4000 K to 10,000 K in Problem 6.13 (as a preliminary to the discussion of chemical equilibrium constants in Chapter 9). Calculate these values for diatomic nitrogen, including chemical energy, assuming the simple internal model.

8.14 The first three electronic levels of diatomic oxygen are:

Term	ϵ/hc, cm^{-1}
${}^3\Sigma_g^-$	0
${}^1\Delta_g$	7,882
${}^1\Sigma_g^+$	13,121

Assume a rigid rotator-harmonic oscillator for which the constants r_ϵ and ω_ϵ are the same for each of the electronic levels and no higher electronic levels are populated.

(*a*) Calculate the fraction of molecules in each of the three electronic levels at 5000 K.

(*b*) Determine the constant-pressure specific heat of O_2 at 5000 K.

8.15 Carry out the development of Eqs. 8.56, 8.57, and 8.58 for the internal partition function for the general internal model.

8.16 The general internal model correction constants were calculated for diatomic hydrogen in Problem 5.18. Use these values to calculate the entropy and enthalpy (with respect to ground level energy) at 5000 K, 1 atm pressure.

8.17 Using the expression for the rotational wave function, Eq. 5.88, show that Eq. 8.75 is correct for a 180° rotation of the molecule.

8.18 List and discuss the possible symmetric and antisymmetric nuclear wave functions for diatomic hydrogen.

8.19 Using the procedure followed in Example 8.5, calculate the equilibrium proportions of ortho and para-hydrogen at 50, 100, and 200 K.

8.20 Calculate the rotational contribution to specific heat for ortho-, para-, normal, and equilibrium hydrogen at 20.4, 50, 100, and 200 K, and plot the results.

8.21 Calculate the enthalpy of conversion of ortho- to para-hydrogen at 20.4 K, and compare the value with the enthalpy of vaporization (218.7 cal/mole) at this temperature. Of what significance is this to the storage of liquid hydrogen?

8.22 By convention, entropies calculated from spectroscopic data do not include the contribution from nuclear spin degeneracy, $R \ln g_n^2$. Calculate the entropy of H_2 at 25 C, 1 atm pressure, using both Eqs. 8.80 and 8.81 for the rotational contribution, correcting in each case for nuclear spin.

8.23 Deuterium is the hydrogen isotope with a mass number of 2 (a nucleus of 1 proton and 1 neutron). The ground level spin number of the nucleus is unity. For the diatomic molecule D_2, the characteristic rotational temperature $\theta_r = 42.7$ K and Ψ_e is symmetric with respect to nuclear exchange.

(*a*) Determine the number of symmetric and antisymmetric nuclear states for the D_2 molecule, and list the wave functions for each in terms of the single-nucleus functions.

(*b*) Express the combined rotational-nuclear partition function for D_2.

(*c*) The D_2 molecules in antisymmetric nuclear states are called para-D_2, and those in symmetric nuclear states ortho-D_2. Determine the high temperature and low temperature percentages of each type at equilibrium.

8.24 Calculate the fraction of molecules in each of the rotational energy levels $\mathbf{j} = 0, 1, 2, 3$ for diatomic oxygen gas at its normal boiling point temperature, 90.1 K. An oxygen atom has a nuclear spin number $\mathbf{m}_n = 0$ and atomic weight of 16. It may be assumed that all the O_2 molecules are in the ground electronic state, for which the electronic wave function Ψ_e is antisymmetric with respect to nuclear exchange.

8.25 Carbon dioxide is heated in a constant-pressure, steady flow process from 300 K, 1 atm, to 1200 K. Determine the heat transfer for the process and the change of entropy of the CO_2.

8.26 Calculate the Gibbs function with respect to ground state energy for nitrous oxide (N_2O) at 800 K, 1 atm pressure.

8.27 Calculate the constant-pressure specific heat and entropy for water at 1000 K, 1 atm.

8.28 Calculate the entropy of H_2O at 25 C, 1 atm pressure. Ideal gas H_2O at 25 C, 1 atm, is a hypothetical state, but it is useful as a reference state for other calculations. Using this value and the steam tables, determine the entropy of liquid water at 25 C (the difference is not equal to s_{fg}).

8.29 Determine the principal moments of inertia for the methane (CH_4) molecule, and calculate the rotational entropy at 400 K.

8.30 Consider the restricted internal rotation in ethane, C_2H_6. The potential shown in Figure 8.8 can be represented in terms of the angle α between methyl groups (groups staggered at $\alpha = 0$) as

$$\Phi(\alpha) = \tfrac{1}{2}\Phi_{\max}(1 - \cos 3\alpha)$$

It has been stated that at low temperature ($kT \ll \Phi_{\max}$), the behavior is essentially that of a harmonic oscillator. Show that in this range

$$\Phi \approx \Phi_{\max}(\tfrac{3}{2}\alpha)^2$$

and that the energy levels of this mode are given by

$$\epsilon_x = h\nu(x + \tfrac{1}{2}); \quad x = 0, 1, 2, \ldots$$

Find an expression for the frequency ν in terms of the significant parameters.

9 Chemical Equilibrium

In studying the properties and behavior of gases at high temperature, we must consider the problems of dissociation, ionization, and chemical equilibrium. These subjects are of particular importance at present, in view of the increased research and applications involving high temperature gases and plasmas. As a gas is heated to high temperature, an appreciable fraction of the molecules possess sufficient energy to dissociate into more elementary compounds or atoms. For example, at the level of energy to which the value of zero is assigned (Figure 5.9) the diatomic molecule is dissociated into two atoms, the distance of separation having become so great that the atoms can no longer be considered to be joined by a bond. Of course, the degree of dissociation (or ionization) is not simply a question of relative energies of the different chemical species. Rather, the problem must be analyzed in terms of the fundamental thermodynamic relations, in order to determine the requirements for equilibrium.

Development of the equilibrium equations for either ordinary chemical reaction or ionization processes can be approached from either the classical or the molecular point of view. First we shall develop the expressions along classical lines and then re-examine the problem from the microscopic viewpoint. The parallel developments from these two viewpoints leads to a greater understanding of chemical equilibrium.

9.1. Chemical Equilibrium Constants

In order to analyze the condition of chemical equilibrium, we consider a general situation in which a tank contains a mixture of four gases, A, B, C, D, at some temperature T and pressure P, as indicated in Figure 9.1. The system is presumed to exist at a condition of chemical equilibrium, according to the reaction equation

$$\nu_A A + \nu_B B \rightleftharpoons \nu_C C + \nu_D D \tag{9.1}$$

where ν_A, ν_B, ν_C, ν_D are the stoichiometric coefficients giving the relative proportions in which the various components take part in the reaction. That is, ν_A moles of substance A combine with ν_B moles of B, resulting in ν_C moles of C and ν_D moles of D, or vice versa, as indicated by the double arrow. At a condition of equilibrium, the reaction is proceeding in both directions at the same rate, so that there is no net change in amount of any substance with time. This is the condition at which the system of Figure 9.1 is presumed to exist. It is assumed that Eq. 9.1 is the only reaction of significance for this system at the given temperature and pressure. A more complex system involving more than one reaction equation will be discussed in Section 9.3.

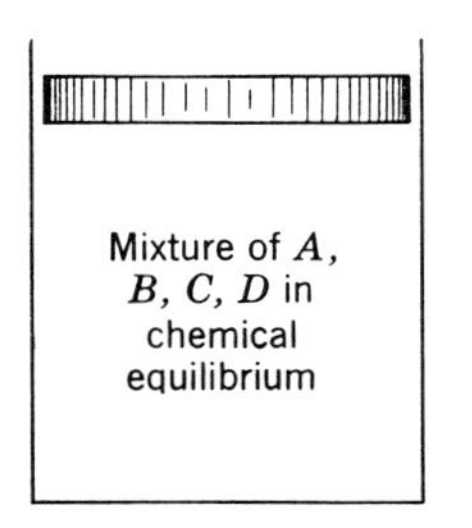

Figure 9.1 An equilibrium mixture at T, P.

It is important to understand that the stoichiometric coefficients are not the amounts of each substance present in the tank but are, instead, the relative proportions in which they take part in the chemical reaction of Eq. 9.1. The amount of each substance present in the tank at any condition is given by n_A, n_B, n_C, n_D. In order to determine the number of moles of each component present at equilibrium, n_{A_e}, n_{B_e}, n_{C_e}, n_{D_e}, let us consider the thermodynamic property relation expressed in terms of the Gibbs function,

$$dG = -S\,dT + V\,dP + \sum_i \bar{G}_i\,dn_i \qquad (9.2)$$

where $\bar{G}_i$ is the partial molal Gibbs function of component i in the mixture. At fixed T and P (as in Figure 9.1), Eq. 9.2 reduces to

$$dG_{T,P} = \sum_i \bar{G}_i\,dn_i \qquad (9.3)$$

From the developments of classical thermodynamics, it is known that the requirement for equilibrium at a given temperature and pressure is that the Gibbs function of the system be a minimum. In other words, at equilibrium,

$$dG_{T,P} = 0 \qquad (9.4)$$

and therefore Eq. 9.3 gives a relation between the number of moles of the components at equilibrium. To determine this minimum point in the Gibbs function for our system, let us consider a small departure from the equilibrium point according to the chemical-reaction equation, Eq. 9.1. We assume that the corresponding change in composition of the system occurs as a result of the reaction proceeding slightly from left to right, as the reaction is written in Eq. 9.1. Then the amounts of A and B present in the tank will decrease slightly, while the amounts of C and D will increase, the change in each being proportional to the stoichiometric coefficients, ν_A, ν_B, ν_C, ν_D. If we call this proportionality constant ρ, also termed the degree of reaction, then the number of moles of each substance present after the small change is given by the number of moles at equilibrium plus the change according to the reaction, or

$$\begin{aligned} n_A &= n_{A_e} - \nu_A \rho \\ n_B &= n_{B_e} - \nu_B \rho \\ n_C &= n_{C_e} + \nu_C \rho \\ n_D &= n_{D_e} + \nu_D \rho \end{aligned} \tag{9.5}$$

The changes in A and B are negative, since we assumed the reaction to proceed from left to right. We are not interested in a finite departure from the equilibrium point—only in determining the composition at which the Gibbs function is a minimum. Therefore, differentiating Eq. 9.5, we obtain

$$\begin{aligned} dn_A &= -\nu_A \, d\rho \\ dn_B &= -\nu_B \, d\rho \\ dn_C &= +\nu_C \, d\rho \\ dn_D &= +\nu_D \, d\rho \end{aligned} \tag{9.6}$$

Substituting these relations into Eq. 9.3, we have

$$dG_{T,P} = (\nu_C \bar{G}_C + \nu_D \bar{G}_D - \nu_A \bar{G}_A - \nu_B \bar{G}_B) \, d\rho = (\Delta G) \, d\rho \tag{9.7}$$

where the quantity

$$\Delta G = \nu_C \bar{G}_C + \nu_D \bar{G}_D - \nu_A \bar{G}_A - \nu_B \bar{G}_B \tag{9.8}$$

has been defined for convenience.

Since the infinitesimal change $d\rho$ is arbitrary, it is apparent from Eqs. 9.4 and 9.7 that the requirement for equilibrium can be written

$$\Delta G = 0 \tag{9.9}$$

in terms of the various reaction equation stoichiometric coefficients and the corresponding partial Gibbs functions of the components in the mixture.

In this chapter, we consider only systems involving gases at high temperature, for which we can accurately assume ideal gas behavior. For a mixture of ideal gases, the partial Gibbs function of each component can be expressed in the form

$$\bar{G}_i = g_i^\circ + RT \ln \frac{y_i P}{P^\circ} \tag{9.10}$$

in which g_i° is the Gibbs function of pure substance i, evaluated at the temperature of the equilibrium mixture and at the standard state pressure P°, which is normally taken as 1 atmosphere. Also in Eq. 9.10, P is the pressure of the equilibrium mixture, and y_i is the mole fraction of component i in the mixture,

$$y_i = \frac{n_i}{\sum_i n_i} \tag{9.11}$$

If now we substitute an expression of the form of Eq. 9.10 for each component A, B, C, D of the mixture into Eq. 9.8, the result is, after collecting terms,

$$\Delta G = \Delta G^\circ + RT \ln \left[\frac{{y_C}^{\nu_C} {y_D}^{\nu_D}}{{y_A}^{\nu_A} {y_B}^{\nu_B}} \left(\frac{P}{P^\circ} \right)^{\nu_C + \nu_D - \nu_A - \nu_B} \right] \tag{9.12}$$

where, for convenience,

$$\Delta G^\circ = \nu_C g_C^\circ + \nu_D g_D^\circ - \nu_A g_A^\circ - \nu_B g_B^\circ \tag{9.13}$$

For a condition of chemical equilibrium in the system, the left side of Eq. 9.12 is equal to zero according to Eq. 9.9. We define the chemical equilibrium constant K by the relation

$$K = \frac{{y_C}^{\nu_C} {y_D}^{\nu_D}}{{y_A}^{\nu_A} {y_B}^{\nu_B}} \left(\frac{P}{P^\circ} \right)^{\nu_C + \nu_D - \nu_A - \nu_B} \tag{9.14}$$

From Eqs. 9.9, 9.12, and 9.14, it is seen that

$$\ln K = -\frac{\Delta G^\circ}{RT} \tag{9.15}$$

The value of the equilibrium constant K is given by Eqs. 9.13 and 9.15 in terms of the Gibbs functions for the pure substances, each evaluated at the temperature of the equilibrium system and the standard state pressure P°. Consequently, the value of K is not a function of the system pressure P and depends only upon the temperature, since P° is fixed (1 atm). It should be pointed out here that several definitions of equilibrium constants appear in the literature. One frequently encountered is our value K divided by the ratio

$$\left(\frac{P}{P^\circ} \right)^{\nu_C + \nu_D - \nu_A - \nu_B}$$

which is, of course, dependent upon system pressure. Another is defined similarly to Eq. 9.14, except that component concentrations are used instead of mole fractions. One must be very careful to note the definition of the equilibrium constant when using values from the many references in this field.

It is the evaluation of the equilibrium constant from Eq. 9.15 that is of particular interest to us in this chapter. Its subsequent use in the determination of equilibrium composition is more properly a topic of classical thermodynamics, and will be discussed only briefly in the following sections. The determination of K at any temperature is accomplished in a straightforward manner by evaluating each of the pure substance Gibbs functions of Eq. 9.13, using the methods and equations of Chapter 8. In doing so, we must be careful to include all the appropriate internal contributions to the Gibbs function. It is necessary, of course, to consider the chemical energy contribution to the Gibbs function given by Eq. 8.46, in order that all the substances present have a common energy reference base. The electronic contributions, even if they are due only to ground level degeneracies, do not in general cancel out of the equation, and must not be overlooked. Nuclear ground level degeneracies, on the other hand, do cancel out and consequently do not contribute to the value of ΔG° for the reaction. These values therefore are omitted from the calculations. These points are clarified in the following example.

Example 9.1

Determine the equilibrium constant for the reaction involving diatomic nitrogen dissociating to the monatomic species at 5000 K.

From Table A.5, for N_2,

$$r_\varepsilon = 1.0976 \text{ Å}$$

$$\omega_\varepsilon = 2357.6 \text{ cm}^{-1}$$

$$g_{e_0} = 1$$

$$D_0 = 9.757 \text{ ev}$$

The reaction equation is written

$$N_2 \rightleftharpoons 2N$$

for which, from Eq. 9.13,

$$\Delta G^\circ = 2g_N{}^\circ - g_{N_2}{}^\circ$$

where the superscript $^\circ$ denotes the value at the standard state pressure P°, namely, 1 atm.

Let us first determine g° for the monatomic nitrogen, for which we need consider only the translational and electronic contributions.

For monatomic nitrogen, the atomic mass is

$$m_N = 1.66 \times 10^{-24}(14.008) = 23.25 \times 10^{-24} \text{ gm}$$

The translational contribution to the partition function at 5000 K and 1 atm is, from Eq. 6.30,

$$\left(\frac{Z_t}{N}\right)^{\circ} = \left(\frac{2\pi m}{h^2}\right)^{3/2} \frac{(kT)^{5/2}}{P^{\circ}}$$

$$= \left[\frac{2\pi \times 23.25 \times 10^{-24}}{(6.625 \times 10^{-27})^2}\right]^{3/2} \frac{(1.3804 \times 10^{-16} \times 5000)^{5/2}}{1.01325 \times 10^{6}}$$

$$= 2.37 \times 10^{9}$$

Therefore the translational Gibbs function for N is, from Eq. 6.18,

$$g_t^{\circ} = -RT \ln \left(\frac{Z_t}{N}\right)^{\circ}$$

$$= -1.987 \times 5000 \ln (2.37 \times 10^{9})$$

$$= -214{,}000 \text{ cal/mole}$$

For the electronic contribution, the ground level term symbol for N is, from Table 5.2,

$$^{4}S_{3/2}$$

Therefore the ground level degeneracy is 4, and the electronic contribution to the Gibbs function is found from Eq. 6.42 to be

$$g_e = -RT \ln Z_e = -1.987 \times 5000 \ln 4 = -13{,}700 \text{ cal/mole}$$

Therefore, for monatomic nitrogen at 5000 K, 1 atm,

$$g_{\mathrm{N}}^{\circ} = -214{,}000 - 13{,}700 = -227{,}700 \text{ cal/mole}$$

Since N_2 is a diatomic gas, we must consider the rotational and vibrational contributions as well as those of translation and electronic ground level. We assume the simple internal model here for diatomic nitrogen.

For diatomic nitrogen, the translational contribution is found in the same manner as for N, except that N_2 has twice the mass. Thus

$$\left(\frac{Z_t}{N}\right)^{\circ} = 2.37 \times 10^{9}(2)^{3/2} = 6.70 \times 10^{9}$$

and

$$g_t^{\circ} = -1.987 \times 5000 \ln (6.70 \times 10^{9}) = -224{,}300 \text{ cal/mole}$$

For the rotational contribution, it was found in Example 8.1 that

$$\theta_r = 2.875 \text{ K}$$

so that, from Eq. 8.13,

$$Z_r = \frac{T}{\sigma \theta_r} = \frac{5000}{2(2.875)} = 870$$

and, from Eq. 8.19,

$$g_r = -RT \ln Z_r = -1.987 \times 5000 \ln 870 = -67{,}150 \text{ cal/mole}$$

For vibration of the molecule,

$$\theta_v = \frac{hc\omega_e}{k} = 1.4388(2357.6) = 3395 \text{ K}, \quad \frac{\theta_v}{T} = 0.679$$

and we find, using Eq. 8.36,

$$\begin{aligned}(g_v - u_{v_0}) &= RT \ln (1 - e^{-(\theta_v/T)}) \\ &= 1.987 \times 5000(-0.7075) \\ &= -7030 \text{ cal/mole}\end{aligned}$$

The electronic contribution for the ground level of the molecule is determined from Eq. 8.40,

$$g_e = -RT \ln Z_{e_0} = -1.987 \times 5000 \ln 1 = 0$$

For the chemical energy, or enthalpy of formation term, we find from Eqs. 8.42 and 8.46 that

$$\begin{aligned}g_{\text{chem}} &= h^\circ = -N_0 D_0 \\ &= -6.023 \times 10^{23} \times 9.757 \times 1.6021 \times 10^{-12} \times \frac{1}{4.184 \times 10^7} \\ &= -225{,}040 \text{ cal/mole}\end{aligned}$$

Thus, for N_2 at 5000 K, 1 atm,

$$\begin{aligned}g_{N_2}{}^\circ &= g_t{}^\circ + g_r + (g_v - u_{v_0}) + g_e + g_{\text{chem}} \\ &= -224{,}300 - 67{,}150 - 7030 + 0 - 225{,}040 \\ &= -523{,}520 \text{ cal/mole}\end{aligned}$$

Substituting these values, we find

$$\begin{aligned}\Delta G^\circ &= 2g_N{}^\circ - g_{N_2}{}^\circ = 2(-227{,}700) - (-523{,}520) \\ &= +68{,}120 \text{ cal/mole}\end{aligned}$$

The equilibrium constant for this reaction is, from Eq. 9.15,

$$\ln K = -\frac{\Delta G^\circ}{RT} = -\frac{68{,}120}{1.987 \times 5000} = -6.86$$

$$K = 0.00103$$

The equilibrium constant in the preceding example was calculated assuming the simple internal model for N_2; namely, rigid rotator, harmonic oscillator, electronic ground level. The assumption of electronic ground level for N_2 is reasonable at this temperature, since the energy of the next electronic term is 49774 cm^{-1}. This is also a reasonable assumption for N at this temperature, since from Table 5.2 we find that for monatomic nitrogen the energy of the first excited electronic level above ground level is 19224 cm^{-1}. For precise computations, however, such terms should be included. The high temperature corrections for centrifugal and vibrational stretching and for anharmonicity are small for N_2 at this temperature, but they should also be considered in accurate calculations (see Example 8.4). Such corrections, although relatively small in magnitude, can assume great importance in the value of ΔG° and K for a reaction, since ΔG° is the result of subtracting one large number from another and is therefore commonly quite small itself.

Example 9.1 points out the importance of the electronic ground level degeneracy contribution to the values ΔG° and K for a reaction. We find from the values in the example that neglecting that contribution would have resulted in an error of 40% in the value of ΔG° for the reaction. On the other hand, the nuclear ground level degeneracies cancel out and do not contribute anything to the result, ΔG°. To demonstrate that this is true, we note that the degeneracy for each nitrogen nucleus, and therefore the partition function also, is given in terms of the nuclear spin quantum number by

$$Z_n = \mathbf{g}_n = 2\mathbf{m}_n + 1$$

This, then, is the degeneracy for the monatomic form of nitrogen. The corresponding contribution to the Gibbs function is

$$(g_n)_N = -RT \ln Z_n = -RT \ln (\mathbf{g}_n)$$

For the homonuclear diatomic molecule, the nuclear ground level degeneracy has previously been found to be $\mathbf{g}_n^{\,2}$, as given by Eq. 8.74. Its Gibbs function contribution is

$$(g_n)_{N_2} = -RT \ln (\mathbf{g}_n)^2 = -2RT \ln (\mathbf{g}_n)$$

The resulting nuclear contribution to ΔG° for the reaction $N_2 \rightleftharpoons 2N$ is then

$$\Delta G_n^{\,\circ} = 2(g_n)_N - (g_n)_{N_2} = 0$$

9.2. Equilibrium Composition

We now turn our attention to the evaluation of the equilibrium composition of a system. The equilibrium equation, Eq. 9.14, specifies composition of the system as a function of T and P, since the equilibrium constant K is a function of temperature according to Eq. 9.15. A knowledge of initial composition in the system along with the equilibrium equation (Eq. 9.14) and $\Sigma_i\, y_i = 1$ gives a sufficient number of relations to evaluate the equilibrium composition. The technique for making this calculation is most easily demonstrated by example problems.

Example 9.2

One mole of N_2 at room temperature is heated to 5000 K, as indicated in Figure 9.2. Determine the equilibrium composition at the outlet, if the pressure is 0.1 atm.

The reaction equation for the system is written

$$N_2 \rightleftharpoons 2N$$

The corresponding equilibrium equation is, from Eq. 9.14,

$$K = \frac{y_N^{\,2}}{y_{N_2}}\left(\frac{P}{P^\circ}\right)^{2-1}$$

The value of K at 5000 K was found in Example 9.1 to be

$$K = 0.00103$$

The reaction equation expresses only the relative proportions in which the substances take part in the reaction, and consequently it gives the relation between the changes in amount of each substance present from initial composition to equilibrium composition. These changes therefore become the unknowns to be determined. In our example, the initial composition is 1 mole of N_2, 0 moles of N. Let the change in number of moles of N_2 during the heating process be $-x$. Then, according to the reaction equation, the change in number of moles of N is $+2x$, and the amount of each substance present at equilibrium is

$$\begin{aligned} n_{N_2} &= 1 - x \\ n_N &= 0 + 2x \\ \hline n_{mix} &= 1 + x \end{aligned}$$

The mole fractions of the components at equilibrium are

$$y_{N_2} = \frac{n_{N_2}}{n_{mix}} = \frac{1-x}{1+x}$$

$$y_N = \frac{n_N}{n_{mix}} = \frac{2x}{1+x}$$

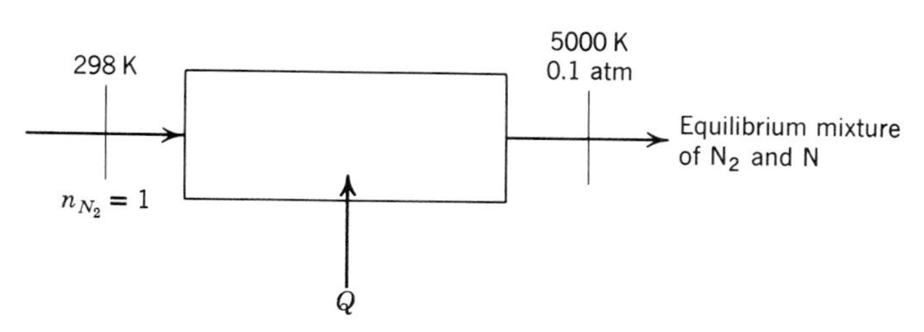

Figure 9.2 Sketch for Example 9.2.

Substituting these values into the equilibrium equation,

$$K = \frac{\left(\dfrac{2x}{1+x}\right)^2}{\left(\dfrac{1-x}{1+x}\right)}\left(\frac{P}{P^\circ}\right) = \left(\frac{4x^2}{1-x^2}\right)\left(\frac{P}{P^\circ}\right)$$

At 5000 K, 0.1 atm, we have

$$x^2 = \frac{\dfrac{K}{4(P/P^\circ)}}{1 + \dfrac{K}{4(P/P^\circ)}} = \frac{\dfrac{0.00103}{4(0.1/1)}}{1 + \dfrac{0.00103}{4(0.1/1)}} = 0.00257$$

$$x = 0.0508$$

We note that -0.0508 is also a mathematical root, but this root is physically meaningless. We find that, at 5000 K, 0.1 atm, we have, under equilibrium conditions,

$$\begin{aligned} n_{N_2} &= 0.9492 \\ n_N &= 0.1016 \\ \hline n_{\text{mix}} &= 1.0508 \end{aligned}$$

The corresponding mole fractions are

$$y_{N_2} = 0.9035, \quad y_N = 0.0965$$

Example 9.3

A mixture of 1 mole of CO_2, $\frac{1}{2}$ mole of CO, and 2 moles of O_2 at 298 K, 1.36 atm pressure, is heated to 3000 K in a constant-pressure process. Determine the resulting equilibrium composition, assuming the only reaction of significance to be

$$CO_2 \rightleftharpoons CO + \tfrac{1}{2}O_2$$

for which $K = 0.378$ at 3000 K.

The equilibrium equation corresponding to the given reaction is

$$K = \frac{y_{CO}\, y_{O_2}^{1/2}}{y_{CO_2}}\left(\frac{P}{P^\circ}\right)^{1+1/2-1}$$

Let $-x$ be the change in number of moles of CO_2 during the heating process. From the reaction equation, the corresponding changes in amount of CO and O_2 are then $+x$ and $+\frac{1}{2}x$, respectively. The number of moles of each component at equilibrium is

$$\begin{aligned} n_{CO_2} &= 1 - x \\ n_{CO} &= \tfrac{1}{2} + x \\ n_{O_2} &= 2 + \tfrac{1}{2}x \\ \hline n_{mix} &= 3\tfrac{1}{2} + \tfrac{1}{2}x \end{aligned}$$

In order that the amount of each substance be greater than zero, the unknown x is restricted to the range

$$-\tfrac{1}{2} \leq x \leq +1$$

The equilibrium mole fraction for each component is the ratio of the number of moles to the number of moles of mixture. Substituting these quantities into the equilibrium equation gives

$$K = \frac{\left(\dfrac{\frac{1}{2} + x}{3\frac{1}{2} + \frac{1}{2}x}\right)\left(\dfrac{2 + \frac{1}{2}x}{3\frac{1}{2} + \frac{1}{2}x}\right)^{1/2}}{\left(\dfrac{1 - x}{3\frac{1}{2} + \frac{1}{2}x}\right)}\left(\frac{P}{P^\circ}\right)^{1/2}$$

or

$$\left(\frac{\frac{1}{2} + x}{1 - x}\right)\left(\frac{4 + x}{7 + x}\right)^{1/2} = \frac{K}{(P/P^\circ)^{1/2}} = \frac{0.378}{(1.36)^{1/2}}$$

The solution is

$$x = -0.048$$

The negative sign in the resulting value of x means only that the amount of CO_2 increased rather than decreased during the heating process. That is, during the heating, the reaction proceeded slightly from right to left, instead of from left to right. Substituting the value for x into the previous expressions, we find that at equilibrium at 3000 K, 1 atm,

$$\begin{aligned} n_{CO_2} &= 1.048 \\ n_{CO} &= 0.452 \\ n_{O_2} &= 1.976 \\ \hline n_{mix} &= 3.476 \end{aligned}$$

and the mole fractions are

$$\begin{aligned} y_{CO_2} &= 0.302 \\ y_{CO} &= 0.130 \\ y_{O_2} &= 0.568 \end{aligned}$$

9.3. Simultaneous Reactions

In developing the equilibrium equation and equilibrium constant expressions of Section 9.1, it was assumed that there was only a single chemical-reaction equation relating the substances present in the system. To demonstrate the more general situation in which there is more than one chemical reaction, we now analyze a system involving two simultaneous reactions by a procedure analogous to that followed in Section 9.1. These results are then readily extended to systems involving several simultaneous reactions.

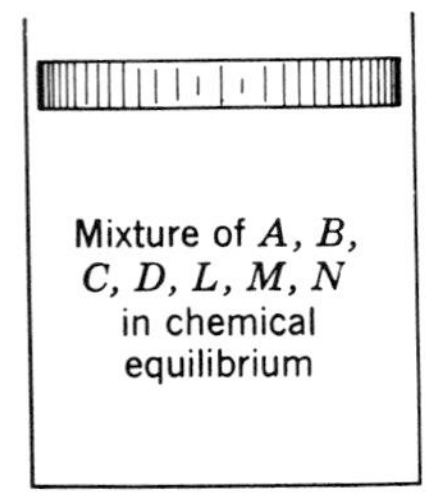

Figure 9.3 A two-reaction equilibrium mixture at T, P.

Consider a mixture of substances A, B, C, D, L, M, and N as indicated in Figure 9.3. These substances are assumed to exist at a condition of chemical equilibrium at temperature T and pressure P, and are related by the two independent reactions

$$(1) \quad \nu_{A_1}A + \nu_B B \rightleftharpoons \nu_C C + \nu_D D \tag{9.16}$$

$$(2) \quad \nu_{A_2}A + \nu_L L \rightleftharpoons \nu_M M + \nu_N N \tag{9.17}$$

We have considered the situation where one of the components (substance A) takes part in each of the reactions, in order to demonstrate the effect of this condition on the resulting equations. As in the preceding section, the changes in amount of the components are related by the stoichiometric coefficients. As already noted, the stoichiometric coefficients are different from the numbers of moles present in the vessel. We also realize that the coefficients ν_{A_1} and ν_{A_2} are not necessarily the same. That is, substance A does not in general take part in each of the reactions to the same extent.

Development of the requirement for equilibrium is completely analogous to that of Section 9.1. We consider that each reaction proceeds an infinitesimal amount toward the right side. This results in a decrease in the number of moles of A, B, and L, and an increase in the moles of C, D, M, and N. Letting the degrees of reaction be ρ_1 and ρ_2 for reactions 1 and 2, respectively, the changes in the number of moles are, for infinitesimal shifts from the equilibrium composition,

$$\begin{aligned}
dn_A &= -\nu_{A_1}\,d\rho_1 - \nu_{A_2}\,d\rho_2 \\
dn_B &= -\nu_B\,d\rho_1 \\
dn_L &= -\nu_L\,d\rho_2 \\
dn_C &= +\nu_C\,d\rho_1 \\
dn_D &= +\nu_D\,d\rho_1 \\
dn_M &= +\nu_M\,d\rho_2 \\
dn_N &= +\nu_N\,d\rho_2
\end{aligned} \tag{9.18}$$

The change in the Gibbs function for the mixture in the vessel at constant temperature and pressure is

$$dG_{T,P} = \bar{G}_A\,dn_A + \bar{G}_B\,dn_B + \bar{G}_C\,dn_C + \bar{G}_D\,dn_D + \bar{G}_L\,dn_L + \bar{G}_M\,dn_M + \bar{G}_N\,dn_N$$

Substituting the expressions of Eq. 9.18 and collecting terms, we have

$$dG_{T,P} = (\nu_C\bar{G}_C + \nu_D\bar{G}_D - \nu_{A_1}\bar{G}_A - \nu_B\bar{G}_B)\,d\rho_1 + (\nu_M\bar{G}_M + \nu_N\bar{G}_N - \nu_{A_2}\bar{G}_A - \nu_L\bar{G}_L)\,d\rho_2 \quad (9.19)$$

It is convenient to express again each of the partial molal Gibbs functions in terms of the relation

$$\bar{G}_i = g_i^\circ + RT\ln\left(\frac{y_iP}{P^\circ}\right)$$

When written in this form, Eq. 9.19 becomes

$$dG_{T,P} = \left[\Delta G_1^\circ + RT\ln\frac{y_C^{\nu_C}y_D^{\nu_D}}{y_A^{\nu_{A_1}}y_B^{\nu_B}}\left(\frac{P}{P^\circ}\right)^{\nu_C+\nu_D-\nu_{A_1}-\nu_B}\right]d\rho_1 + \left[\Delta G_2^\circ + RT\ln\frac{y_M^{\nu_M}y_N^{\nu_N}}{y_A^{\nu_{A_2}}y_L^{\nu_L}}\left(\frac{P}{P^\circ}\right)^{\nu_M+\nu_N-\nu_{A_2}-\nu_L}\right]d\rho_2 \quad (9.20)$$

In this equation, the standard state change in the Gibbs function for each reaction is defined as

$$\Delta G_1^\circ = \nu_C g_C^\circ + \nu_D g_D^\circ - \nu_{A_1} g_A^\circ - \nu_B g_B^\circ \quad (9.21)$$

$$\Delta G_2^\circ = \nu_M g_M^\circ + \nu_N g_N^\circ - \nu_{A_2} g_A^\circ - \nu_L g_L^\circ \quad (9.22)$$

Equation 9.20 expresses the change in Gibbs function of the system at constant T, P, for infinitesimal degrees of reaction of both reactions 1 and 2 (Eqs. 9.16 and 9.17). The requirement for equilibrium is that $dG_{T,P} = 0$. Therefore, since reactions 1 and 2 are independent, $d\rho_1$ and $d\rho_2$ can be independently varied. It follows that at equilibrium each of the bracketed terms of Eqs. 9.20 must be zero. Defining equilibrium constants for the two reactions by

$$\ln K_1 = -\frac{\Delta G_1^\circ}{RT} \quad (9.23)$$

and

$$\ln K_2 = -\frac{\Delta G_2^\circ}{RT} \quad (9.24)$$

we find that, at equilibrium,

$$K_1 = \frac{y_C^{\nu_C}y_D^{\nu_D}}{y_A^{\nu_{A_1}}y_B^{\nu_B}}\left(\frac{P}{P^\circ}\right)^{\nu_C+\nu_D-\nu_{A_1}-\nu_B} \quad (9.25)$$

and

$$K_2 = \frac{y_M{}^{\nu_M} y_N{}^{\nu_N}}{y_A{}^{\nu_{A_2}} y_L{}^{\nu_L}} \left(\frac{P}{P^\circ}\right)^{\nu_M+\nu_N-\nu_{A_2}-\nu_L} \tag{9.26}$$

The equilibrium constants K_1 and K_2 are functions of temperature only, and they can be evaluated from Eqs. 9.23 and 9.24 by the procedures followed in Section 9.1. The equilibrium equations (Eqs. 9.25 and 9.26) must be solved simultaneously for the equilibrium composition of the mixture. The following example is presented to demonstrate and clarify this procedure.

Example 9.4

One mole of H_2O vapor is heated to 3000 K, 1 atm pressure. Determine the equilibrium composition, assuming that H_2O, H_2, O_2, and OH are present.

There are two independent reactions relating the four components of the mixture at equilibrium. These can be written as

$$H_2O \rightleftharpoons H_2 + \tfrac{1}{2}O_2 \qquad (1)$$

$$H_2O \rightleftharpoons \tfrac{1}{2}H_2 + OH \qquad (2)$$

Let a be the number of moles of H_2O dissociating according to reaction (1) during the heating, and b the number of moles of H_2O dissociating according to reaction (2). Since the initial composition is 1 mole of H_2O, the changes according to the two reactions are

$$\begin{array}{lcccl} & H_2O & \rightleftharpoons H_2 & + \tfrac{1}{2}O_2 & \quad (1) \\ \text{Change:} & -a & +a & +\tfrac{1}{2}a & \\ & H_2O & \rightleftharpoons \tfrac{1}{2}H_2 & + OH & \quad (2) \\ \text{Change:} & -b & +\tfrac{1}{2}b & +b & \end{array}$$

Therefore the number of moles of each component at equilibrium is its initial number plus the change, so that at equilibrium

$$\begin{aligned} n_{H_2O} &= 1 - a - b \\ n_{H_2} &= a + \tfrac{1}{2}b \\ n_{O_2} &= \tfrac{1}{2}a \\ n_{OH} &= b \\ \hline n_{mix} &= 1 + \tfrac{1}{2}a + \tfrac{1}{2}b \end{aligned}$$

The overall chemical reaction that occurs during the heating process can be written

$$H_2O \rightarrow (1 - a - b)H_2O + (a + \tfrac{1}{2}b)H_2 + \tfrac{1}{2}aO_2 + bOH$$

The right side of this expression is the equilibrium composition of the system. Since the number of moles of each substance must necessarily be greater than zero, we find that the possible values of a and b are restricted to

$$a \geq 0$$

$$b \geq 0$$

$$(a + b) \leq 1$$

The equilibrium equations corresponding to the two reactions are

$$K_1 = \frac{y_{H_2} y_{O_2}^{1/2}}{y_{H_2O}} \left(\frac{P}{P^\circ} \right)^{1+1/2-1}$$

$$K_2 = \frac{y_{H_2}^{1/2} y_{OH}}{y_{H_2O}} \left(\frac{P}{P^\circ} \right)^{1/2+1-1}$$

Since the mole fraction of each component is the ratio of the number of moles of the component to the total number of moles of the mixture, these equations can be written in the form

$$K_1 = \frac{\left(\dfrac{a + \frac{1}{2}b}{1 + \frac{1}{2}a + \frac{1}{2}b} \right) \left(\dfrac{\frac{1}{2}a}{1 + \frac{1}{2}a + \frac{1}{2}b} \right)^{1/2}}{\left(\dfrac{1 - a - b}{1 + \frac{1}{2}a + \frac{1}{2}b} \right)} \left(\frac{P}{P^\circ} \right)^{1/2}$$

$$= \left(\frac{a + \frac{1}{2}b}{1 - a - b} \right) \left(\frac{\frac{1}{2}a}{1 + \frac{1}{2}a + \frac{1}{2}b} \right)^{1/2} \left(\frac{P}{P^\circ} \right)^{1/2}$$

and

$$K_2 = \frac{\left(\dfrac{a + \frac{1}{2}b}{1 + \frac{1}{2}a + \frac{1}{2}b} \right)^{1/2} \left(\dfrac{b}{1 + \frac{1}{2}a + \frac{1}{2}b} \right)}{\left(\dfrac{1 - a - b}{1 + \frac{1}{2}a + \frac{1}{2}b} \right)} \left(\frac{P}{P^\circ} \right)^{1/2}$$

$$= \left(\frac{a + \frac{1}{2}b}{1 + \frac{1}{2}a + \frac{1}{2}b} \right)^{1/2} \left(\frac{b}{1 - a - b} \right) \left(\frac{P}{P^\circ} \right)^{1/2}$$

Since $P = 1$ atm and the values of K_1, K_2 can be determined as in Section 9.1, we have two equations in the two unknowns a and b. At 3000 K, the values of K are found to be (by calculations similar to that performed in Example 9.1),

$$K_1 = 0.0457, \quad K_2 = 0.0543$$

Therefore the equations can be solved simultaneously for a and b. The values satisfying the equations are

$$a = 0.1074, \quad b = 0.1110$$

Substituting these values into the expressions for the number of moles of each component and of the mixture, we find the equilibrium mole fractions to be

$$y_{H_2O} = 0.7045$$
$$y_{H_2} = 0.1468$$
$$y_{O_2} = 0.0485$$
$$y_{OH} = 0.1002$$

The methods used in this section can readily be extended to equilibrium systems involving more than two independent reactions. In each case, the result is a number of simultaneous equilibrium equations equal to the number of independent reactions. Solution of a large set of non-linear simultaneous equations naturally becomes quite difficult and is not readily accomplished by hand calculations. Instead, solution of these problems is normally made using iterative procedures on a digital computer.

9.4. Developments from Statistical Mechanics

In Section 9.1, we developed the requirement for chemical equilibrium from the macroscopic, or classical, viewpoint and obtained the equilibrium equation, Eq. 9.14. This relation expresses equilibrium composition in terms of pressure and the equilibrium constant, a temperature-dependent quantity defined by Eq. 9.15. In the present section, we repeat this development, in this case from the molecular viewpoint, using the methods and relations developed in earlier chapters. A consideration of the problem of chemical equilibrium from these two different points of view leads to a greater insight and understanding of the subject.

In the present development, we consider a particular type of chemical reaction process, the dissociation of the diatomic species AB to the two monatomic species A and B. The results of the analysis of this problem can then be readily extended to the more general situation. The reaction considered here is written

$$AB \rightleftharpoons A + B \tag{9.27}$$

We assume as in previous developments that the mixture behaves as an ideal gas mixture of the three components, AB, A, and B. The first part of the analysis of this three-component mixture is analogous to the determination of the most probable distribution for the components in an ideal gas mixture made in Section 3.7 (in that case, for a binary mixture). Each species has its own set of energy levels,

$$\begin{matrix} \epsilon_{AB_1}, & \epsilon_{AB_2}, \ldots, & \epsilon_{AB_j} \\ \epsilon_{A_1}, & \epsilon_{A_2}, \ldots, & \epsilon_{A_j} \\ \epsilon_{B_1}, & \epsilon_{B_2}, \ldots, & \epsilon_{B_j} \end{matrix} \tag{9.28}$$

the values of which are fixed for a given system volume V and which have the corresponding degeneracies $g_{AB_1}, g_{AB_2}, \ldots, g_{AB_j}, g_{A_1}, g_{A_2}, \ldots, g_{A_j}, g_{B_1}, g_{B_2}, \ldots, g_{B_j}$.

Let us consider first a mixture comprised of a given total number of particles of each species N_{AB}, N_A, N_B (not necessarily the number of each at a condition of chemical equilibrium) contained in volume V at some temperature T. These particles are distributed in some manner among the various energy levels, the distribution being specified by the number of particles of each species in each of its energy levels,

$$\begin{matrix} N_{AB_1}, & N_{AB_2}, \ldots, & N_{AB_j} \\ N_{A_1}, & N_{A_2}, \ldots, & N_{A_j} \\ N_{B_1}, & N_{B_2}, \ldots, & N_{B_j} \end{matrix} \tag{9.29}$$

with the restrictions

$$\begin{aligned} N_{AB} &= \sum_{AB_j} N_{AB_j} \\ N_A &= \sum_{A_j} N_{A_j} \\ N_B &= \sum_{B_j} N_{B_j} \end{aligned} \tag{9.30}$$

For any given distribution of particles among levels, the thermodynamic probability w for the mixture is

$$\begin{aligned} w &= (w_{AB})(w_A)(w_B) \\ &= \left(\prod_{AB_j} \frac{g_{AB_j}^{N_{AB_j}}}{N_{AB_j}!}\right)\left(\prod_{A_j} \frac{g_{A_j}^{N_{A_j}}}{N_{A_j}!}\right)\left(\prod_{B_j} \frac{g_{B_j}^{N_{B_j}}}{N_{B_j}!}\right) \end{aligned} \tag{9.31}$$

As discussed in Section 3.7, each state for component A can be associated with any state of B or of AB, etc. Therefore the value of w for the mixture is the product of those for the individual constituents, as indicated in Eq. 9.31. The energy corresponding to this given distribution is expressed by

$$U = \sum_{AB_j} N_{AB_j}\epsilon_{AB_j} + \sum_{A_j} N_{A_j}\epsilon_{A_j} + \sum_{B_j} N_{B_j}\epsilon_{B_j} \tag{9.32}$$

We seek the most probable distribution for this system, subject to the given constraints, Eqs. 9.30 and 9.32, a fixed number of particles of each type, and fixed energy of the system. For this purpose, it is convenient to express Eq. 9.31 in the logarithmic form, which becomes

$$\ln w = \sum_{AB_j} N_{AB_j} \ln \frac{g_{AB_j}}{N_{AB_j}} + \sum_{A_j} N_{A_j} \ln \frac{g_{A_j}}{N_{A_j}} + \sum_{B_j} N_{B_j} \ln \frac{g_{B_j}}{N_{B_j}} + N_{AB} + N_A + N_B \quad (9.33)$$

The most probable distribution is that having the maximum value of w. Therefore, differentiating Eq. 9.33, and setting the result equal to zero, we have

$$d \ln w = -\sum_{AB_j} \ln \frac{N_{AB_j}}{g_{AB_j}}\, dN_{AB_j} - \sum_{A_j} \ln \frac{N_{A_j}}{g_{A_j}}\, dN_{A_j} - \sum_{B_j} \ln \frac{N_{B_j}}{g_{B_j}}\, dN_{B_j} = 0 \quad (9.34)$$

subject to the constraints

$$dN_{AB} = \sum_{AB_j} dN_{AB_j} = 0 \quad (9.35)$$

$$dN_A = \sum_{A_j} dN_{A_j} = 0 \quad (9.36)$$

$$dN_B = \sum_{B_j} dN_{B_j} = 0 \quad (9.37)$$

$$dU = \sum_{AB_j} \epsilon_{AB_j}\, dN_{AB_j} + \sum_{A_j} \epsilon_{A_j}\, dN_{A_j} + \sum_{B_j} \epsilon_{B_j}\, dN_{B_j} = 0 \quad (9.38)$$

Using the method of undetermined multipliers to find the most probable distribution, we multiply Eqs. 9.35 to 9.38 by α_{AB}, α_A, α_B, and β, respectively. Adding the resulting expressions to the negative (for convenience) of Eq. 9.34 and collecting terms, we conclude that, for the most probable distribution,

$$\ln \frac{N_{AB_j}}{g_{AB_j}} + \alpha_{AB} + \beta\epsilon_{AB_j} = 0 \quad \text{for all } AB_j \quad (9.39)$$

$$\ln \frac{N_{A_j}}{g_{A_j}} + \alpha_A + \beta\epsilon_{A_j} = 0 \quad \text{for all } A_j \quad (9.40)$$

$$\ln \frac{N_{B_j}}{g_{B_j}} + \alpha_B + \beta\epsilon_{B_j} = 0 \quad \text{for all } B_j \quad (9.41)$$

If $\beta = 1/kT$ is substituted as in previous developments, Eq. 9.39 can be written in the form

$$N_{AB_j} = g_{AB_j} e^{-\alpha_{AB}} e^{-\epsilon_{AB_j}/kT} \quad (9.42)$$

Summing over all AB_j, we have

$$N_{AB} = \sum_{AB_j} N_{AB_j} = e^{-\alpha_{AB}} \sum_{AB_j} g_{AB_j} e^{-\epsilon_{AB_j}/kT} = e^{-\alpha_{AB}} Z_{AB} \tag{9.43}$$

where the partition function Z_{AB} for substance AB is defined in the usual manner. Therefore Eq. 9.39 for the most probable distribution can be expressed as

$$N_{AB_j} = \frac{N_{AB}}{Z_{AB}} g_{AB_j} e^{-\epsilon_{AB_j}/kT} \tag{9.44}$$

By a similar procedure,

$$N_{A_j} = \frac{N_A}{Z_A} g_{A_j} e^{-\epsilon_{A_j}/kT} \tag{9.45}$$

and

$$N_{B_j} = \frac{N_B}{Z_B} g_{B_j} e^{-\epsilon_{B_j}/kT} \tag{9.46}$$

Equations 9.44 to 9.46 express the most probable distribution of particles of the various species among their respective energy levels for given N_{AB}, N_A, N_B, and U. The thermodynamic probability for this most probable distribution is found by substituting these expressions into Eq. 9.33; the result is

$$\ln w_{\text{mp}} = N_{AB}\left[\ln\left(\frac{Z_{AB}}{N_{AB}}\right) + 1\right] + N_A\left[\ln\left(\frac{Z_A}{N_A}\right) + 1\right] + N_B\left[\ln\left(\frac{Z_B}{N_B}\right) + 1\right] + \frac{U}{kT} \tag{9.47}$$

To review briefly, we began our development by considering a mixture comprised of given N_{AB}, N_A, and N_B, but not necessarily the proportions for chemical equilibrium, and we found that the thermodynamic probability for the most probable distribution in the mixture of these proportions is given by Eq. 9.47. There will be an equation of the same form for each mixture of different proportions (different values of N_{AB}, N_A, N_B), but of course the value of w_{mp} will be different for each different mixture. The particular mixture with the largest value of w_{mp} is the chemical equilibrium mixture and can be found by determining the maximum of Eq. 9.47 while varying N_{AB}, N_A, N_B. The possible values of N_{AB}, N_A, N_B are, however, restricted according to the initial composition in the system. The total number of atoms of A must remain the same, whether in the diatomic AB or in the monatomic form, and the total number of atoms of B must similarly remain constant. Since there is one atom of A

and one of B in AB,

$$\text{total atoms of type } A = N_{AB} + N_A = \text{constant} \tag{9.48}$$

$$\text{total atoms of type } B = N_{AB} + N_B = \text{constant} \tag{9.49}$$

To find the maximum in Eq. 9.47 subject to these interrelations, we differentiate Eq. 9.47 with respect to N_{AB}, N_A, N_B, at fixed U, V and set the result equal to zero. Upon performing this differentiation, it is found that, for the chemical equilibrium composition,

$$d \ln w_{\text{mp}} = \ln \frac{Z_{AB}}{N_{AB}} dN_{AB} + \ln \frac{Z_A}{N_A} dN_A + \ln \frac{Z_B}{N_B} dN_B = 0 \tag{9.50}$$

subject to the constraints

$$dN_{AB} + dN_A = 0 \tag{9.51}$$

$$dN_{AB} + dN_B = 0 \tag{9.52}$$

We now multiply Eq. 9.51 by γ and Eq. 9.52 by δ and add the result to Eq. 9.50. By the method of undetermined multipliers, we conclude that, for the equilibrium composition N_{AB_e}, N_{A_e}, N_{B_e},

$$\gamma + \delta + \ln \frac{Z_{AB}}{N_{AB_e}} = 0 \tag{9.53}$$

$$\gamma + \ln \frac{Z_A}{N_{A_e}} = 0 \tag{9.54}$$

$$\delta + \ln \frac{Z_B}{N_{B_e}} = 0 \tag{9.55}$$

Eliminating the multipliers γ and δ from these equations, we have

$$\frac{N_{A_e} N_{B_e}}{N_{AB_e}} = \frac{Z_A Z_B}{Z_{AB}} \tag{9.56}$$

which is the equilibrium equation and is analogous to the expression (Eq. 9.14) developed from the classical viewpoint. To demonstrate this analogy, recall that the mole fraction of component i is

$$y_i = \frac{n_i}{n} = \frac{N_i}{N} \tag{9.57}$$

and also that the partition function, including all contributions, can be written for component i as

$$Z_i = V \times f_i(T) \tag{9.58}$$

Therefore, substituting these relations into Eq. 9.56 for the three species, we have

$$\frac{y_A y_B}{y_{AB}} = \frac{V}{N} f(T) = \frac{kT}{P} f(T) = K\left(\frac{P^\circ}{P}\right) \tag{9.59}$$

which is the classical result (Eq. 9.14 applied to the reaction Eq. 9.27) in terms of the equilibrium constant K.

9.5. Ionization

In this final section of this chapter, we consider the equilibrium of systems involving ionized gases, or plasmas, a field that has gained in significance in recent years. In previous sections we have discussed chemical equilibrium, with a particular emphasis on molecular dissociation. One reaction that has been considered is

$$N_2 \rightleftharpoons 2N$$

Dissociation occurs to an appreciable extent for most molecules only at high temperature (of the order of magnitude 3000–10,000 K). At still higher temperatures, such as found in electric arcs, some of the atoms become ionized. That is, they lose an electron, according to the reaction

$$N \rightleftharpoons N^+ + e^-$$

where N^+ denotes a singly ionized atom (one that has lost one electron and consequently has a positive charge), and e^- denotes the free electron. As temperature is increased still further, many of the ionized atoms lose another electron, according to the reaction

$$N^+ \rightleftharpoons N^{++} + e^-$$

and thus become doubly ionized. At still higher temperature, triply ionized atoms are formed, and eventually a temperature is reached at which all the electrons have been stripped from the nucleus.

Ionization generally is appreciable only at high temperature. However, at a given temperature dissociation and ionization both tend to increase as the pressure is decreased. Consequently, dissociation and ionization may be appreciable in such environments as the upper atmosphere, even at moderate temperature. Other effects, such as radiation, also cause ionization of atoms, but these effects are not considered here.

The problems encountered in analyzing the composition in a plasma are more difficult than those for an ordinary chemical reaction, since the free electrons in the mixture do not exchange energy with the positive ions and neutral atoms at the same rate as they do with one another. Consequently, the electron gas in a plasma may persist for long periods of time at a temperature different from that of the heavy particles. However, if we assume a condition of thermodynamic equilibrium in the plasma, we can treat the ionization equilibrium in exactly the same manner as an ordinary chemical equilibrium analysis.

At these extremely high temperatures, we may assume that the plasma behaves as an ideal gas mixture of neutral atoms, positive ions, and electron gas. Thus, for the ionization of some atomic species A,

$$A \rightleftharpoons A^+ + e^- \tag{9.60}$$

we may write the ionization equilibrium equation in the form

$$K = \frac{(y_{A^+})(y_{e^-})}{(y_A)}\left(\frac{P}{P^\circ}\right)^{1+1-1} \tag{9.61}$$

The ionization equilibrium constant K is defined in the ordinary manner as

$$\ln K = -\frac{\Delta G^\circ}{RT} \tag{9.62}$$

and is a function of temperature only. The standard state change in Gibbs function for the reaction in Eq. 9.60 is found from

$$\Delta G^\circ = g_{A^+}{}^\circ + g_{e^-}{}^\circ - g_A{}^\circ \tag{9.63}$$

The standard state Gibbs function for each component at the given plasma temperature and 1 atm can be calculated using the procedures of statistical thermodynamics. In doing so, there are two problems, chemical energy (or reference level) and electronic states for the neutral atoms, positive ions, and electron gas. The chemical energy for the neutral atoms and electron gas is zero, and that for the positive ions is the ionization energy of the atom. This value is determined from the atomic spectra, and it is commonly called the limit.

The electronic partition function for the electron gas is a constant value of 2, which results from the two possible orientations of the spin vector. For the neutral atom and positive ion, however, evaluation of the electronic partition function becomes very complex at high temperature. The summation

$$Z_e = \sum_j g_{e_j} e^{-\epsilon_{ej}/kT}$$

tends to diverge at high temperature because of the extremely large number of possible electronic states that must be included. Several theories from quantum mechanics give the maximum electronic state for which the electron can still be considered bound, thereby giving a cut-off level for the summation. An alternative method frequently employed uses an empirical equation having several terms to approximate the electronic levels, instead of all the observed energy values.

Once the various standard state Gibbs functions have been determined, the equilibrium constant can be calculated from Eq. 9.62. Solution of the ionization equilibrium equation (Eq. 9.61) is then accomplished in the same manner as discussed in Section 9.2 for an ordinary chemical reaction equilibrium.

Example 9.5

Calculate the ionization equilibrium constant for the reaction

$$Ar \rightleftharpoons Ar^{+} + e^{-}$$

at 10,000 K, using the accompanying data for electronic levels.†

	ϵ_i/k, °K	g_i
	0	1
Ar	162,500	60
	.	.
	.	.
	.	.
	183,000	Limit (ionization)
	0	4
	2,062	2
Ar^+	156,560	2
	.	.
	.	.
	.	.
	320,800	Limit
e^-	0	2

The equilibrium constant K is given by Eq. 9.62 in terms of the standard state Gibbs functions (according to Eq. 9.63).

For the neutral argon atoms,

$$m = 39.944(1.66 \times 10^{-24}) = 66.3 \times 10^{-24} \text{ gm}$$

† Data from A. B. Cambel, *Plasma Physics and Magnetofluidmechanics*, McGraw-Hill Book Co., New York, 1963, p. 119.

Therefore, evaluating Eq. 6.30 at 1 atm pressure and 10,000 K gives

$$\left(\frac{Z_t}{N}\right)^\circ = \left(\frac{2\pi m}{h^2}\right)^{3/2} \frac{(kT)^{5/2}}{P^\circ}$$

$$= \left(\frac{2\pi \times 66.3 \times 10^{-24}}{(6.625 \times 10^{-27})^2}\right)^{3/2} \frac{(1.3804 \times 10^{-16} \times 10^4)^{5/2}}{1.01325 \times 10^6} = 6.45 \times 10^{10}$$

From Eq. 6.18, the translational contribution to the Gibbs function is

$$\begin{aligned} g_t^\circ &= -RT \ln \left(\frac{Z_t}{N}\right)^\circ \\ &= -1.987 \times 10^4 \ln (6.45 \times 10^{10}) \\ &= -495{,}000 \text{ cal/mole} \end{aligned}$$

The electronic partition function is found from Eq. 6.37:

$$Z_e = 1 + 60e^{-16.25} + \cdots = 1 + 5.4 \times 10^{-6}$$

so that, from Eq. 6.42,

$$g_e = -RT \ln Z_e = -1.987 \times 10^4 \ln (5.4 \times 10^{-6}) = -0.107 \text{ cal/mole}$$

which is negligible compared with the translational contribution. Therefore, for Ar at 10,000 K,

$$g_{\text{Ar}}^\circ = g_t^\circ + g_e = -495{,}000 \text{ cal/mole}$$

For singly ionized argon, we proceed in exactly the same manner, but now we must include a chemical (ionization energy) contribution. Since

$$M_{\text{Ar}^+} \approx M_{\text{Ar}}$$

it follows that, for Ar^+,

$$g_t^\circ = -495{,}000 \text{ cal/mole}$$

The electronic contribution is found to be

$$\begin{aligned} Z_e &= 4 + 2e^{-0.2062} + 2e^{-15.656} + \cdots \\ &= 5.628 \end{aligned}$$

so that

$$g_e = -1.987 \times 10^4 \ln 5.628 = -34{,}300 \text{ cal/mole}$$

From Eq. 8.46,

$$g_{\text{chem}} = h^\circ = +183{,}000(1.987) = +363{,}700 \text{ cal/mole}$$

Therefore, for Ar^+ at 10,000 K,

$$g^{\circ}_{Ar^+} = g_t^{\circ} + g_e + g_{chem} = -495{,}000 - 34{,}300 + 363{,}700$$
$$= -165{,}600 \text{ cal/mole}$$

Finally, we must consider the electron gas, for which, from Table A.1, $m = 9.1086 \times 10^{-28}$ gm. Using this value and Eq. 6.30, we have

$$\left(\frac{Z_t}{N}\right)^{\circ} = 3290$$

From Eq. 6.18,

$$g_t^{\circ} = -1.987 \times 10^4 \ln (3290) = -161{,}000 \text{ cal/mole}$$

Since the electron has a degeneracy of 2 (spin),

$$g_e = -1.987 \times 10^4 \ln 2 = -13{,}700 \text{ cal/mole}$$

Thus, for e^- at 10,000 K,

$$g_{e^-}^{\circ} = g_t^{\circ} + g_e = -161{,}000 - 13{,}700 = -174{,}700 \text{ cal/mole}$$

Now, from Eq. 9.63,

$$\Delta G^{\circ} = g^{\circ}_{Ar^+} + g_{e^-}^{\circ} - g_{Ar}^{\circ} = -165{,}600 - 174{,}700 - (-495{,}000)$$
$$= +154{,}700 \text{ cal/mole}$$

Using Eq. 9.62,

$$\ln K = \frac{-\Delta G^{\circ}}{RT} = \frac{-154{,}700}{1.987 \times 10^4} = -7.78$$

or

$$K = 0.00042$$

We conclude that, at this temperature, argon would become appreciably ionized only at very low pressures.

We can analyze simultaneous reactions, such as simultaneous molecular dissociation and ionization processes or multiple ionization reactions, in the same manner as the ordinary simultaneous chemical reactions of Section 9.3. In doing so, we again make the assumption of thermal equilibrium in the plasma, which, as mentioned above, is a reasonable approximation. The necessary equilibrium constants are evaluated by the procedures discussed previously, after calculation of the standard state Gibbs functions for the components. Figure 9.4 shows the equilibrium composition of air at high temperature and low density, and indicates the overlapping regions of the various dissociation and ionization processes.

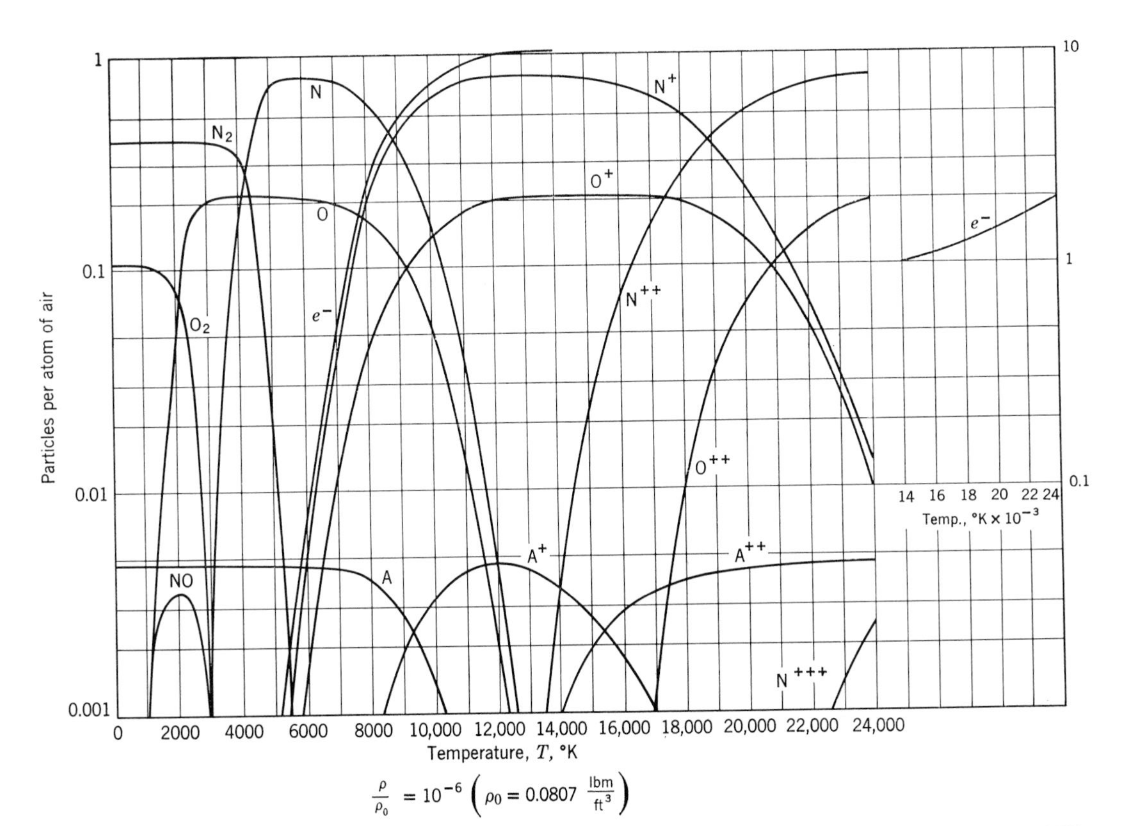

Figure 9.4 Equilibrium composition of air. From W. E. Moeckel and K. C. Weston NACA TN 4265 (1958).

PROBLEMS

9.1 Enthalpies of formation of various chemical species are frequently tabulated relative to a base in which the enthalpy of the common form of elements (O_2, N_2, H_2, C(s), etc.) is arbitrarily taken as zero at 25 C. Using data in Tables 5.2 and A.5, calculate the enthalpy of formation of monatomic nitrogen at 25 C relative to this base.

9.2 Calculate the equilibrium constant for the reaction

$$\tfrac{1}{2}N_2 + \tfrac{1}{2}O_2 \rightleftharpoons NO$$

at 4000 K. See Problem 8.14 for the electronic states of oxygen.

9.3 The equilibrium constant in Example 9.3 was given as $K = 0.378$. Using the various sources of data for CO_2, CO, and O_2, show that this is the correct value.

9.4 A mixture of 1 mole of N_2 and $\frac{1}{2}$ mole O_2 is heated to 4000 K, 1 atm, and results in an equilibrium mixture of N_2, O_2, and NO.

(*a*) Using the equilibrium constant determined in Problem 9.2, find the equilibrium composition.

(*b*) If the heating process was steady flow, how much heat transfer was required?

9.5 Repeat Example 9.3, assuming that the initial mixture also includes 2 moles of N_2, which does not dissociate during the process.

9.6 (*a*) Repeat Example 9.2 for a pressure of 0.01 atm.

(*b*) Find the distribution of nitrogen molecules among the various vibrational levels at 5000 K, and compare this distribution with the fraction of molecules dissociated at the conditions given in part (*a*). This illustrates the important point that dissociation is not a function only of energy.

9.7 Diatomic fluorine (F_2) is heated to 1500 K, 2 atm pressure, during which process some of the fluorine dissociates according to the reaction $\frac{1}{2}F_2 \rightleftharpoons F$.

(*a*) Calculate the equilibrium constant for this reaction at 1500 K. Assume the simple internal model for F_2.

(*b*) Determine the equilibrium composition at 1500 K, 2 atm.

9.8 (*a*) Using the electronic level data for diatomic oxygen given in Problem 8.14, calculate the equilibrium constant for the reaction $O_2 \rightleftharpoons 2O$ at 4000 K.

(*b*) One mole of oxygen is heated to 4000 K, 1.5 atm. Find the equilibrium composition.

9.9 Gibbs functions for monatomic and diatomic nitrogen have been calculated in Problems 6.13 and 8.13 from 4000 K to 10,000 K.

(*a*) Calculate the equilibrium constant for the reaction $N_2 \rightleftharpoons 2N$ over this range of temperature, and plot ln K vs. $(1/T)$.

(*b*) For an initial composition of 1 mole of N_2, plot equilibrium composition vs. temperature for 100 atm, 1 atm, and 0.01 atm from 4000 K to 10,000 K.

9.10 The equilibrium constant in Example 9.1 was determined using the simple internal model for N_2. Using the result of Example 8.4, recalculate K

for this reaction at 5000 K, assuming the general internal model for N_2. Make a similar comparison at 10,000 K.

9.11 (*a*) Assuming the general internal model, calculate the Gibbs function for H_2 at 5000 K, 1 atm pressure. Include the chemical energy of H_2 in the result. (see Problems 5.18 and 8.16 regarding the general internal model for H_2).

(*b*) Calculate the equilibrium constant at 5000 K for the reaction $H_2 \rightleftharpoons 2H$.

9.12 Repeat Problem 9.11, assuming the simple internal model for H_2. Is the difference in K significant?

9.13 The equilibrium constant for a particular reaction can be found from those for other reactions. Consider the reactions

$$(1) \quad H_2O \rightleftharpoons H_2 + \tfrac{1}{2}O_2$$

$$(2) \quad CO_2 \rightleftharpoons CO + \tfrac{1}{2}O_2$$

At 100 K:

$$\ln K_1 = -23.169$$

$$\ln K_2 = -23.535$$

(*a*) Use this information to calculate K at 1000 K for the water-gas reaction

$$CO + H_2O \rightleftharpoons CO_2 + H_2$$

(*b*) One mole of CO_2, 1 mole of CO, and 2 moles of H_2O are fed to a catalytic reactor maintained at 1000 K. An equilibrium mixture of CO, H_2O, CO_2, H_2 leaves at this temperature and 1 atm pressure. Find the composition at the outlet.

9.14 Repeat Example 9.4, assuming that the initial composition is 1 mole of H_2O, 1 mole of O_2, and that the equilibrium pressure is 2 atm.

9.15 One mole of air (assume 78% N_2, 21% O_2, 1% Ar) at room temperature is heated to 7200 R, 30 lbf/in.^{2}. Find the equilibrium composition at this state, assuming that N_2, O_2, NO, O, and Ar are present. (The necessary equilibrium constants have been calculated in Problems 9.2 and 9.8.)

9.16 Show that, if Eq. 9.47 is maximized at constant U, V, the result is Eq. 9.50.

9.17 Carry out a maximization at constant T, V, analogous to that of Eqs. 9.47 to 9.50. (In terms of the independent variables T, V, it is necessary to utilize the Helmholtz function instead of entropy.)

9.18 Consider the molecular dissociation $A_2 \rightleftharpoons 2A$, where A is an arbitrary substance. Develop an expression for the chemical equilibrium constant in terms of the significant fundamental parameters. Assume ground electronic levels only, with degeneracies g_{e_A}, $g_{e_{A_2}}$, and use the simple internal model for A_2.

9.19 Consider the ionization of argon discussed in Example 9.5. Compare the fraction of neutral argon atoms in excited electronic levels at this temperature and 0.01 atm pressure with the fraction of atoms ionized. Discuss this result (see Problem 9.6 for the analogous result in molecular dissociation).

9.20 At temperatures above 10,000 K, nitrogen is essentially completely dissociated to the atomic species (see Problem 9.9), and ionization begins to be of significance.

(*a*) Determine the ionization equilibrium constant at 15,000 K for the reaction

$$N \rightleftharpoons N^+ + e^-$$

For atomic N, use Table 5.2 for electronic levels, and the ionization energy of 14.54 ev. For the positive ion N^+,

Term	ϵ/hc, cm^{-1}
$^3P_0$	0
$^3P_1$	49.1
$^3P_2$	131.3
$^1D_2$	15,315.7

(*b*) For an initial composition of one mole of N_2 at room temperature, what is the equilibrium composition at 15,000 K, 0.1 atm?

9.21 One mole of argon at room temperature is heated to 20,000 K, 1 atm pressure. Assume that the plasma at this condition consists of an equilibrium mixture of Ar, Ar^+, Ar^{++}, e^- according to the simultaneous reactions

$$(1) \quad Ar \rightleftharpoons Ar^+ + e^-$$

$$(2) \quad Ar^+ \rightleftharpoons Ar^{++} + e^-$$

(*a*) Calculate the Gibbs function for each constituent at 20,000 K, 1 atm.

(*b*) Calculate the ionization equilibrium constant for each reaction at this temperature.

(*c*) Evaluate the equilibrium composition. Use the data from Example 9.5 and the following electronic terms for Ar^{++}:

Term	ϵ/hc, cm^{-1}
$^3P_2$	0
$^3P_1$	1,112.1
$^3P_0$	1,570.2
$^1D_2$	14,010
$^1S_0$	33,266

10 Statistical Mechanics for Systems of Dependent Particles

In previous chapters, we have dealt with systems of independent particles. These systems possess no energy that results from the configuration of the particles and intermolecular forces, and we are correct in speaking of the energy of a particle, since the energy of any particle is not dependent upon the location or energy of the remainder of the particles in the system. In this chapter, we consider the possibility of a system that has configurational energy. In such a system, it is not possible to analyze the behavior by consideration of the single-particle energies. Instead, we introduce the concepts of the canonical and grand canonical ensembles, and, by using methods and procedures analogous to those already developed, we find that it is possible to determine the equilibrium behavior and thermodynamic properties for the system of dependent particles.

10.1 The Canonical Ensemble

Let us consider a thermodynamic system consisting of a substance which is at relatively high pressure, is contained in a tank of given volume V, and is free to exchange energy with its surroundings. The system under consideration is one containing billions of particles having the various modes of energy discussed previously, and, because it is at high pressure, the system also possesses a configurational energy that is dependent upon the intermolecular forces, positions, and possibly even the relative orientations of the molecules.

For the system described, we cannot correctly speak of the energy per particle, because the particles are not independent, and consequently our earlier analyses are not valid. The problem can be approached, however, by a procedure analogous to that followed previously. Let us suppose that this system of volume V and containing N particles is but one member of an extremely large collection or ensemble of such systems, η in number, each having the same volume V and same number of particles,

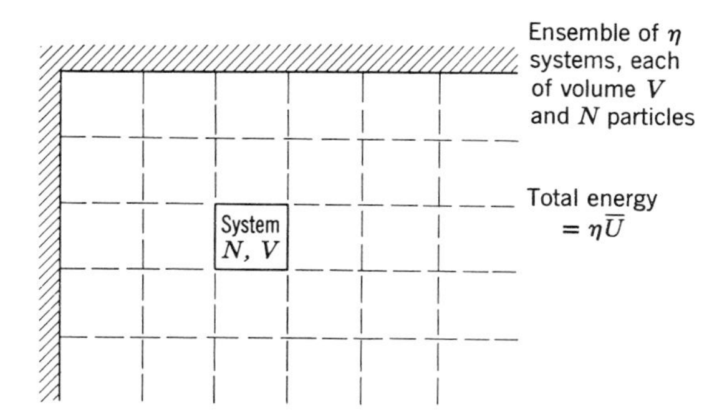

Figure 10.1 Representation of a canonical ensemble.

N. Such an ensemble is shown in Figure 10.1. This concept was developed by J. Willard Gibbs; he called the collection of systems a canonical ensemble. We shall, in general, refer to it simply as an ensemble. The systems of the ensemble are distinguishable from one another, because each has a fixed location in the ensemble. The systems are free to exchange energy with one another, but not mass. That is, the number of particles in each of the η systems remains fixed. We assume that the overall energy of the ensemble is fixed, although that of the individual systems may vary.

At any instant of time, each system is in some system quantum state. The various system states are functions not only of the volume and any other boundary parameter of importance, but of the locations of the particles as well. We could describe, at least in principle, the state of the ensemble by specifying the state of each of the η distinguishable systems at this instant of time. However, as with previous developments, our interest lies in the ensemble distribution. That is, we are not concerned with which system is in which system state; we are concerned about the number of systems that are in each of the states at any time. Consider a particular ensemble distribution for which there are a number of systems η_1 in system quantum state 1, a number of systems η_2 in system state 2, . . . , a number η_i in system state i. There are w ways of achieving this particular ensemble distribution, and, since the systems are distinguishable from one another, w is given by the expression

$$w = \frac{\eta!}{\prod \eta_i!} \tag{10.1}$$

As in the development for a system of independent particles, we seek the most probable distribution of systems among the various system states. The possible distributions are, however, subject to two constraints.

The number of systems in the ensemble is fixed, or

$$\eta = \sum_i \eta_i \tag{10.2}$$

where η_i is the number of systems in system state i, and the total energy of the ensemble is fixed, or

$$\eta \bar{U} = \sum_i \eta_i U_i \tag{10.3}$$

where U_i is the energy of system state i, and $\bar{U}$ is the average energy per system.

Before proceeding to determine the most probable distribution, we should make a few clarifying remarks regarding average values of properties. Consider some arbitrary thermodynamic property E which is a function of energy. The average value of this property over all the systems for some particular distribution of systems, d, is, from Chapter 2,

$$\bar{E}_d = \frac{1}{\eta} \sum_i \eta_{i_d} E_i \tag{10.4}$$

in which E_i is the average value of E in system state i. The number of ways that this distribution d can occur is w_d, while the total number of ways that all possible distributions can occur is the sum of the thermodynamic probability over all distributions. That is,

$$w_{\text{tot}} = \sum_d w_d \tag{10.5}$$

Therefore the so-called ensemble average of the property E is the average of $\bar{E}_d$ over all possible distributions, or

$$\bar{E} = \frac{1}{w_{\text{tot}}} \sum_d w_d \bar{E}_d \tag{10.6}$$

It is a basic assumption of statistical mechanics that, as the number of systems making up the ensemble becomes very large, the time-average value of E for our system of interest becomes equal to the ensemble average $\bar{E}$ as given by Eq. 10.6. Thus, if we are able to calculate ensemble averages of various thermodynamic properties where the ensemble consists of a large number of systems, the results will be equivalent to determining time-average properties of our single thermodynamic system of interest. It is apparent that this gives us a most important and powerful tool for analysis.

We make one additional observation at this point, this concerning the nature of the thermodynamic equilibrium state. We recall from earlier discussions that the most probable distribution is representative of a group of distributions all differing from the most probable by a negligible amount; that is, all have values of $\bar{E}_d$ identical (for all practical purposes) with $\bar{E}_{d_{\text{mp}}}$. There are, of course, many other distributions consistent with the system constraints (Eqs. 10.2 and 10.3), these having measurably different values of $\bar{E}_d$. However, because the thermodynamic probabilities, w_d, for these distributions are negligibly small in comparison to w_{mp}, they do not contribute appreciably to Eqs. 10.5 and 10.6. Thus we conclude that Eq. 10.5 may be replaced by

$$\ln w_{\text{tot}} \approx \ln w_{\text{mp}} \tag{10.7}$$

We also realize that the ensemble average of E given by Eq. 10.6 reduces to the form

$$\bar{E} \approx \bar{E}_{d_{\text{mp}}} = \frac{1}{\eta} \sum_i \eta_{i_{\text{mp}}} E_i \tag{10.8}$$

where the $\eta_{i_{\text{mp}}}$'s in the summation are the numbers of systems in each of the system states for the most probable distribution. Equation 10.8 is seen to be a very important expression, because, from our basic hypothesis, the ensemble average $\bar{E}$ is the same as the time-average of E for our thermodynamic system. In order to evaluate this average from Eq. 10.8, however, we must first determine the most probable ensemble distribution.

To find the most probable distribution of systems among the various system quantum states, we proceed as in our earlier development. Consider some particular distribution, for which the thermodynamic probability w is given by Eq. 10.1. In logarithmic form, and using Stirling's formula and Eq. 10.2, we have

$$\ln w = \eta \ln \eta - \eta - \sum_i (\eta_i \ln \eta_i - \eta_i) = \eta \ln \eta - \sum_i \eta_i \ln \eta_i \tag{10.9}$$

Therefore, for the most probable distribution,

$$d \ln w = -\sum_i (\ln \eta_i \, d\eta_i + d\eta_i) = 0 \tag{10.10}$$

But, from Eq. 10.2,

$$d\eta = \sum_i d\eta_i = 0 \tag{10.11}$$

so that, for the most probable distribution, Eq. 10.10 becomes

$$d \ln w = -\sum_i \ln \eta_i \, d\eta_i = 0 \tag{10.12}$$

subject to the two constraints, namely, Eq. 10.11 and

$$dU = \sum_i U_i \, d\eta_i = 0 \tag{10.13}$$

Using the method of undetermined multipliers, we multiply Eq. 10.11 by α and Eq. 10.13 by β, and add to the negative of Eq. 10.12. The result is

$$\sum_i (\ln \eta_i + \alpha + \beta U_i)\, d\eta_i = 0 \tag{10.14}$$

By the method of undetermined multipliers, we conclude that

$$\ln \eta_i + \alpha + \beta U_i = 0$$

for all i, or

$$\eta_i = e^{-\alpha} e^{-\beta U_i} \tag{10.15}$$

This is the equilibrium distribution equation for the systems of the canonical ensemble.

In order to eliminate the undetermined multipliers α and β, let us first substitute the distribution equation (Eq. 10.15) into Eq. 10.2:

$$\eta = \sum_i \eta_i = e^{-\alpha} \sum_i e^{-\beta U_i} \tag{10.16}$$

Therefore the multiplier α can be eliminated from the distribution equation, which becomes

$$\eta_i = \frac{\eta}{Z_N} e^{-\beta U_i} \tag{10.17}$$

where Z_N, which is called the system or canonical partition function, is defined as,

$$Z_N = \sum_i e^{-\beta U_i} \tag{10.18}$$

We use the subscript N in the symbol for the system partition function as a reminder that Z_N is a function of the number of particles in a system (since U_i depends upon the particles) as well as being a function of β and V.

We now utilize Eq. 10.3 to express the energy of a system in terms of Z_N. Dividing Eq. 10.3 by η (or alternatively from Eq. 10.8), we have

$$\bar{U} = \frac{1}{\eta} \sum_i \eta_i U_i \tag{10.19}$$

For the most probable distribution, we substitute Eq. 10.17 into Eq. 10.19; the result is

$$\bar{U} = \frac{1}{Z_N} \sum_i U_i e^{-\beta U_i} = \frac{1}{Z_N}\left[-\left(\frac{\partial Z_N}{\partial \beta}\right)_{V,N}\right] = -\left(\frac{\partial \ln Z_N}{\partial \beta}\right)_{V,N} \tag{10.20}$$

The thermodynamic property entropy is defined by the expression

$$S = \eta \bar{S} = k \ln w_{\text{tot}} \tag{10.21}$$

where $\bar{S}$ is the average entropy per system. However, for our application, Eq. 10.7 is valid, so that, using Eqs. 10.9 and 10.17, we obtain

$$\bar{S} = \frac{1}{\eta} k \ln w_{\text{mp}} = \frac{k}{\eta}\Big[\eta \ln \eta - \sum_i \eta_{i_{\text{mp}}} \ln \eta_{i_{\text{mp}}}\Big]$$
$$= \frac{k}{\eta}\Big[\eta \ln \eta - \sum_i \eta_{i_{\text{mp}}} \ln \Big(\frac{\eta}{Z_N} e^{-\beta U_i}\Big)\Big]$$

When we use Eq. 10.19, this reduces to the form

$$\bar{S} = k \ln Z_N + k\beta\bar{U} \tag{10.22}$$

which is a general expression for the average entropy of a system.

As a point of interest, we note that the equation for entropy can alternatively be written in the form

$$\bar{S} = \frac{k}{\eta}\Big[\eta \ln \eta - \sum_i \eta_{i_{\text{mp}}} \ln \eta_{i_{\text{mp}}}\Big] = \frac{k}{\eta}\Big[- \sum_i \eta_{i_{\text{mp}}} \ln \Big(\frac{\eta_{i_{\text{mp}}}}{\eta}\Big)\Big]$$
$$= -k \sum_i \Big(\frac{\eta_{i_{\text{mp}}}}{\eta}\Big) \ln \Big(\frac{\eta_{i_{\text{mp}}}}{\eta}\Big) = -k \sum_i P_{i_{\text{mp}}} \ln P_{i_{\text{mp}}} \tag{10.23}$$

in which $P_{i_{\text{mp}}}$ is the mathematical probability that for the most probable distribution a given system will be found in state i. It is seen that, in this form, the entropy has the same mathematical expression and characteristics as the entropy of communication theory, discussed in Chapter 4.

In order that the multiplier β may be identified, let us now differentiate Eq. 10.22 with respect to $\bar{U}$. Making use of Eq. 10.20, this becomes

$$\Big(\frac{\partial \bar{S}}{\partial \bar{U}}\Big)_{V,N} = k\Big(\frac{\partial \ln Z_N}{\partial \bar{U}}\Big)_{V,N} + k\beta + k\bar{U}\Big(\frac{\partial \beta}{\partial \bar{U}}\Big)_{V,N}$$
$$= k\Big(\frac{\partial \ln Z_N}{\partial \bar{U}}\Big)_{V,N} + k\beta - k\Big(\frac{\partial \ln Z_N}{\partial \beta}\Big)_{V,N}\Big(\frac{\partial \beta}{\partial \bar{U}}\Big)_{V,N}$$
$$= k\beta \tag{10.24}$$

The property relation of classical thermodynamics is written

$$dU = T\,dS - P\,dV + \mu\,dN$$

where, from the basic assumption of classical thermodynamics, the properties are understood to be time-average values. From this relation, it follows that

$$\Big(\frac{\partial S}{\partial U}\Big)_{V,N} = \frac{1}{T} \tag{10.25}$$

Comparing the two results (Eqs. 10.23 and 10.25), we find the relation

$$\beta = \frac{1}{kT} \tag{10.26}$$

This relation of β to the temperature has previously been developed for a classical monatomic ideal gas, and later it was presumed to be more generally applicable. We now find that Eq. 10.26 is indeed a general expression, following from the relation of the microscopic and macroscopic developments of thermodynamics.

By the use of Eq. 10.26, our general relations developed in terms of β can more meaningfully be written in terms of the temperature. The partition function, Eq. 10.18, is

$$Z_N = \sum_i e^{-U_i/kT} \tag{10.27}$$

From Eq. 10.20, the average internal energy is

$$\bar{U} = kT^2\left(\frac{\partial \ln Z_N}{\partial T}\right)_{V,N} \tag{10.28}$$

and, from Eq. 10.22, the average entropy is

$$\bar{S} = k \ln Z_N + \frac{\bar{U}}{T} \tag{10.29}$$

Therefore the Helmholtz function is

$$\bar{A} = \bar{U} - T\bar{S} = -kT \ln Z_N \tag{10.30}$$

From the property relation,

$$d\bar{A} = -\bar{S}\,dT - \bar{P}\,dV + \bar{\mu}\,dN$$

and, from Eq. 10.30, it follows that

$$\left(\frac{\partial \bar{A}}{\partial V}\right)_{T,N} = -\bar{P} = -kT\left(\frac{\partial \ln Z_N}{\partial V}\right)_{T,N}$$

or

$$\bar{P} = kT\left(\frac{\partial \ln Z_N}{\partial V}\right)_{T,N} \tag{10.31}$$

Equation 10.31 is an expression for the equation of state of the substance. Expressions for other thermodynamic properties of interest follow directly from their definitions and the relations above.

10.2. The System of Independent Particles as a Special Case

In order to complete our development of the canonical ensemble, we must show that the earlier development assuming a system of independent particles is merely a special case of the canonical ensemble model. That is, we must evaluate the system partition function (Eq. 10.27)

$$Z_N = \sum_i e^{-U_i/kT}$$

for the special case of independent particles (i.e., an ideal gas).

Consider a system of N independent particles contained in volume V. The assumption of a system of independent particles implies the freedom to speak of single-particle quantum or energy states, which are fixed by specification of the boundary parameters, V in this case. Let the index j refer to the jth particle of the system, this index being restricted to values between 1 and N. Let C denote a single-particle quantum cell, so that C_j is the quantum cell occupied by the jth particle, and ϵ_{C_j} is its energy. Therefore, for the system in system state i, the particles are distributed in some manner among the various single-particle states, resulting in the system energy

$$U_i = \sum_j \epsilon_{C_j} \tag{10.32}$$

There is a problem, however, in substituting Eq. 10.32 into Eq. 10.27 in order to sum over all possible system states. The particles are indistinguishable, and we cannot tell which particle is in which single-particle state. Thus the system state i is given merely by the number of particles, N_C, in each single-particle cell C, rather than by specification of the list of all C_j for the individual particles. As discussed in previous chapters, some systems of particles obey Fermi-Dirac statistics while others obey Bose-Einstein statistics. For a Fermi-Dirac system, the various occupation numbers N_C are either 0 or 1. That is, a single-particle cell is either empty or contains 1 particle, but not more than 1. Thus each system state i corresponds to $N!$ permutations of the N numbers C_j. For a Bose-Einstein system, on the other hand, occupation numbers greater than 1 are permitted, so that the corresponding total number of permutations would consequently be somewhat less than the $N!$ of the Fermi-Dirac system. However, we recall that at moderately low density the number of states available far exceeds the number of particles. Therefore, the number of cells for which N_C exceeds 1 is negligibly small, and the total number of permutations approaches $N!$. Thus, we may use this factor for Bose-Einstein as well as for Fermi-Dirac systems.

Now we substitute Eq. 10.32 into Eq. 10.27 and divide by $N!$ to correct for the indistinguishability of the particles; Eq. 10.27 becomes

$$Z_N = \frac{1}{N!} \sum_{C_j \text{ for each } j} e^{-\epsilon_{C_j}/kT} \tag{10.33}$$

In this equation, the summation must be taken over all the C_j (cells) for each j (particle); that is, the product of N summations over all the cells. Therefore Eq. 10.33 can also be written as

$$Z_N = \frac{1}{N!}\left[\sum_C e^{-\epsilon_C/kT}\right]^N = \frac{Z^N}{N!} \tag{10.34}$$

where Z is the single-particle partition function defined in Chapter 3; it is written here as a summation over all the single-particle cells or states.

In logarithmic form, the system partition function is, applying Stirling's formula to $N!$,

$$\begin{aligned}\ln Z_N = \ln\left(\frac{Z^N}{N!}\right) &= N \ln Z - N \ln N + N \\ &= N\left[\ln \frac{Z}{N} + 1\right]\end{aligned} \tag{10.35}$$

From Eqs. 10.28 and 10.35, the internal energy for a system of independent particles is found to be

$$\bar{U} = NkT^2\left(\frac{\partial \ln Z}{\partial T}\right)_V \tag{10.36}$$

and, from Eqs. 10.29 and 10.35, the entropy is

$$\bar{S} = Nk\left[\ln \frac{Z}{N} + 1\right] + \frac{\bar{U}}{T} \tag{10.37}$$

both of which are identical with the expressions developed in Chapter 4 for independent particles. It follows that other thermodynamic properties for a system of independent particles are then given in terms of these equations as in our earlier development.

10.3. Fluctuations in Internal Energy

In classical thermodynamics we assume a condition of thermodynamic equilibrium, for which the thermodynamic properties U, P, . . . are considered to have fixed and determinable values. We realize that on a microscopic level these properties are not fixed at all, but are instead continually fluctuating about some average values. In Section 10.1 we determined expressions for these average values of the thermodynamic properties for a system assumed to exist in a state of equilibrium or, in other words, for the most probable distribution.

We now ask a most important question, namely, what is the magnitude of the instantaneous fluctuation in a given thermodynamic property about its equilibrium value? This question has tremendous implications.

If fluctuations are large, even macroscopically observable, then the assumption of a classical equilibrium state and the application and validity of classical thermodynamics may not be reasonable.

Let us now examine the fluctuations in internal energy of a substance. To evaluate the magnitude of the fluctuations, we utilize the standard deviation in U, in a manner similar to that described in Chapter 2. The standard deviation σ_U is defined as

$$\sigma_U = [\overline{(U - \bar{U})^2}]^{1/2} = [\overline{U^2} - \bar{U}^2]^{1/2} \tag{10.38}$$

where $\bar{U}$, the average of U, is, from Eq. 10.8 and the distribution equation,

$$\begin{aligned}\bar{U} &= \frac{1}{\eta}\sum_i \eta_{i_{\text{mp}}} U_i = \frac{1}{\eta}\sum_i \frac{\eta}{Z_N} U_i e^{-U_i/kT} \\ &= \frac{1}{Z_N}\sum_i U_i e^{-U_i/kT}\end{aligned} \tag{10.39}$$

Similarly, $\overline{U^2}$, the average of U^2, is found to be

$$\overline{U^2} = \frac{1}{\eta}\sum_i \eta_{i_{\text{mp}}} U_i^2 = \frac{1}{Z_N}\sum_i U_i^2 e^{-U_i/kT} \tag{10.40}$$

Differentiating Eq. 10.39 with respect to T at constant V, N, we have

$$\begin{aligned}\left(\frac{\partial \bar{U}}{\partial T}\right)_{V,N} &= -\frac{1}{Z_N}\left(\frac{\partial \ln Z_N}{\partial T}\right)_{V,N}\sum_i U_i e^{-U_i/kT} + \frac{1}{Z_N}\sum_i \frac{U_i^2}{kT^2} e^{-U_i/kT} \\ &= -\frac{\bar{U}^2}{kT^2} + \frac{\overline{U^2}}{kT^2}\end{aligned} \tag{10.41}$$

However, the left side of Eq. 10.41 is

$$\left(\frac{\partial \bar{U}}{\partial T}\right)_{V,N} = n\left(\frac{\partial \bar{u}}{\partial T}\right)_{V,N} = n\bar{C}_v \tag{10.42}$$

where n is the number of moles and $\bar{C}_v$ is the average value of the constant-volume specific heat per mole. Comparing Eqs. 10.41 and 10.42 and substituting into Eq. 10.38, the standard deviation in energy is found to be

$$\sigma_U = [kT^2(n\bar{C}_v)]^{1/2} \tag{10.43}$$

Example 10.1

Develop an expression for the fractional standard deviation in internal energy of a classical monatomic ideal gas.

For a classical monatomic ideal gas, it was determined in Chapter 6 that

$$\bar{U} = \tfrac{3}{2}NkT$$

and

$$(n\bar{C}_v) = \tfrac{3}{2}Nk$$

Therefore, using Eq. 10.43, we have

$$\frac{\sigma_U}{\bar{U}} = \frac{[kT^2(\frac{3}{2}Nk)]^{1/2}}{\frac{3}{2}NkT} = \left(\frac{2}{3N}\right)^{1/2}$$

We find that the fractional standard deviation in U is inversely dependent upon the number of particles in the system. For a cubic centimeter of gas at room temperature and pressure, the ratio $\sigma_U/\bar{U}$ is of the order of magnitude of 10^{-10}, a quantity much too small to be detected experimentally. For a real substance, the deviation in U is also negligibly small in nearly every region of temperature and pressure. However, in the immediate vicinity of the critical point, the specific heat becomes very large. From the general expression, Eq. 10.43, we see that in such a region fluctuations in U become appreciable, regardless of the size of the system, so that the existence of a classical equilibrium state in this region is not strictly correct but is only a close approximation. Fluctuations in the vicinity of the critical point are discussed again in Section 10.5.

10.4. The Grand Canonical Ensemble

In this section we consider the completely general case, a thermodynamic system of dependent particles that is permitted to exchange mass as well as energy with its surroundings. The procedure is analogous to that followed for the canonical ensemble in Section 10.1, but is, however, much more complex because of the additional generality. We imagine an ensemble comprised of a very large number of distinguishable systems η of the same type as our system of interest, each of volume V. The total energy of the ensemble is fixed as in the previous case, as is the total number of particles. The number of particles per system is a variable in this ensemble, however, since the systems are allowed to exchange mass with one another. This type of ensemble was termed the grand canonical ensemble by Gibbs; it is represented in Figure 10.2.

Permitting N to be a variable complicates the analysis tremendously, since the various system energy states are now functions of the number of particles in the system, as well as of their location and the system boundary parameters. Therefore we must consider the double variable Ni, the ith state of the system when it contains N particles. The energy of this state is then designated U_{Ni}. As the number of particles in a system changes, the system states i change accordingly. Hence there is a complete set of states i for each value of N.

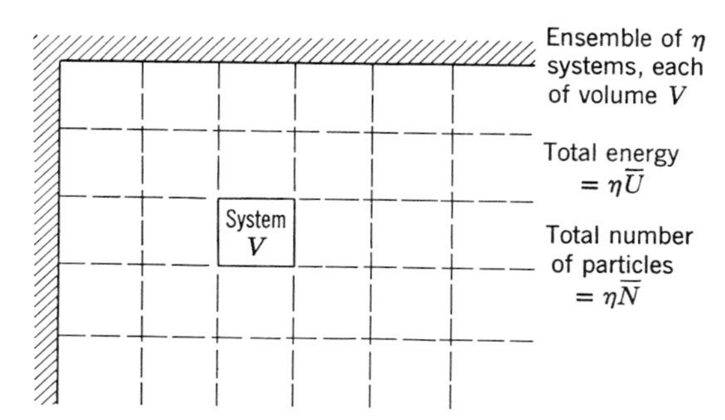

Figure 10.2 Representation of a grand canonical ensemble.

Let us now consider a particular distribution of systems among the various possible states. We use the symbol η_{Ni} to designate the number of systems containing N particles which are found in state i for N particles. Thus, at some instant of time, there is some number of systems η_{11} having one particle and in system state 1 for one particle, some number of systems η_{12} having one particle and in system state 2 for one particle, . . . , some number of systems η_{Ni} having N particles and in system state i for N particles. The thermodynamic probability of such a distribution is given by

$$w = \frac{\eta!}{\prod_N \prod_i \eta_{Ni}!} \tag{10.44}$$

where the product must be taken over all values of N and i.

For the grand canonical ensemble, the possible distributions are subject to three constraints. The total number of systems is fixed, or

$$\eta = \sum_N \sum_i \eta_{Ni} \tag{10.45}$$

Also, the total energy of the ensemble is fixed, or

$$\eta\bar{U} = \sum_N \sum_i \eta_{Ni} U_{Ni} \tag{10.46}$$

where $\bar{U}$ is the average energy per system. Finally, the total number of particles in the ensemble is fixed, or

$$\eta\bar{N} = \sum_N \sum_i \eta_{Ni} N \tag{10.47}$$

where $\bar{N}$ is the average number of particles per system.

In order to determine the most probable distribution, we first write the expression for thermodynamic probability, Eq. 10.44, in logarithmic form and apply Stirling's formula. The result is

$$\ln w = \eta \ln \eta - \eta - \sum_N \sum_i (\eta_{Ni} \ln \eta_{Ni} - \eta_{Ni}) = \eta \ln \eta - \sum_N \sum_i \eta_{Ni} \ln \eta_{Ni} \tag{10.48}$$

Differentiating Eq. 10.48 and setting the result equal to zero for the maximum of w, we have

$$d \ln w = - \sum_N \sum_i (\ln \eta_{Ni}\, d\eta_{Ni} + d\eta_{Ni}) = 0 \tag{10.49}$$

But, from Eq. 10.45,

$$d\eta = \sum_N \sum_i d\eta_{Ni} = 0 \tag{10.50}$$

so that Eq. 10.49 for the maximum of w becomes

$$d \ln w = - \sum_N \sum_i \ln \eta_{Ni}\, d\eta_{Ni} = 0 \tag{10.51}$$

Equation 10.51 specifies the most probable distribution, subject to the given constraints, Eq. 10.50 and also

$$\sum_N \sum_i U_{Ni}\, d\eta_{Ni} = 0 \tag{10.52}$$

and

$$\sum_N \sum_i N\, d\eta_{Ni} = 0 \tag{10.53}$$

Let us now multiply Eq. 10.50 by α, Eq. 10.52 by β, and Eq. 10.53 by $-\ln \lambda$ (for convenience) and add these to the negative of Eq. 10.51. The result is

$$\sum_N \sum_i [\ln \eta_{Ni} + \alpha + \beta U_{Ni} - N \ln \lambda]\, d\eta_{Ni} = 0 \tag{10.54}$$

By the method of undetermined multipliers, the term inside the brackets must be zero for all N and i. Thus, for the most probable distribution,

$$\eta_{Ni} = \lambda^N e^{-\alpha} e^{-\beta U_{Ni}} \quad \text{for all possible } N \text{ and } i \tag{10.55}$$

To eliminate the multiplier α from the distribution equation, Eq. 10.55, we substitute Eq. 10.55 into Eq. 10.45 and obtain the relation

$$\eta = \sum_N \sum_i \eta_{Ni} = e^{-\alpha} \sum_N \sum_i \lambda^N e^{-\beta U_{Ni}} \tag{10.56}$$

If we use this expression to eliminate α, the distribution equation becomes

$$\eta_{Ni} = \frac{\eta}{Z_G} \lambda^N e^{-\beta U_{Ni}} \tag{10.57}$$

where Z_G, which is called the grand partition function, is defined as

$$Z_G = \sum_N \sum_i \lambda^N e^{-\beta U_{Ni}} \tag{10.58}$$

Proceeding as in the development for the canonical ensemble, we now express the average internal energy for a system of the grand canonical ensemble. From Eqs. 10.46 and 10.57, the average internal energy is

$$\bar{U} = \frac{1}{\eta}\sum_N \sum_i \eta_{Ni} U_{Ni} = \frac{1}{Z_G}\sum_N \sum_i \lambda^N U_{Ni} e^{-\beta U_{Ni}}$$

$$= \frac{1}{Z_G}\left[-\left(\frac{\partial Z_G}{\partial \beta}\right)_{V,\lambda}\right] = -\left(\frac{\partial \ln Z_G}{\partial \beta}\right)_{V,\lambda} \qquad (10.59)$$

The entropy is defined by Eq. 10.21. However, for our application, Eq. 10.7,

$$\ln w_{\text{tot}} \approx \ln w_{\text{mp}}$$

is a valid assumption, so that, from Eqs. 10.48 and 10.57, the average entropy is

$$\bar{S} = \frac{1}{\eta} k \ln w_{\text{mp}}$$

$$= \frac{k}{\eta}\left[\eta \ln \eta - \sum_N \sum_i \eta_{Ni_{\text{mp}}} \ln\left(\frac{\eta}{Z_G} \lambda^N e^{-\beta U_{Ni}}\right)\right]$$

$$= \frac{k}{\eta}\left[\eta \ln \eta - \eta \ln \eta + \eta \ln Z_G\right.$$

$$\left. - \ln \lambda \sum_N \sum_i \eta_{Ni_{\text{mp}}} N + \beta \sum_N \sum_i \eta_{Ni_{\text{mp}}} U_{Ni}\right]$$

This reduces to

$$\bar{S} = k \ln Z_G - \bar{N} k \ln \lambda + k\beta \bar{U} \qquad (10.60)$$

Now, substituting the distribution equation, Eq. 10.57, into Eq. 10.47, we have

$$\bar{N} = \frac{1}{\eta}\sum_N \sum_i \eta_{Ni} N = \frac{1}{Z_G}\sum_N \sum_i N \lambda^N e^{-\beta U_{Ni}}$$

$$= \frac{\lambda}{Z_G}\left(\frac{\partial Z_G}{\partial \lambda}\right)_{\beta,V} = \left(\frac{\partial \ln Z_G}{\partial \ln \lambda}\right)_{\beta,V} \qquad (10.61)$$

In order to eliminate the multiplier β, we proceed as in the development for the canonical ensemble. Differentiating Eq. 10.60 with respect to $\bar{U}$ at constant V, N, we obtain

$$\left(\frac{\partial \bar{S}}{\partial \bar{U}}\right)_{V,N} = k\left(\frac{\partial \ln Z_G}{\partial \bar{U}}\right)_{V,N} - \bar{N} k\left(\frac{\partial \ln \lambda}{\partial \bar{U}}\right) + k\beta + k\bar{U}\left(\frac{\partial \beta}{\partial \bar{U}}\right)_{V,N} \qquad (10.62)$$

Substituting Eq. 10.59 for $\bar{U}$ and Eq. 10.61 for $\bar{N}$ into Eq. 10.62 and collecting terms, we find

$$\left(\frac{\partial \bar{S}}{\partial \bar{U}}\right)_{V,N} = k\beta + k\left[\left(\frac{\partial \ln Z_G}{\partial \bar{U}}\right)_{V,N} - \left(\frac{\partial \ln Z_G}{\partial \ln \lambda}\right)_{V,\beta}\left(\frac{\partial \ln \lambda}{\partial \bar{U}}\right)_{V,N} - \left(\frac{\partial \ln Z_G}{\partial \beta}\right)_{V,\lambda}\left(\frac{\partial \beta}{\partial \bar{U}}\right)_{V,N}\right] \tag{10.63}$$

Examination of Eq. 10.63 reveals that the term inside the brackets is identically zero. Thus a comparison of Eqs. 10.63 and 10.25 results once again in the relation

$$\beta = \frac{1}{kT} \tag{10.64}$$

as for the previous development.

Using Eq. 10.64, we can now write the grand partition function in the form

$$Z_G = \sum_N \sum_i \lambda^N e^{-U_{Ni}/kT} \tag{10.65}$$

which is related to the system partition function Z_N of Eq. 10.27 by

$$Z_G = \sum_N \lambda^N\left(\sum_i e^{-U_{Ni}/kT}\right) = \sum_N \lambda^N Z_N \tag{10.66}$$

The average internal energy per system of the grand canonical ensemble given by Eq. 10.59 now can be expressed as

$$\bar{U} = kT^2\left(\frac{\partial \ln Z_G}{\partial T}\right)_{V,\lambda} \tag{10.67}$$

and the average entropy, Eq. 10.60, is

$$\bar{S} = k \ln Z_G - \bar{N}\, k \ln \lambda + \frac{\bar{U}}{T} \tag{10.68}$$

Therefore the Helmholtz function $\bar{A}$ is

$$\bar{A} = \bar{U} - T\bar{S} = \bar{N}\, kT \ln \lambda - kT \ln Z_G \tag{10.69}$$

Also, from the property relation the chemical potential per particle, μ, is

$$\mu = \left(\frac{\partial \bar{A}}{\partial \bar{N}}\right)_{T,V} = kT \ln \lambda + \bar{N}\, kT\left(\frac{\partial \ln \lambda}{\partial \bar{N}}\right)_{T,V} - kT\left(\frac{\partial \ln Z_G}{\partial \bar{N}}\right)_{T,V}$$

But, using Eq. 10.61 for $\bar{N}$, we have

$$\begin{aligned}\mu &= kT \ln \lambda + kT\left(\frac{\partial \ln Z_G}{\partial \ln \lambda}\right)_{T,V}\left(\frac{\partial \ln \lambda}{\partial \bar{N}}\right)_{T,V} - kT\left(\frac{\partial \ln Z_G}{\partial \bar{N}}\right)_{T,V} \\ &= kT \ln \lambda\end{aligned} \tag{10.70}$$

so that the multiplier λ is related to the chemical potential μ by the expression

$$\lambda = e^{\mu/kT} \tag{10.71}$$

As a result of this relation, the variable λ is termed the absolute activity of the substance.

The Gibbs function $\bar{G}$ is expressed as

$$\bar{G} = \bar{N}\mu = \bar{N}\, kT \ln \lambda \tag{10.72}$$

Therefore

$$\bar{P}V = \bar{G} - \bar{A} = kT \ln Z_G \tag{10.73}$$

This is the equation of state in terms of the grand partition function. It is from this expression, Eq. 10.73, that the virial equation of state will be developed in Chapter 11.

10.5. Fluctuations in Density

In this section we use the results and developments of the grand canonical ensemble to determine fluctuations in density about the mean, or equilibrium, value. To do so, we must evaluate the standard deviation in the number of particles in a system of the grand canonical ensemble, which is

$$\sigma_N = [\overline{(N - \bar{N})^2}]^{1/2} = [\overline{N^2} - \bar{N}^2]^{1/2} \tag{10.74}$$

Consider the distribution equation for the grand canonical ensemble, given by Eq. 10.57. Substituting Eq. 10.71 for λ and Eq. 10.64 for β, we have

$$\frac{\eta_{Ni}}{\eta} = \frac{1}{Z_G} e^{N\mu/kT} e^{-U_{Ni}/kT} \tag{10.75}$$

In this section, we are concerned only with the distribution of particles among the systems of the ensemble. Therefore, if the expression given by Eq. 10.75 is summed over all values of i for a given N, the result specifies the distribution of N for any state i. The equation is

$$\frac{\eta_N}{\eta} = \frac{1}{\eta} \sum_i \eta_{Ni} = \frac{1}{Z_G} e^{N\mu/kT} \sum_i e^{-U_{Ni}/kT} = \frac{1}{Z_G} e^{N\mu/kT} Z_N \tag{10.76}$$

The average value of N required in Eq. 10.74 is given by

$$\begin{aligned} \bar{N} &= \frac{1}{\eta} \sum_N \eta_N N = \frac{1}{Z_G} \sum_N N e^{N\mu/kT} Z_N \\ &= \frac{1}{Z_G}\, kT \left(\frac{\partial Z_G}{\partial \mu}\right)_{T,V} \end{aligned} \tag{10.77}$$

Similarly,

$$\overline{N^2} = \frac{1}{\eta} \sum_N \eta_N N^2 = \frac{1}{Z_G} \sum_N N^2 e^{N\mu/kT} Z_N$$
$$= \frac{1}{Z_G} (kT)^2 \left(\frac{\partial^2 Z_G}{\partial \mu^2} \right)_{T,V} \tag{10.78}$$

But, utilizing Eq. 10.77, we can also write this expression as

$$\overline{N^2} = \frac{(kT)^2}{Z_G} \left[\frac{\partial}{\partial \mu} \left(\frac{\bar{N} Z_G}{kT} \right) \right]_{T,V} = \frac{kT}{Z_G} \left[\bar{N} \left(\frac{\partial Z_G}{\partial \mu} \right)_{T,V} + Z_G \left(\frac{\partial \bar{N}}{\partial \mu} \right)_{T,V} \right]$$
$$= (\bar{N})^2 + kT \left(\frac{\partial \bar{N}}{\partial \mu} \right)_{T,V} \tag{10.79}$$

Thus, from Eqs. 10.74 and 10.79 the standard deviation in N is found to be

$$\sigma_N = \left[kT \left(\frac{\partial \bar{N}}{\partial \mu} \right)_{T,V} \right]^{1/2} \tag{10.80}$$

Also of interest is the fractional standard deviation in N,

$$\frac{\sigma_N}{\bar{N}} = \left[\frac{kT}{\bar{N}^2} \left(\frac{\partial \bar{N}}{\partial \mu} \right)_{T,V} \right]^{1/2} \tag{10.81}$$

The expressions in the form of Eqs. 10.80 and 10.81 are not very convenient because of the unfamiliar partial derivative in the equations. This derivative can, however, be expressed in more familiar and useful terms by the methods of classical thermodynamics. For equilibrium values (omitting the bars),

$$\left(\frac{\partial N}{\partial \mu} \right)_{T,V} = \frac{-1}{(\partial V/\partial N)_{T,\mu} (\partial \mu/\partial V)_{T,N}}$$
$$= \frac{-1}{(\partial V/\partial N)_{T,P} (\partial \mu/\partial P)_{T,N} (\partial P/\partial V)_{T,N}} \tag{10.82}$$

Holding T, μ constant in the partial derivative above is, of course, equivalent to requiring T, P to be constant, as all three are intensive properties. Now, from the property relation

$$dG = -S\,dT + V\,dP + \mu\,dN$$

we find

$$\left(\frac{\partial \mu}{\partial P} \right)_{T,N} = \left(\frac{\partial V}{\partial N} \right)_{T,P} \tag{10.83}$$

Substituting Eq. 10.83 into Eq. 10.82, we have

$$\left(\frac{\partial N}{\partial \mu} \right)_{T,V} = \frac{-1}{(\partial V/\partial N)_{T,P}^2 (\partial P/\partial V)_{T,N}} = \frac{-1}{(V/N)^2 (\partial P/\partial V)_{T,N}} \tag{10.84}$$

The last step of Eq. 10.84 follows from Euler's theorem on homogeneous functions, since at constant T, P the volume is a homogeneous function of first degree in the mass.

We therefore find that the fractional standard deviation in the number of particles given by Eq. 10.81 is more conveniently expressed as

$$\frac{\sigma_N}{\bar{N}} = \left[\frac{-kT}{V^2(\partial \bar{P}/\partial V)_{T,N}}\right]^{1/2} \tag{10.85}$$

Example 10.2

Develop an expression for the fractional standard deviation in N for an ideal gas.

For an ideal gas, the equation of state is

$$\bar{P}V = \bar{N}kT$$

Therefore

$$\left(\frac{\partial \bar{P}}{\partial V}\right)_{T,N} = -\frac{\bar{N}kT}{V^2}$$

Substituting into Eq. 10.85, we have

$$\frac{\sigma_N}{\bar{N}} = \left[\frac{-kT}{V^2(-\bar{N}kT/V^2)}\right]^{1/2} = \left(\frac{1}{\bar{N}}\right)^{1/2}$$

We find that for an ideal gas the fractional standard deviation in N is inversely dependent upon the average number of particles in the system, and it is seen to be negligibly small for normal macroscopic systems. Fluctuations would become appreciable only for an extremely rarefied gas. For a real substance at nearly any condition of temperature and pressure, the fluctuation in number or number density is similarly found, by using Eq. 10.85, to be totally negligible. However, near the critical point of a substance the partial derivative

$$\left(\frac{\partial P}{\partial V}\right)_{T,N}$$

approaches zero, so that density fluctuations become appreciable. This causes the phenomenon of critical opalescence, a scattering of light near the critical point making the substance appear cloudy in this region. These large scale fluctuations also make very difficult the measurement of critical density for a substance.

In this chapter, we have considered as special examples the energy fluctuations in a system of a canonical ensemble and density fluctuations in a system of a grand canonical ensemble, both for an ideal gas only. To determine similar fluctuations for real substances from the expressions developed here requires a knowledge of the thermodynamic behavior of the substance. This is the subject of the next chapter.

PROBLEMS

10.1 Calculate the fractional standard deviation in internal energy for a 1 cm^3 volume of carbon dioxide at 0 C and a pressure of

(*a*) 1 atm.

(*b*) 1 micron Hg.

10.2 Consider a system comprised of a monatomic ideal gas represented as a member of a canonical ensemble. For some purposes, it is convenient to group together those system states having a common value of energy, U_i. If this energy is essentially the same value as the mean energy $\bar{U}$, show that the number of states, w_{U_i}, of energy U_i is given by

$$w_{U_i} = \left(\frac{V}{N}\right)^N \left(\frac{4\pi m}{3Nh^2}\right)^{3N/2} e^{5N/2} U_i^{3N/2}$$

10.3 (*a*) Using the result of Problem 10.2, show that the probability for observing a value of energy U_i compared to that for the mean energy $\bar{U}$ is

$$\frac{P(U_i)}{P(\bar{U})} = \frac{w_{U_i}}{w_{\text{mp}}} e^{-(U_i - \bar{U})/kT}$$

(*b*) If $U_i = \bar{U}(1 + x)$, where x is very small, show that the relation of part (*a*) reduces to

$$e^{-3Nx^2/4}$$

Evaluate this ratio for a 1 mole system and $x = 10^{-6}, 10^{-9}, 10^{-12}$.

10.4 Find an expression for the fractional standard deviation in energy, $\sigma_u/(U - U_0)$, for a Debye solid at very low temperature, and calculate the value for 1 gm of carbon at 20 K.

10.5 Show that, by mathematical considerations, the term inside the brackets of Eq. 10.63 is identically zero.

10.6 Calculate the fractional standard deviation in N for a 1 mm^3 volume in a helium system at 15 C and a pressure of

(*a*) 1 atm.

(*b*) 10^{-4} micron Hg.

10.7 The van der Waals equation of state is written as

$$P = \frac{RT}{v - b} - \frac{a}{v^2}$$

where

$$a = \frac{27}{64} \frac{R^2 T_c^2}{P_c}$$

$$b = \frac{v_c}{3} = \frac{RT_c}{8P_c}$$

subscript c denoting a property at the critical point.

(*a*) Develop an expression for the fractional standard deviation for a van der Waals gas in terms of the significant parameters.

(*b*) If $v = v_c$, find $\sigma_N/\bar{N}$ in terms of the reduced temperature $T_r = T/T_c$.

10.8 The isothermal compressibility β_T is defined as

$$\beta_T = -\frac{1}{V}\left(\frac{\partial V}{\partial P}\right)_T$$

(*a*) Develop an expression for $\sigma_N/\bar{N}$ in terms of β_T.

(*b*) For ethyl alcohol, (C_2H_5OH), $\beta_T = 1.12 \times 10^{-10}$ cm^2/dyne at 20 C. Calculate the fractional standard deviation in number density for a 1 cm^3 sample at this temperature.

10.9 Estimate the fractional standard deviation in number density for a 0.1 in.^{3} sample of liquid water at 100 F, 500 lbf/in^2. From the steam tables at 100 F,

$$P_{sat} = 0.9492 \text{ lbf/in.}^2, \quad v_f = 0.016132 \text{ ft}^3\text{/lbm}$$

P, lbf/in.^{2}	$v - v_f$, ft^3/lbm
200	-1.1×10^{-5}
400	-2.1×10^{-5}
600	-3.2×10^{-5}

10.10 In Section 10.2, it was shown that for a system of independent particles the system partition function reduces to the form in expression 10.34, in terms of the single-particle partition function. Beginning with Eq, 10.66, show that for this case the grand partition function Z_G reduces to $e^{Z\lambda}$, resulting in the ideal gas equation of state. Also demonstrate that Eqs. 10.67 and 10.68 reduce to their familiar independent-particle forms.

10.11 A highly sensitive galvanometer consists of a coil suspended in a magnetic field. A lamp and scale are located at a distance of 1 meter from the coil. Light reflected from a mirror fixed to the coil is focused on the scale in such a way that the deflection resulting from a small flow of current through the coil can be detected. One factor limiting the sensitivity of this device is Brownian motion, the random and unequal bombardment of the delicate coil suspension by molecules of air, which results in a random angular displacement θ of the coil. The associated mean potential energy is $\frac{1}{2}K\,\overline{\theta^2}$, also equal to one degree of freedom of thermal energy $\frac{1}{2}kT$. If the suspension torsion constant $K = 10^{-4}$ dyne-cm/rad, what is the nature of the corresponding fluctuation in light-spot location on the scale at 25 C?

10.12 One factor limiting the sensitivity in electronic detectors and amplifiers is Johnson, or thermal, noise. Because of thermal agitation velocity, the free electrons in a conductor tend to move in random directions. Thus, instantaneous fluctuations result in a noise current and, consequently, a noise voltage. It can be shown that, in a frequency interval ν to $\nu + d\nu$, the mean-square thermal noise EMF is

$$d(\overline{\Delta\mathscr{E}^2}) = 4\mathscr{R}kT\,d\nu$$

where $\mathscr{R}$ is the resistance. Calculate the mean-square thermal noise EMF at 300 K and at 4.2 K in the frequency range 0 to 10^3 sec^{-1} for a resistance of 1000 ohms.

10.13 Show that the fluctuation σ_U in internal energy for a grand canonical ensemble can be expressed as

$$\sigma_U{}^2 = (kT^2 n\, \bar{C}_v) + \left(\frac{\partial \bar{U}}{\partial \bar{N}}\right)_{T,V} \sigma_{N^2}$$

where σ_N is the standard deviation in number density, given by Eq. 10.80.

10.14 Carry out the development of the grand canonical ensemble for a mixture of two substances, A and B. Show that the grand partition function in this case is given by

$$Z_{G_{AB}} = \sum_{N_A} \sum_{N_B} \lambda_A^{N_A} \lambda_B^{N_B} Z_{N_{AB}}$$

where

$$\lambda_A = e^{\mu_A/kT}, \quad \lambda_B = e^{\mu_B/kT}$$

and $Z_{N_{AB}}$ is the system partition function for a given set of N_A and N_B,

$$Z_{N_{AB}} = \sum_i e^{-U_{N_A N_B i}/kT}$$

Show also that the equivalent of Eq. 10.61 for a mixture is the pair of expressions

$$\bar{N}_A = \lambda_A \left(\frac{\partial \ln Z_{G_{AB}}}{\partial \lambda_A}\right)_{T,V,\lambda_B}, \quad \bar{N}_B = \lambda_B \left(\frac{\partial \ln Z_{G_{AB}}}{\partial \lambda_B}\right)_{T,V,\lambda_A}$$

The relations discussed here will be basic to the development, in Section 11.4, of the equation of state for a mixture.

10.15 Using the results of Problem 10.14 and the relation for $Z_{N_{AB}}$ found in Problem 4.14,

$$Z_{N_{AB}} = \left(\frac{Z_A^{N_A}}{N_A!}\right)\left(\frac{Z_B^{N_B}}{N_B!}\right)$$

show that the grand canonical ensemble equations reduce to the familiar ideal gas mixture relations.

11 The Behavior of Real Gases and Liquids

In this chapter, we shall utilize the expressions for the canonical and grand canonical ensembles, which were developed in Chapter 10, to examine the behavior of real substances. In particular, we shall develop the theoretical equation of state for a substance in terms of the intermolecular potential energy, and also discuss several models for the potential function that have been used to correlate experimental data.

11.1. The Virial Equation of State

An equation of state for a substance is the functional relation between the properties pressure, specific volume, and temperature. The availability of such an expression is of great importance, since, by using the methods of classical thermodynamics, other thermodynamic properties of interest can be evaluated for a substance in terms of the equation of state and the specific heat. Most of these equations presently in use are empirical in form; that is, they are mathematical expressions that are so fitted as to best represent experimental P-V-T data.

The equation of state can, however, be derived on a theoretical basis from the partition function, either the system partition function given by Eq. 10.27, expressing P-V-T behavior according to Eq. 10.31, or in terms of the grand partition function, Eq. 10.66, which expresses P-V-T behavior according to Eq. 10.73. The latter is the more convenient to use, because it does not require taking a partial derivative of $\ln Z_G$ with respect to volume.

We therefore utilize Eq. 10.73 as the basic relation

$$PV = kT \ln Z_G$$

where the grand partition function Z_G is defined by Eq. 10.66,

$$Z_G = \sum_N \lambda^N Z_N$$

The absolute activity λ is related to the chemical potential μ by Eq. 10.71,

$$\lambda = e^{\mu/kT}$$

and the system partition function Z_N is expressed by Eq. 10.27 as

$$Z_N = \sum_i e^{-U_i/kT}$$

The system state energy U_i includes intermolecular potential energy due to the configuration and possibly the relative orientations of the particles constituting the system (although the latter will not be discussed here) in addition to the various single-particle energy modes discussed earlier. Our task, then, is to develop Eq. 10.73 to obtain a relation between P, V, and T in terms of the intermolecular potential energy.

Let us first substitute Eq. 10.66 into Eq. 10.73 and expand. The result is

$$\frac{PV}{kT} = \ln Z_G = \ln\left[\sum_N \lambda^N Z_N\right] = \ln\left[1 + \sum_{N\geq 1} \lambda^N Z_N\right] \tag{11.1}$$

The last step in this relation follows from the fact that, for $N = 0$, $U = 0$ and therefore $Z_0 = 1$. Using the logarithmic expansion, $\ln(1 + x) = x - x^2/2 + x^3/3 - \cdots$ in Eq. 11.1, we have

$$\frac{PV}{kT} = \ln Z_G = \lambda Z_1 + \lambda^2(Z_2 - \tfrac{1}{2}Z_1^2) + \lambda^3(Z_3 - Z_1Z_2 + \tfrac{1}{3}Z_1^3) + \cdots \tag{11.2}$$

We note that Z_1 is the ordinary single-particle partition function Z used in the analysis of systems of independent particles. Equation 11.2 reduces to the first term, λZ_1, as λ becomes very small. For convenience, let us now define a variable $\mathfrak{z}$, which is termed the active number density, as

$$\mathfrak{z} = \frac{\lambda Z_1}{V} \tag{11.3}$$

which from Eq. 11.2 is seen to reduce to the number density N/V or P/kT for very small λ (ideal gas). This is also noted from an evaluation of Eq. 11.3 for an ideal gas:

$$\ln \mathfrak{z} = \ln \lambda + \ln \frac{Z_1}{V} = \frac{G}{NkT} + \ln \frac{Z_1}{V}$$

$$= -\ln \frac{Z_1}{N} + \ln \frac{Z_1}{V} = \ln \frac{N}{V}$$

Let us now express Eq. 11.1 in terms of the active number density $\mathfrak{z}$. This relation becomes

$$\frac{PV}{kT} = \ln Z_G = \ln\left[1 + \sum_{N\geq 1}\left(\frac{Z_N V^N}{Z_1^{\,N}}\right)\mathfrak{z}^N\right] \tag{11.4}$$

In order to develop the last expression in terms of the configurational or intermolecular potential energy, we note that, when considering any number of particles, N, greater than 1, the energy U_i in the summation for Z_N can be separated into two parts:

$$U_i = U_{i(\text{ind part})} + \Phi(\vec{r}_1, \vec{r}_2, \ldots, \vec{r}_N) \tag{11.5}$$

in which the first term is the system energy for independent particles, and $\Phi(\vec{r}_1, \vec{r}_2, \ldots, \vec{r}_N)$ is the configurational energy, a function of the position of each of the N particles in the system. We do not consider the possibility of energy resulting from the relative orientations of the particles; it is reasonable to neglect this contribution for fairly light non-polar molecules. In addition, we also assume that quantum effects are negligible, and that we can consider Φ to be a continuous function of the particle positions. Therefore, substituting Eq. 11.5 into Eq. 10.27, utilizing Eq. 10.34 for the independent particle summation, and integrating the continuous function Φ over the limits of the container, we obtain

$$\begin{aligned} Z_N &= \sum_{\text{all states}} \exp\left[\frac{-U_{i(\text{ind part})}}{kT}\right] \exp\left[\frac{-\Phi(\vec{r}_1, \vec{r}_2, \ldots, \vec{r}_N)}{kT}\right] \\ &= \left(\frac{Z_1^{\,N}}{N!}\right)\frac{1}{V^N}\iint\cdots\int \exp\left[\frac{-\Phi(\vec{r}_1, \vec{r}_2, \ldots, \vec{r}_N)}{kT}\right] d\vec{r}_1\, d\vec{r}_2 \cdots d\vec{r}_N \end{aligned} \tag{11.6}$$

in which

$$d\vec{r}_1 = dx_1\, dy_1\, dz_1, \ldots$$

This relation is more conveniently written in the form

$$Z_N = \left(\frac{Z_1^{\,N}}{N!}\right)\left(\frac{Q_N}{V^N}\right) \tag{11.7}$$

where the quantities Q_N for the various values of N, termed the configurational integrals, are defined by

$$Q_N = \iint\cdots\int \exp\left[\frac{-\Phi(\vec{r}_1, \vec{r}_2, \ldots, \vec{r}_N)}{kT}\right] d\vec{r}_1\, d\vec{r}_2 \cdots d\vec{r}_N \tag{11.8}$$

It should be pointed out that, in those cases where quantum effects are important, the quantity Q_N is defined by Eq. 11.7 for each N but is not expressed by the integral of Eq. 11.8.

We now substitute Eq. 11.7 into Eq. 11.4 and, again using the logarithmic expansion, obtain the relation

$$\frac{PV}{kT} = \ln Z_G = \ln\left[1 + \sum_{N\geq 1}\frac{Q_N}{N!}\mathfrak{z}^N\right] = \sum_{l\geq 1} V b_l \mathfrak{z}^l \tag{11.9}$$

The coefficients b_l, which are functions of the configurational integrals, are termed cluster integrals, the first few of which are seen to be

$$
\begin{aligned}
b_1 &= \frac{Q_1}{V} = 1 \\
b_2 &= \frac{Q_2 - Q_1^2}{2!V} \\
b_3 &= \frac{Q_3 - 3Q_1Q_2 + 2Q_1^3}{3!V} \\
&\cdots\cdots\cdots\cdots\cdots\cdots
\end{aligned}
\tag{11.10}
$$

The equation of state has now been developed to the form of Eq. 11.9, in which the b_l are functions of the Q_N, which in turn are functions of the intermolecular energy. In order to complete the development, we must now find an expression for the active number density $\mathfrak{z}$ in terms of the P-V-T behavior of the substance. In the development of the grand canonical ensemble, it was found that the average number of particles per system is given by Eq. 10.61 (omitting the bar in $\bar{N}$)

$$N = \lambda \left(\frac{\partial \ln Z_G}{\partial \lambda} \right)_{T,V}$$

However, differentiating Eq. 11.3 at constant T, V, we obtain

$$d\mathfrak{z}_{T,V} = \left(\frac{Z_1}{V} \right) d\lambda_{T,V} = \frac{\mathfrak{z}}{\lambda}\, d\lambda_{T,V} \tag{11.11}$$

and, substituting into Eq. 10.61, we find

$$N = \mathfrak{z} \left(\frac{\partial \ln Z_G}{\partial \mathfrak{z}} \right)_{T,V} = \mathfrak{z} \left[\frac{\partial (PV/kT)}{\partial \mathfrak{z}} \right]_{T,V} \tag{11.12}$$

If we substitute Eq. 11.9 into Eq. 11.12 and carry out the differentiation, the result is

$$N = \sum_{l \geq 1} V l b_l \mathfrak{z}^l \tag{11.13}$$

or

$$\frac{N}{V} = b_1 \mathfrak{z} + 2b_2 \mathfrak{z}^2 + 3b_3 \mathfrak{z}^3 + \cdots \tag{11.14}$$

which is a series for the number density N/V in terms of the active number density $\mathfrak{z}$. It is noted that at small $\mathfrak{z}$ the number density reduces to $\mathfrak{z}$ in agreement with its definition. For our purpose, however, we need the inverse of the series, Eq. 11.14; that is, $\mathfrak{z}$ as a function of N/V. This expression is written as

$$\mathscr{z} = a_1\left(\frac{N}{V}\right) + a_2\left(\frac{N}{V}\right)^2 + a_3\left(\frac{N}{V}\right)^3 + \cdots \tag{11.15}$$

where the relations between the unknown coefficients $a_1, a_2, \ldots$, and the cluster integrals $b_1, b_2, \ldots$, must be found. If we substitute Eq. 11.15 for $\mathscr{z}$ into Eq. 11.14 and collect terms, the result is an identity in N/V, from which the desired relations can be determined. This identity is found to be

$$\frac{N}{V} = a_1 b_1\left(\frac{N}{V}\right) + (a_2 b_1 + 2a_1^2 b_2)\left(\frac{N}{V}\right)^2 + (a_3 b_1 + 4a_1 a_2 b_2 + 3a_1^3 b_3)\left(\frac{N}{V}\right)^3 + \cdots \tag{11.16}$$

from which we conclude that

$$a_1 b_1 = 1$$
$$a_2 b_1 + 2a_1^2 b_2 = 0$$
$$a_3 b_1 + 4a_1 a_2 b_2 + 3a_1^3 b_3 = 0$$
$$\cdots\cdots\cdots\cdots\cdots$$

Therefore, using $b_1 = 1$ from Eq. 11.10, we have

$$\begin{aligned} a_1 &= \frac{1}{b_1} = 1 \\ a_2 &= -2b_2 \\ a_3 &= 8b_2^2 - 3b_3 \\ &\cdots\cdots\cdots \end{aligned} \tag{11.17}$$

so that the desired series, Eq. 11.15, can be written in the form

$$\mathscr{z} = \left(\frac{N}{V}\right) - 2b_2\left(\frac{N}{V}\right)^2 + (8b_2^2 - 3b_3)\left(\frac{N}{V}\right)^3 + \cdots \tag{11.18}$$

The development of the equation of state is now complete. We substitute Eq. 11.18 for $\mathscr{z}$ into Eq. 11.9 and collect terms; after we divide by N, the result is

$$\frac{PV}{NkT} = 1 + (-b_2)\left(\frac{N}{V}\right) + (4b_2^2 - 3b_3)\left(\frac{N}{V}\right)^2 + \cdots \tag{11.19}$$

But

$$\frac{N}{V} = \frac{N/n}{V/n} = \frac{N_0}{v}$$

Therefore, in terms of the molal specific volume v, the equation of state is

$$\frac{Pv}{RT} = 1 + \frac{B(T)}{v} + \frac{C(T)}{v^2} + \cdots \tag{11.20}$$

This form of the equation of state is termed the virial form, which is an infinite series for the compressibility factor Pv/RT in terms of the reciprocal volume. The temperature-dependent coefficient $B(T)$ is called the second virial coefficient and, by comparison with Eq. 11.19, is given in terms of the cluster integral b_2 as

$$B(T) = -b_2 N_0 \tag{11.21}$$

We recall from Eq. 11.10 that b_2 is a result only of binary interactions, in terms of the configurational integral Q_2. At moderately low density, only the second virial coefficient term contributes significantly to the non-ideality of the gas, and we find that this is a consequence of two-particle interactions only.

Similarly, $C(T)$ is termed the third virial coefficient, which, by comparison of Eqs. 11.19 and 11.20, is seen to be

$$C(T) = (4b_2^2 - 2b_3)N_0^2 \tag{11.22}$$

We find from Eqs. 11.10 that the third virial coefficient is a consequence of binary and tertiary interactions on the molecular level, in terms of the configurational integrals Q_2 and Q_3. Thus, whereas at relatively low density only two-particle interactions are of importance in explaining departure from ideal gas behavior [the $B(T)/v$ term], as density is increased three-particle interactions also become significant [the $C(T)/v^2$ term in the equation of state].

Had more terms been carried in the series expansions throughout our development, we could also have obtained expressions for higher virial coefficients, and we could, at least in principle, express these temperature-dependent parameters as functions of higher order molecular interactions. Thus Eq. 11.20 is a quite general equation of state for gases, for which the various virial coefficients are represented in terms of the intermolecular potential energy. The problem of expressing and evaluating the potential energy Φ becomes at least partially empirical and depends upon the selection of an analytical model to represent intermolecular forces.

11.2. The Virial Coefficients

In this section we further develop the equations for the second and third virial coefficients (for an angle-independent potential) as defined by Eqs. 11.21 and 11.22, respectively. In those expressions, the virial coefficients are given as functions of the cluster integrals, which in turn are defined as combinations of the configurational integrals by the set of Eqs. 11.10. The definition of the configurational integrals is given by Eq. 11.7, and where quantum effects can be neglected these are expressed according to Eq. 11.8.

In order to evaluate the configurational integrals for various values of N, it is necessary to integrate Eq. 11.8, which contains the intermolecular potential energy $\Phi(\vec{r}_1, \vec{r}_2, \ldots, \vec{r}_N)$ for the number of particles under consideration. In order to utilize a suitable analytical model for the potential function between pairs of particles, we assume that the various pair potentials are additive and depend only upon the distance of separation. That is,

$$\Phi(\vec{r}_1, \vec{r}_2, \ldots, \vec{r}_N) = \sum_{i \geq 1}^{N-1} \sum_{j>1}^{N} \varphi(r_{ij}) \tag{11.23}$$

in which $\varphi(r_{ij})$ is the potential between a pair of molecules i and j. The first few configurational integrals are found to be, from Eq. 11.8,

$$\begin{aligned} Q_1 &= \int_V d\vec{r}_1 = V \\ Q_2 &= \iint\limits_V \exp\left[-\varphi(r_{12})/kT\right] d\vec{r}_1\, d\vec{r}_2 \\ Q_3 &= \iiint\limits_V \exp\left\{-[\varphi(r_{12}) + \varphi(r_{13}) + \varphi(r_{23})]/kT\right\} d\vec{r}_1\, d\vec{r}_2\, d\vec{r}_3 \end{aligned} \tag{11.24}$$

The second virial coefficient $B(T)$ was defined in terms of the cluster integral b_2 by Eq. 11.21. For b_2, from Eq. 11.10,

$$\begin{aligned} b_2 &= \frac{1}{2V}(Q_2 - Q_1^2) = \frac{1}{2V} \iint\limits_V \left[e^{-\varphi(r_{12})/kT} - 1\right] d\vec{r}_1\, d\vec{r}_2 \\ &= \frac{1}{2V} \iint\limits_V f_{12}\, d\vec{r}_1\, d\vec{r}_2 \end{aligned} \tag{11.25}$$

in which the f function has been defined for convenience by the relation

$$f_{ij} = e^{-\varphi(r_{ij})/kT} - 1 \tag{11.26}$$

Since the potential energy $\varphi(r_{ij})$ for a pair of molecules approaches zero (and consequently $f_{ij} \to 0$ also) very rapidly as the distance of separation, r_{ij}, increases, it is appropriate to change variables from $\vec{r}_1, \vec{r}_2$, to $\vec{r}_1$ and

$$\vec{r}_{12} = \vec{r}_2 - \vec{r}_1 \tag{11.27}$$

Using Eq. 11.27, we find that Eq. 11.25 becomes

$$b_2 = \frac{1}{2V}\iint\limits_V f_{12}\, d\vec{r}_1\, d\vec{r}_{12} = \frac{1}{2V}\int_V d\vec{r}_1 \int_V f_{12}\, d\vec{r}_{12}$$

$$= \frac{1}{2}\int_V f_{12}\, d\vec{r}_{12} \tag{11.28}$$

The separation of the integrals is reasonable inasmuch as f_{12} is independent of the location of the volume element $d\vec{r}_1$ (except for a small region very close to the wall of the container, which is a negligible portion of the total volume V). Therefore the resulting integral in Eq. 11.28 depends only upon the distance of separation, r, and can be written as

$$b_2 = \frac{1}{2}\int_V f_r 4\pi r^2\, dr = 2\pi \int_0^\infty f_r r^2\, dr \tag{11.29}$$

The integrand of Eq. 11.29 approaches zero rapidly for large r. Thus no error is introduced by integrating r over the convenient limits 0 to ∞ instead of over the dimensions of the container. If Eq. 11.29 is substituted into Eq. 11.21, the second virial coefficient becomes

$$B(T) = -2\pi N_0 \int_0^\infty f_r r^2\, dr \tag{11.30}$$

which can be integrated and evaluated if a suitable model for the pair potential $\varphi(r)$ is available.

The third virial coefficient $C(T)$, defined by Eq. 11.22, requires evaluation of the cluster integral b_3, which from Eq. 11.10 and the expressions for the configurational integrals, Eqs. 11.24, is

$$b_3 = \frac{1}{6V}[Q_3 - 3Q_1Q_2 + 2Q_1^3]$$

$$= \frac{1}{6V}\iiint\limits_V [e^{-\varphi(r_{12})/kT} e^{-\varphi(r_{13})/kT} e^{-\varphi(r_{23})/kT} - e^{-\varphi(r_{12})/kT}$$

$$- e^{-\varphi(r_{13})/kT} - e^{-\varphi(r_{23})/kT} + 2]\, d\vec{r}_1\, d\vec{r}_2\, d\vec{r}_3 \tag{11.31}$$

However, from the definition of the f functions (Eq. 11.26), this can alternatively be expressed as

$$b_3 = \frac{1}{6V}\iiint\limits_V [f_{12}f_{13}f_{23} + f_{12}f_{13} + f_{12}f_{23} + f_{13}f_{23}]\, d\vec{r}_1\, d\vec{r}_2\, d\vec{r}_3 \tag{11.32}$$

To obtain the expression required for b_2^2 in the third virial coefficient equation, we note that, from Eq. 11.28,

$$\iiint_V f_{12}f_{13}\, d\vec{r}_1\, d\vec{r}_2\, d\vec{r}_3 = \left[\int_V d\vec{r}_1\right]\left[\int_V f_{12}\, d\vec{r}_{12}\right]\left[\int_V f_{13}\, d\vec{r}_{13}\right]$$

$$= [V][2b_2][2b_2] = 4Vb_2^2 \qquad (11.33)$$

Of course, similar expressions result from the use of the pairs 12 and 23 or 13 and 23 as indices.

Using Eq. 11.32 and three equations of the form of Eq. 11.33, we find that the third virial coefficient defined by Eq. 11.22 becomes

$$C(T) = -\frac{N_0^2}{3V}[6Vb_3 - 3(4Vb_2^2)] = -\frac{N_0^2}{3V}\iiint_V f_{12}f_{13}f_{23}\, d\vec{r}_1\, d\vec{r}_2\, d\vec{r}_3 \qquad (11.34)$$

which is found to be a considerably more difficult expression to integrate in terms of a pair potential function $\varphi(r_{ij})$ than is the corresponding equation (Eq. 11.30) for the second virial coefficient.

11.3. Intermolecular Potential Functions and Evaluation of the Virial Coefficients

Under the assumptions of an angle-independent intermolecular potential function between molecules and the additivity of forces, the second and third virial coefficients have been found in the preceding section to be given by Eqs. 11.30 and 11.34, respectively. To integrate these equations and thereby calculate the virial coefficients, we must have an analytical model for the potential energy between pairs of molecules. The form for the potential function can be deduced at least in part from quantum mechanics, but a quantitative representation remains at least partially empirical. Thus we require experimental values for the virial coefficients in order to calculate potential function constants and, in turn, to evaluate proposed potential functions and to correlate and predict other thermodynamic and transport properties.

In order to determine virial coefficients from experimental data, we first rearrange the virial equation of state to the form

$$\left(\frac{Pv}{RT} - 1\right)v = B(T) + C(T)\frac{1}{v} + \cdots$$

If at any given temperature, low density experimental data (P vs. v) are plotted as

$$\left(\frac{Pv}{RT} - 1\right)v \text{ vs. } \frac{1}{v}$$

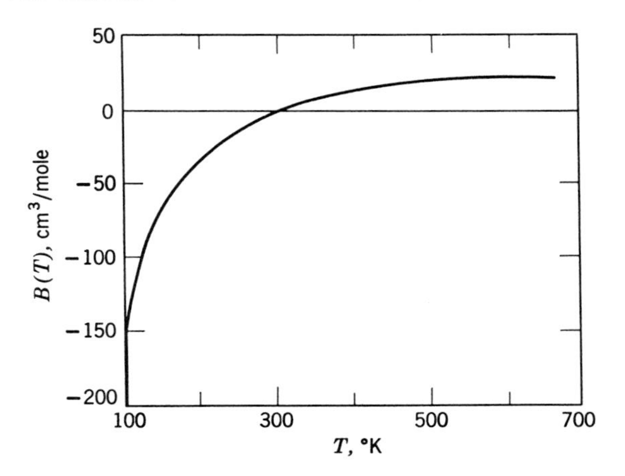

Figure 11.1 The second virial coefficient for nitrogen.

then the second virial coefficient B at this temperature is found as the intercept as $(1/v) \to 0$ ($1/v \to 0$ as the pressure approaches zero). Similarly, the third virial coefficient C at this temperature is found as the slope of the curve as $(1/v) \to 0$. Obviously, very precise low density P-V-T data are required in order to evaluate the virial coefficients by this procedure. Such data are available for a number of substances over fairly extensive ranges of temperature, and from these data accurate virial coefficients have been determined. As an example, the second virial coefficient for nitrogen is shown in Figure 11.1 as a function of temperature. From the virial equation, it is seen that the $B(T)/v$ term is the first-order contribution to non-ideal behavior of a substance. At low temperature, where B is negative, the compressibility factor Pv/RT is therefore less than unity at low to moderate densities. On the other hand, at higher temperature B becomes positive,

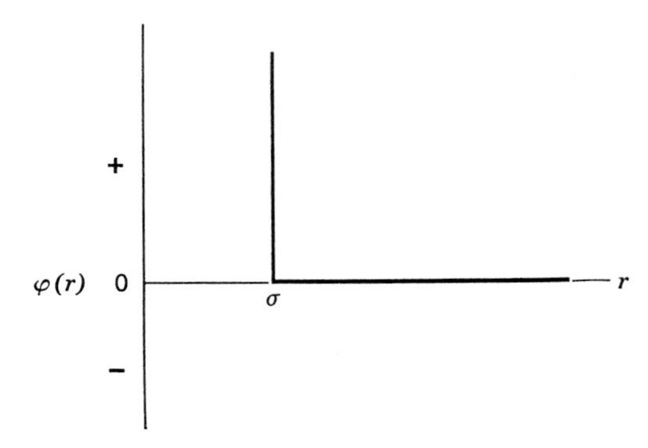

Figure 11.2 The rigid spheres potential.

and consequently Pv/RT becomes greater than unity. At very high temperature, beyond the range of Figure 11.1, B approaches zero and the gas behaves as an ideal gas at even moderately high density. When an analytical model for the intermolecular potential is used to evaluate $B(T)$ from Eq. 11.30, the second virial coefficient temperature dependency should be as shown in Figure 11.1. Otherwise, the model cannot be suitable for subsequent accurate calculations over a significant range of temperature.

Although numerous analytical potential functions have been proposed and utilized in thermodynamic calculations, only a few will be discussed here. The most elementary model is that of the rigid spheres, the potential of which is shown in Figure 11.2. The analytical representation of this model is

$$\begin{aligned} \varphi(r) &= \infty, \quad r < \sigma \\ \varphi(r) &= 0, \quad r > \sigma \end{aligned} \tag{11.35}$$

in which the constant σ is termed the collision diameter. This model gives a simplified representation of the strong forces of repulsion between molecules at small distances of separation. However, it makes no attempt to consider the longer range forces of attraction between molecules, and for this reason it cannot be considered a realistic potential function. The sole advantage of this potential is its simplicity. The second virial coefficient is easily evaluated from Eq. 11.30, and it is found to be

$$B = -2\pi N_0\left[\int_0^\sigma (-1)r^2\,dr + \int_\sigma^\infty (0)r^2\,dr\right] = \tfrac{2}{3}\pi N_0\sigma^3 = b_0 \tag{11.36}$$

The resulting second virial coefficient is found to be a positive constant, which is not at all in agreement with the behavior shown in Figure 11.1.

By a similar but more difficult integration of Eq. 11.34, the third virial coefficient for the rigid spheres potential is found to be

$$C = \tfrac{5}{8}b_0^2 \tag{11.37}$$

which is also a positive constant. Thus the compressibility factor Pv/RT is always greater than unity for this model; hence the rigid spheres potential can be reasonable (even qualitatively) only at high temperature. This model has frequently been used for preliminary calculations in the high temperature region when more accurate information is not available.

A somewhat more realistic intermolecular potential is the point centers of repulsion model, shown in Figure 11.3. This model is represented analytically as

$$\varphi(r) = \beta r^{-\delta} \tag{11.38}$$

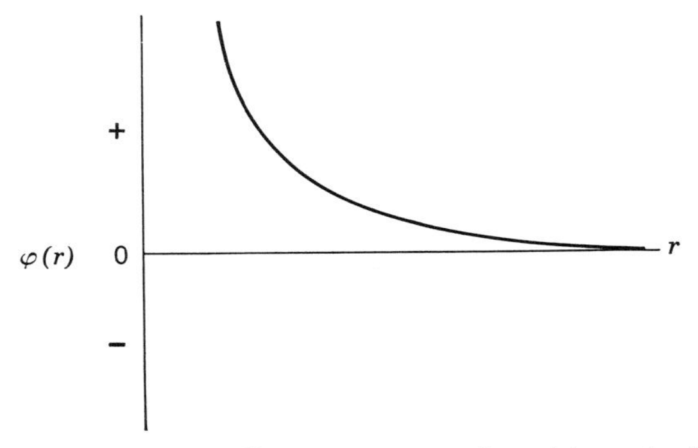

Figure 11.3 The point centers of repulsion potential.

where δ is the index of repulsion, usually in the range from 9 to 15, and β is a constant. This representation of the forces of repulsion between molecules is quite reasonable, but, as with the previous model, there is no consideration of the important longer range forces of attraction. Upon substitution of Eq. 11.38 into Eq. 11.30 and integrating, we find that, for $\delta > 3$, the result is given in terms of the gamma function as

$$B(T) = \tfrac{2}{3}\pi N_0 \left(\frac{\beta}{kT}\right)^{3/\delta} \Gamma\left(\frac{\delta - 3}{\delta}\right) \tag{11.39}$$

which is temperature-dependent. This result, although a significant improvement over that for the rigid spheres potential, is suitable only for high temperature computations and is not reasonable at low temperature, where attractive forces are dominant.

The Lennard-Jones potential, perhaps the best known of the many intermolecular potential functions, includes a repulsive term of the form used

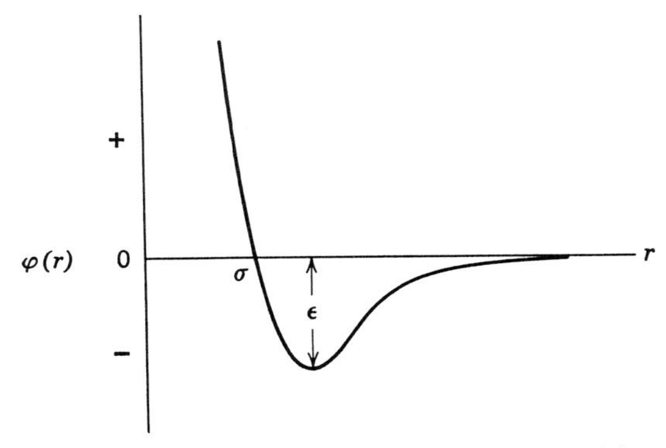

Figure 11.4 The Lennard-Jones potential.

in the point centers of repulsion model and also has a term to account for forces of attraction. This potential function is represented in Figure 11.4; in its more general form it is expressed analytically as

$$\varphi(r) = \beta r^{-\delta} - \alpha r^{-\gamma} \tag{11.40}$$

where $\delta > \gamma$, and both are greater than one. The first term in this equation accounts for repulsion and, since $\delta > \gamma$, this term dominates at small r. The second term is the potential resulting from forces of attraction and dominates at large r. Thus there is an attractive "potential well" as seen in Figure 11.4, the potential approaching zero at large distances of separation and becoming very large for molecules in close proximity to one another, as a result of the strong short range repulsive forces. This behavior is qualitatively very reasonable.

It has been shown from quantum mechanical calculations regarding the forces of attraction that the index of attraction γ should be taken as 6. This value has also been found to give good agreement with experimental data. Furthermore, if we use the mathematical requirement for the minimum point in the potential well, at which distance

$$\left[\frac{\partial \varphi(r)}{\partial r}\right] = 0 \quad \text{and} \quad \varphi(r) = -\epsilon$$

and also the fact that $\varphi(r) = 0$ at $r = \sigma$, then the constants α and β of Eq. 11.40 can be eliminated in terms of ϵ and σ. The resulting expression, using the theoretical value $\gamma = 6$, is found to be

$$\varphi(r) = \frac{\epsilon}{\delta - 6}\left(\frac{\delta^{\delta}}{6^{6}}\right)^{1/(\delta-6)}\left[\left(\frac{\sigma}{r}\right)^{\delta} - \left(\frac{\sigma}{r}\right)^{6}\right] \tag{11.41}$$

where $\delta > 6$.

It has not been possible to predict the index of repulsion, δ, from quantum mechanical calculations. The value $\delta = 12$ is normally used with this potential, as it results in the best correlation with experimental data for most substances. In this form the potential is called the Lennard-Jones (6–12) potential, which from Eq. 11.41 with $\delta = 12$ is

$$\varphi(r) = 4\epsilon\left[\left(\frac{\sigma}{r}\right)^{12} - \left(\frac{\sigma}{r}\right)^{6}\right] \tag{11.42}$$

To evaluate $B(T)$ for the Lennard-Jones (6–12) potential, we use the dimensionless or reduced variables

$$r^* = \frac{r}{\sigma}$$
$$T^* = \frac{kT}{\epsilon} \tag{11.43}$$

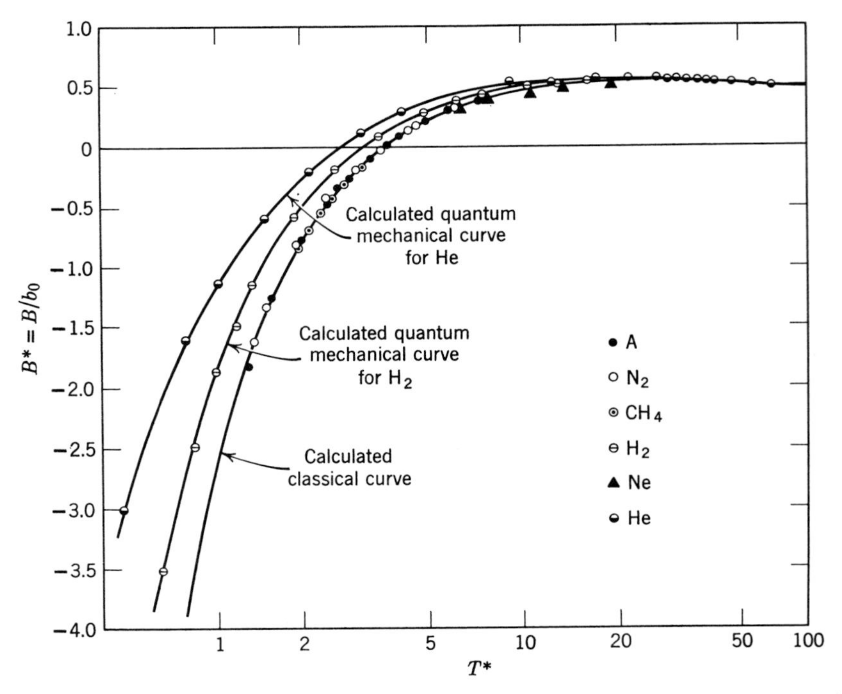

Figure 11.5 The reduced second virial coefficient for the Lennard-Jones potential. (From J. O. Hirschfelder, C. F. Curtiss, and R. B. Bird, *Molecular Theory of Gases and Liquids*, John Wiley and Sons, New York, second printing, corrected, with notes added, March, 1964.

in terms of which Eq. 11.42 becomes

$$\frac{\varphi(r)}{kT} = \frac{4}{T^*}\left[r^{*-12} - r^{*-6}\right] \tag{11.44}$$

Substituting Eq. 11.44 into Eq. 11.30, we find that

$$B(T) = b_0 B^*(T^*) \tag{11.45}$$

where

$$b_0 = \tfrac{2}{3}\pi N_0 \sigma^3 \tag{11.46}$$

the same as the result for the rigid spheres potential, while the reduced second virial coefficient $B^*(T^*)$ is

$$B^*(T^*) = 3\int_0^\infty \left\{1 - \exp\left[-\frac{4}{T^*}\left(r^{*-12} - r^{*-6}\right)\right]\right\} r^{*2}\, dr^* \tag{11.47}$$

If the exponential in the integrand of Eq. 11.47 is expanded in a series, the result can then be integrated term by term. The final result is a series expression for B^* as a function only of T^*, values of which are tabulated in the Appendix, Table A.7. Determination of the second virial coefficient $B(T)$ at any temperature for a particular substance then requires suitable values for the constants ϵ and σ for that substance. These constants are determined for each substance by trial and error, to give the best representation of experimentally determined second virial coefficients. The Lennard-Jones force constants ϵ and σ are given for a number of substances in Table A.8.

Figure 11.5 shows a plot of the reduced second virial coefficient B^* as a function of reduced temperature T^*; it indicates the degree of accuracy with which this potential function represents the experimental data for most simple substances. It is noted, however, that the values for helium and hydrogen, both of which exhibit large quantum effects, cannot be accurately correlated by the result of Eq. 11.47, which assumes classical behavior. For these substances and also for neon, the configurational integrals Q_N are defined according to Eq. 11.7 and are not correctly given in integral form by Eq. 11.8, so that the second virial coefficient for these substances must include the appropriate quantum corrections.

The third virial coefficient for the Lennard-Jones (6–12) potential is evaluated in a manner similar to that presented for $B(T)$. The result is an expression of the form

$$C(T) = b_0^2 C^*(T^*) \tag{11.48}$$

with b_0 defined by Eq. 11.46. The reduced third virial coefficient $C^*(T^*)$ is, however, a complicated series expression in which the coefficient of each term includes a double integral that can be only approximately determined by numerical integration. The resulting values of C^* are tabulated as a function of T^* in the Appendix, Table A.7, and are shown in Figure 11.6. In this figure, $C^*(T^*)$ is compared with experimental values found from low density P-V-T data as discussed previously. The agreement is not of the accuracy found for the second virial coefficient; nevertheless it is seen to be quite reasonable.

Many other potential functions have been used to represent virial coefficient behavior, some of which are more complex than those considered here. Considerations made in some of these potentials include exponential repulsive terms, models of molecules with hard cores surrounded by softer penetrable cores, non-spherical molecules with angle-dependent potential functions, etc., for which the mathematical development and computations become quite involved.†

† The reader is referred to the very complete reference, *Molecular Theory of Gases and Liquids*, by J. O. Hirschfelder, C. F. Curtiss, and R. B. Bird (John Wiley and Sons, New York, 1964) for an extensive discussion of intermolecular potential functions and virial coefficients.

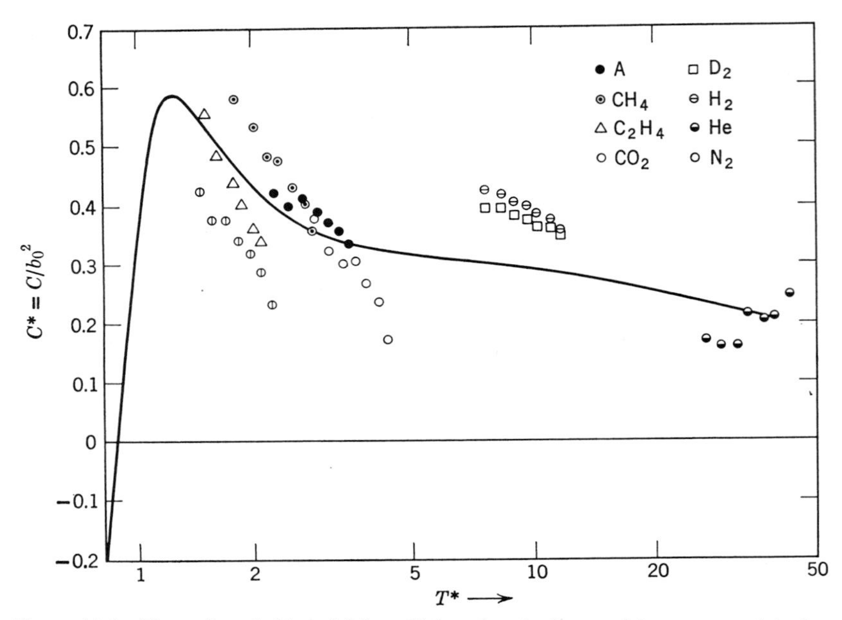

Figure 11.6 The reduced third virial coefficient for the Lennard-Jones potential. From J. O. Hirschfelder, C. F. Curtiss, and R. B. Bird, *Molecular Theory of Gases and Liquids*, John Wiley and Sons, New York, second printing, corrected, with notes added, March, 1964.

11.4. The Virial Equation of State for a Mixture

The development of the equation of state from a theoretical standpoint in Section 11.1 assumes a pure substance, but for many applications it is necessary to know the thermodynamic properties and behavior for a mixture. The mixture equation of state must necessarily be more complicated, as it must include composition as a variable in addition to pressure, specific volume, and temperature. Because of the added complexity, in this section we limit our development to that for a binary mixture of components A and B and include only the first-order terms (the second virial coefficient terms).

The basic relation for the equation of state development is again Eq. 10.73, in which the grand partition function $Z_{G_{AB}}$ is now given for a mixture of A and B by

$$
\begin{aligned}
Z_{G_{AB}} &= \sum_{N_A} \sum_{N_B} \lambda_A{}^{N_A} \lambda_B{}^{N_B} Z_{N_{AB}} \\
&= 1 + \lambda_A Z_{1_A} + \lambda_B Z_{1_B} + \lambda_A{}^2 Z_{2_A} + \lambda_A \lambda_B Z_{2_{AB}} + \lambda_B{}^2 Z_{2_B} + \cdots
\end{aligned}
\tag{11.49}
$$

In this expression, λ_A and λ_B are defined in terms of the component chemical potentials μ_A and μ_B as

$$\begin{aligned} \lambda_A &= e^{\mu_A/kT} \\ \lambda_B &= e^{\mu_B/kT} \end{aligned} \tag{11.50}$$

The system partition function $Z_{N_{AB}}$ includes all combinations of numbers of molecules, N_A and N_B. Thus, in Eq. 11.49, Z_{1_A} and Z_{1_B} are the normal single-particle partition functions for pure A and B, respectively, each at the temperature and volume of the system. Similarly, Z_{2_A} is the partition function for two molecules of A in the system, Z_{2_B} is that for two molecules of B, and $Z_{2_{AB}}$ is the partition function for two molecules, one of A and one of B, in the system.

Using Eq. 10.71 and the logarithmic series for Eq. 11.49, we find that at low density

$$\frac{PV}{kT} = \ln Z_{G_{AB}} = \lambda_A Z_{1_A} + \lambda_B Z_{1_B} + \cdots \tag{11.51}$$

all other terms being of second degree or higher in λ. Let us define the active number densities $\mathfrak{z}_A$ and $\mathfrak{z}_B$ for the components as

$$\begin{aligned} \mathfrak{z}_A &= \frac{\lambda_A Z_{1_A}}{V} \\ \mathfrak{z}_B &= \frac{\lambda_B Z_{1_B}}{V} \end{aligned} \tag{11.52}$$

and, since at very small λ_A, λ_B, Eq. 11.51 reduces to the indicated first-order terms, we find that

$$\mathfrak{z}_A \rightarrow \frac{N_A}{V}, \quad \mathfrak{z}_B \rightarrow \frac{N_B}{V}$$

as $\lambda_A, \lambda_B \rightarrow 0$ (that is, as the pressure approaches zero and the system behaves as an ideal gas).

From the definitions of $\mathfrak{z}_A$, $\mathfrak{z}_B$ in Eq. 11.52, the equation of state for a mixture using Eq. 11.49 can be written in the form

$$\frac{PV}{kT} = \ln Z_{G_{AB}} = \ln\left[\sum_{N_A}\sum_{N_B}\left(\frac{Z_{N_{AB}} V^{N_A+N_B}}{Z_{1_A}{}^{N_A} Z_{1_B}{}^{N_B}}\right)\mathfrak{z}_A{}^{N_A}\mathfrak{z}_B{}^{N_B}\right] \tag{11.53}$$

which is analogous to Eq. 11.4 for a pure substance. As in the development for a pure substance, we wish to re-express the quantity in the parentheses of Eq. 11.53 as a function of the system configurational energy. For any given N_A, N_B, the system state energy U_i can be separated into two parts:

$$U_{i_{N_{AB}}} = U_{i_{N_{AB}}(\text{ind part})} + \Phi_{N_{AB}}(\vec{r}_{A_1}, \ldots, \vec{r}_{A_{N_A}}, \vec{r}_{B_1}, \ldots, \vec{r}_{B_{N_B}}) \tag{11.54}$$

in which the first term is the system energy for independent particles, and the second term is the configurational energy, assumed to be a function of the positions (but not orientation) of all the $N_A + N_B$ particles under consideration. We make the additional assumption that $\Phi_{N_{AB}}$ is a continuous function of the particle positions, so that the system partition function $Z_{N_{AB}}$ for given N_A, N_B becomes

$$Z_{N_{AB}} = \left(\frac{Z_{1_A}{}^{N_A}}{N_A!}\right)\left(\frac{Z_{1_B}{}^{N_B}}{N_B!}\right)\frac{1}{V^{N_A+N_B}}\iint\cdots\int$$
$$\times \exp\left[\frac{-\Phi_{N_{AB}}(\vec{r}_{A_1},\ldots,\vec{r}_{A_{N_A}},\vec{r}_{B_1},\ldots,\vec{r}_{B_{N_B}})}{kT}\right]$$
$$\times\, d\vec{r}_{A_1}\cdots d\vec{r}_{A_{N_A}}\, d\vec{r}_{B_1}\cdots d\vec{r}_{B_{N_B}} \tag{11.55}$$

in which $d\vec{r}_{A_1} = dx_{A_1}\,dy_{A_1}\,dz_{A_1}$, etc. Equation 11.55 is conveniently written as

$$Z_{N_{AB}} = \left(\frac{Z_{1_A}{}^{N_A}}{N_A!}\right)\left(\frac{Z_{1_B}{}^{N_B}}{N_B!}\right)\frac{Q_{N_{AB}}}{V^{N_A+N_B}} \tag{11.56}$$

for which the various configurational integrals $Q_{N_{AB}}$ are given by

$$Q_{N_{AB}} = \iint\cdots\int\exp\left[\frac{-\Phi_{N_{AB}}(\vec{r}_{A_1},\ldots,\vec{r}_{A_{N_A}},\vec{r}_{B_1},\ldots,\vec{r}_{B_{N_B}})}{kT}\right]$$
$$\times\, d\vec{r}_{A_1}\cdots d\vec{r}_{A_{N_A}}\, d\vec{r}_{B_1}\cdots d\vec{r}_{B_{N_B}} \tag{11.57}$$

We note at this point that should the assumption of classical mechanics (continuous function $\Phi_{N_{AB}}$) not be valid, then the $Q_{N_{AB}}$ are defined according to Eq. 11.56 and are not given in integral form by Eq. 11.57. Evaluation of the $Q_{N_{AB}}$ for such a case then must include quantum corrections, which are not considered here.

If we now substitute Eq. 11.56 into Eq. 11.53, we have

$$\frac{PV}{kT} = \ln Z_{G_{AB}} = \ln\left[\sum_{N_A}\sum_{N_B}\left(\frac{Q_{N_{AB}}}{N_A!N_B!}\right)\mathfrak{z}_A{}^{N_A}\mathfrak{z}_B{}^{N_B}\right]$$
$$= \ln\left[1 + Q_{1_A}\mathfrak{z}_A + Q_{1_B}\mathfrak{z}_B + \tfrac{1}{2}Q_{2_A}\mathfrak{z}_A{}^2 + Q_{2_{AB}}\mathfrak{z}_A\mathfrak{z}_B + \tfrac{1}{2}Q_{2_B}\mathfrak{z}_B{}^2 + \cdots\right] \tag{11.58}$$

where the subscripts are to be understood in the same manner as those of Eq. 11.49. That is, Q_{2_A} is the configurational integral as given by Eq. 11.57 for two molecules of A and no B in the system at T, V, while $Q_{2_{AB}}$ is that for one molecule of A and one of B, etc. Using the logarithmic series to expand Eq. 11.58 and collecting terms, we write the equation of state as

$$\frac{PV}{kT} = V[b_{1_A}\mathfrak{z}_A + b_{1_B}\mathfrak{z}_B + b_{2_A}\mathfrak{z}_A{}^2 + 2b_{2_{AB}}\mathfrak{z}_A\mathfrak{z}_B + b_{2_B}\mathfrak{z}_B{}^2 + \cdots] \tag{11.59}$$

In this expression, the various cluster integrals b_{1_A}, etc., are defined in terms of the configurational integrals by the relations

$$\begin{aligned} b_1 &= \frac{Q_{1A}}{V} = 1, \qquad b_{1_B} = \frac{Q_{1B}}{V} = 1 \\ b_{2_A} &= \frac{Q_{2_A} - Q_{1_A}{}^2}{2!V}, \qquad b_{2_B} = \frac{Q_{2_B} - Q_{1_B}{}^2}{2!V} \\ b_{2_{AB}} &= \frac{Q_{2_{AB}} - Q_{1_A}Q_{1_B}}{2!V} \end{aligned} \tag{11.60}$$

The problem remaining in our development is that of relating the active number densities $\mathfrak{z}_A$ and $\mathfrak{z}_B$ to number densities as was done for the pure substance in Section 11.1. Substitution into Eq. 11.59 will then yield the desired equation of state for the mixture. The expression for a mixture component analogous to the pure substance relation of Eq. 10.61 is

$$N_A = \lambda_A\left(\frac{\partial \ln Z_{G_{AB}}}{\partial \lambda_A}\right)_{T,V,\lambda_B} \tag{11.61}$$

But, differentiating Eq. 11.52 at constant T, V, we have

$$d\mathfrak{z}_A = \frac{Z_{1A}}{V}\, d\lambda_{A_{T,V}} = \frac{\mathfrak{z}_A}{\lambda_A}\, d\lambda_{A_{T,V}} \tag{11.62}$$

and, substituting into Eq. 11.61,

$$N_A = \mathfrak{z}_A\left(\frac{\partial \ln Z_{G_{AB}}}{\partial \mathfrak{z}_A}\right)_{T,V,\mathfrak{z}_B} = \mathfrak{z}_A\left[\frac{\partial(PV/kT)}{\partial \mathfrak{z}_A}\right]_{T,V,\mathfrak{z}_B} \tag{11.63}$$

We now substitute Eq. 11.59 for PV/kT into Eq. 11.63, perform the differentiation, and divide by V. The result is

$$\frac{N_A}{V} = b_{1_A}\mathfrak{z}_A + 2b_{2_A}\mathfrak{z}_A{}^2 + 2b_{2_{AB}}\mathfrak{z}_A\mathfrak{z}_B + \cdots \tag{11.64}$$

By an identical procedure for component B, we find

$$\frac{N_B}{V} = b_{1_B}\mathfrak{z}_B + 2b_{2_B}\mathfrak{z}_B{}^2 + 2b_{2_{AB}}\mathfrak{z}_A\mathfrak{z}_B + \cdots \tag{11.65}$$

However, for our purposes, the inverses of these series, Eqs. 11.64 and 11.65, are required. These are obtained algebraically in the same manner as for the pure substance in Eqs. 11.15 to 11.18, from which we obtain the series

$$\mathfrak{z}_A = \left(\frac{N_A}{V}\right) - 2b_{2_A}\left(\frac{N_A}{V}\right)^2 - 2b_{2_{AB}}\left(\frac{N_A}{V}\right)\left(\frac{N_B}{V}\right) + \cdots \tag{11.66}$$

$$\mathfrak{z}_B = \left(\frac{N_B}{V}\right) - 2b_{2_B}\left(\frac{N_B}{V}\right)^2 - 2b_{2_{AB}}\left(\frac{N_A}{V}\right)\left(\frac{N_B}{V}\right) + \cdots \tag{11.67}$$

In Eqs. 11.66 and 11.67, the cluster integrals b_{1_A} and b_{1_B} have been set equal to unity in accordance with Eq. 11.60.

The series expressions, Eqs. 11.66 and 11.67, are now substituted into Eq. 11.59 for $\mathfrak{z}_A$ and $\mathfrak{z}_B$, respectively. After collecting terms and dividing by $N = N_A + N_B$, we have the equation of state in the form

$$\frac{PV}{NkT} = 1 - \frac{b_{2_A}N_A{}^2 + 2b_{2_{AB}}N_AN_B + b_{2_B}N_B{}^2}{NV} + \cdots \tag{11.68}$$

This equation is preferably written in terms of the molal specific volume of the mixture, v, and the component mole fractions y_A and y_B. From the relations

$$\frac{N_A{}^2}{NV} = \frac{N_A{}^2}{N^2}\left(\frac{N_0}{v}\right) = \frac{y_A{}^2N_0}{v}, \quad \text{etc.}$$

Eq. 11.68 becomes

$$\frac{Pv}{RT} = 1 + \frac{B_M(T, y)}{v} + \cdots \tag{11.69}$$

the desired virial equation of state for a mixture. In Eq. 11.69, the second virial coefficient of the mixture, $B_M(T, y)$, is a function of composition as well as of temperature and is, by comparison with Eq. 11.68,

$$B_M(T, y) = y_A{}^2B_A(T) + 2y_Ay_BB_{AB}(T) + y_B{}^2B_B(T) \tag{11.70}$$

where

$$\begin{aligned} B_A(T) &= -b_{2_A}N_0 \\ B_{AB}(T) &= -b_{2_{AB}}N_0 \\ B_B(T) &= -b_{2_B}N_0 \end{aligned} \tag{11.71}$$

We have found that the mixture coefficient $B_M(T, y)$ is expressed by Eq. 11.70 in terms of the composition and the three temperature functions of Eq. 11.71. Each of the latter is a consequence of two-particle interactions; $B_A(T)$ and $B_B(T)$ are the ordinary second virial coefficients for pure A and pure B, respectively. The coefficient $B_A(T)$ is, of course, a result of interactions between two molecules of A, while $B_B(T)$ results from interactions between two molecules of B. The remaining coefficient of Eq. 11.71, $B_{AB}(T)$, is a result of the two-particle interactions of unlike molecules A and B and is termed the interaction coefficient for the mixture. By a procedure similar to that followed in Section 11.2, we utilize Eqs. 11.71, 11.60, and 11.57 for $B_{AB}(T)$, b_{AB}, and $Q_{2_{AB}}$ to obtain

$$B_{AB}(T) = -\frac{N_0}{2V}\iint_V [e^{-\varphi(r_{AB})/kT} - 1]\, d\vec{r}_A\, d\vec{r}_B$$

$$= -\frac{N_0}{2V}\int_V d\vec{r}_A \int_V f_{r_{AB}}\, d\vec{r}_{AB}$$

$$= -2\pi N_0 \int_0^\infty f_{r_{AB}} r_{AB}^2\, dr_{AB} \tag{11.72}$$

This equation is of the same form as that found for a pure substance in Eq. 11.30. Consequently, for a given potential function, the resulting expression for $B_{AB}(T)$ will be of the same form as that found for $B(T)$ in Section 11.3. The one important difference, however, is that the significant parameters are in this case the distance and energy between unlike molecules. For example, in using the Lennard-Jones (6–12) potential, the interaction coefficient $B_{AB}(T)$ depends upon the force constants ϵ_{AB}, σ_{AB}. From a study of low density experimental data for various mixtures, it has been found that reasonable values to use for these constants are the combinations of the pure substance constants:

$$\begin{aligned} \epsilon_{AB} &= (\epsilon_A \epsilon_B)^{1/2} \\ \sigma_{AB} &= \sigma_A + \sigma_B \end{aligned} \tag{11.73}$$

from which the interaction coefficient $B_{AB}(T)$ can be evaluated at any temperature.

11.5. The Liquid Phase

The theory of the behavior of gases at moderate densities is well developed, and has been discussed in some detail in the earlier sections of this chapter. The behavior of the solid phase, the simple theories of which have been discussed in Chapter 7, is also quite well understood. It has been possible to analyze these phases of matter rather extensively, because their analytical models lend themselves readily to analysis. In the gas of moderate density, the molecules are distributed at random throughout the system volume and are, on the average, relatively widely dispersed. Thus only two-particle or two- and three-particle intermolecular interactions are of importance in describing the behavior of the gaseous system. On the other hand, the atoms in a solid are held quite rigidly in a lattice structure by strong interatomic forces, and vibrate only moderately about their respective equilibrium positions.

Between these two extreme cases are the very highly compressed gas and liquid phases, for which the analytical models, although very numerous, are not nearly so satisfactory. The basic problem is that of evaluating the system partition function given by Eq. 10.27,

$$Z_N = \sum_i e^{-U_i/kT}$$

If this could be done, the equation of state could then be determined from Eq. 10.31,

$$P = kT\left(\frac{\partial \ln Z_N}{\partial V}\right)_{T,N}$$

However, the system state energy U_i (and therefore Z_N and the equation of state) is strongly dependent upon the intermolecular potential energy and cannot be evaluated exactly. The virial expansion approach used for gases is not valid for the liquid phase (even if high-order virial coefficients could be evaluated), since the series becomes divergent in the liquid density range. Therefore most theories treat the liquid phase as a pseudo-solid, in which the molecules are held in a lattice structure, although much more loosely than those of a solid. The theories become very complicated mathematically, and the discussion here will be primarily qualitative.

Restriction of a molecule to a volume (V/N) by its neighbors in the lattice leads to a theory of the liquid termed a cell theory, of which there are many modifications; it also presents an immediate difficulty in the analytical model. To point out this difficulty, let us consider the atoms of a monatomic ideal gas, for which, at a given T, V,

$$Z_{N_{\text{gas}}} = \frac{Z^N}{N!} = \frac{(CV)^N}{N!} = \frac{C^N V^N}{N!} \tag{11.74}$$

where

$$C = \left(\frac{2\pi mkT}{h^2}\right)^{3/2} \tag{11.75}$$

Now suppose that these atoms are not permitted access to the entire system volume V, but that each is restricted to a particular cell of volume (V/N), as in a solid. The atoms now become effectively distinguishable from one another because the cells are distinguishable and the atoms are not permitted to change cells. Therefore, in this case, there is no factor $N!$ in the system partition function to discount for particle distinguishability, so that, for atoms restricted to cells of volume (V/N),

$$Z_{N_{\text{cell}}} = Z^N = \left(C\frac{V}{N}\right)^N = \frac{C^N V^N}{N^N} \tag{11.76}$$

The entropy difference between these two models is, from Eq. 10.29 with the same U,

$$S_{\text{gas}} - S_{\text{cell}} = k \ln Z_{N_{\text{gas}}} - k \ln Z_{N_{\text{cell}}} = k \ln \frac{N^N}{N!} = Nk = nR \tag{11.77}$$

The first model is correct for a monatomic ideal gas, and the second is correct for a solid. The inherent difference in viewpoint leads to an entropy difference of R per mole, which is called communal entropy, and is presumed to be present in a gas but not in a solid. Viewed in another manner, the cell model would give the correct entropy for a gas if the single-particle partition function Z in Eq. 11.76 were multiplied by e. Let us then include in Eq. 11.76 such a multiplying factor ν which we term the communal entropy parameter, equal to unity for a solid and to e if Eq. 11.76 is used for a gas.

One basic problem of applying the cell model to the liquid phase is the question whether the communal entropy is present in the liquid. Some models assume that it is; that is, $\nu = e$ throughout the liquid range. Other cell theories assume that ν varies continuously from 1 at the melting point to e at the boiling point. The different theories are in fundamental disagreement on this point, and the question has not been resolved.

In the cell theory of the liquid, the system partition function must include the lattice energy contribution $(\varphi(0)/2)$, where $\varphi(0)$ is the potential energy, so that Eq. 11.76 is written

$$Z_N = (C\nu v_F e^{-\varphi(0)/2kT})^N \tag{11.78}$$

In this expression, the cell volume (V/N) is replaced by the "free" volume v_F, since a molecule is not free to move throughout the entire volume of its cell. The molecules themselves occupy a certain volume, and the intermolecular potential of a molecule with its neighbors gives different probabilities of occupancy to different portions of the cell. Even for rigid spheres, for which we assume a constant potential except during collision, the exact free volume is quite complex geometrically. A simplified free volume for this case can, however, be found in the following manner. Consider only one dimension x and rigid molecules of diameter σ, as shown

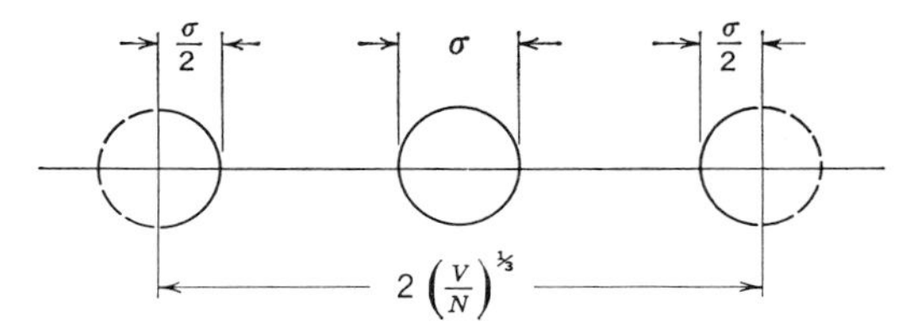

Figure 11.7 One-dimensional movement for a rigid sphere confined to a cell.

in Figure 11.7. The maximum x-directional movement possible for the central molecule is

$$2\left[\left(\frac{V}{N}\right)^{1/3} - \sigma\right]$$

Generalizing this argument to three dimensions, we find that free volume is

$$v_F = 8\left[\left(\frac{V}{N}\right)^{1/3} - \sigma\right]^3 \tag{11.79}$$

Using Eq. 11.79 for the free volume, $\nu = e$, and a lattice energy equal to the negative of the vaporization energy of the liquid molecule in Eq. 11.78 results in the most elementary model of the liquid phase, which is fairly reasonable, at least in a qualitative sense.

When we introduce a more realistic intermolecular potential $\varphi(\vec{r})$ that varies over the cell, the free volume is then expressed in the form

$$v_F = \int_{\text{cell}} e^{-[\varphi(\vec{r})-\varphi(0)]/kT}\, d\vec{r} \tag{11.80}$$

In order to evaluate Eq. 11.80, it is necessary to have an expression for $\varphi(\vec{r})$, which should in reality be angle-dependent, as the potential is a function of the positions of the neighboring molecules. In order to circumvent this extremely difficult problem, it is common to use an equivalent spherical potential for which the potential resulting from neighboring molecules is "smeared" uniformly over the surface of a sphere. If this approximation is made, and the Lennard-Jones potential used for $\varphi(\vec{r})$ in Eq. 11.80, the equation of state can then be evaluated from Eqs. 11.78 and 10.31. This expression is called the Lennard-Jones and Devonshire equation, which is considerably more realistic than the elementary model described previously.

There are many other theories of the liquid phase, more refined and more complex than those discussed here. If some cells are allowed to be vacant, the resulting theories are termed hole theories of the liquid. These theories are more reasonable than basic cell theories in that they eliminate the communal entropy problem completely. Still more advanced treatments utilize radial distribution functions to approximate the system partition function, and a completely different approach to the problem, which may prove valuable, is the direct computation of the partition function using a digital computer and a statistical sampling procedure. However, because of the computational difficulties, the latter method has thus far been applied to systems of only a few hundred particles.

PROBLEMS

11.1 Carry through the development indicated in inverting the series of Eq. 11.14 to obtain the series for $\mathfrak{z}$ in terms of number density, Eq. 11.18.

11.2 Extend the development of Section 11.1 to find the fourth virial coefficient $D(T)$ (the coefficient of $1/v^3$ in Eq. 11.20) in terms of the appropriate cluster integrals.

11.3 Show that Eq. 11.32, the expression for the cluster integral b_3, follows from Eqs. 11.26 and 11.31.

11.4 The procedure by which second and third virial coefficients are obtained graphically from experimental data has been described in Section 11.3. The following data have recently been measured † for argon at -70 C:

P, atm	Pv/RT
36.9839	0.89541
31.6923	0.91031
24.1428	0.93184
20.5770	0.94209
15.5444	0.95635
13.1974	0.96299

Use these data to find B and C for argon at -70 C.

11.5 Integrate the equation for the second virial coefficient, using the point centers of repulsion potential, Eq. 11.38. Show that the result is given by Eq. 11.39.

11.6 A simple potential that includes forces of both attraction and repulsion is the Sutherland potential, shown in Figure 11.8. Show that the second virial coefficient for this potential is of the form

$$B(T) = b_0 B^*(T^*)$$

where b_0 is the same as Eq. 11.46, $T^* = (kT/\epsilon)$, and $B^*(T^*)$ is a series in T^*. Determine the first three terms in the series for $B^*(T^*)$.

11.7 The square-well potential is shown in Figure 11.9. Show that the second virial coefficient is

$$B(T) = b_0[1 - (\mathscr{R}^3 - 1)(e^{\epsilon/kT} - 1)]$$

where b_0 is given by Eq. 11.46.

† R. W. Crain, Jr.; Ph.D. Thesis, University of Michigan, 1965.

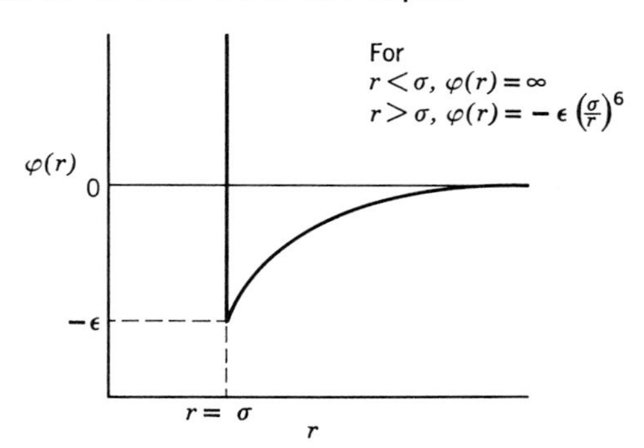

Figure 11.8 Sketch for Problem 11.6.

11.8 Consider the Lennard-Jones potential.

(*a*) Show that Eq. 11.41 follows from Eq. 11.40 as a result of the requirement for the minimum point in the potential well.

(*b*) For the Lennard-Jones (6–12) potential, plot the pair-potential $\varphi(r)$ and the individual attractive and repulsive terms vs. r for nitrogen.

11.9 Expand the exponential term of Eq. 11.47 for the Lennard-Jones (6–12) potential, and find the terms in the resulting series for the reduced second virial coefficient $B^*(T^*)$.

11.10 Oxygen is contained in a rigid 1-liter tank at 0 C, 100 atm pressure. If the tank is then cooled to -100 C, find the final pressure. Use second and third virial coefficients calculated from the Lennard-Jones potential.

11.11 Using the force constants for argon given in Table A.8 and the Lennard-Jones (6–12) potential, calculate the second and third virial coefficients at -70 C. Compare these results with those of Problem 11.4.

11.12 Problem 11.4 illustrated the determination of virial coefficients from experimental PVT data. Such values can be used to determine the best values of force constants for an intermolecular potential function. Experimental values

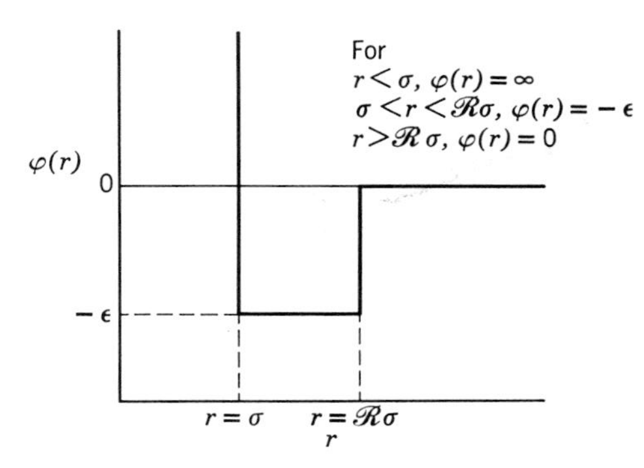

Figure 11.9 Sketch for Problem 11.7.

for argon are

T, °C	B, cm^3/mole
0	−21.18
−70	−46.35
−110	−73.20
−130	−94.69

(*a*) Determine ϵ/k and σ for the Lennard-Jones (6–12) potential from the 0 C and −110 C values of B.

(*b*) Repeat this calculation using the −70 C and −130 C values of B. What conclusion can be drawn from these results?

11.13 In order to calculate thermodynamic properties of a system (by the statistical thermodynamics approach), it is necessary to evaluate the system partition function.

(*a*) Show that Z_N can be expressed as the ideal gas partition function plus corrections in the form

$$\ln Z_N = \ln\left(\frac{Z^N}{N!}\right) - \frac{NB(T)}{v} - \frac{NC(T)}{2v^2} - \cdots$$

(*b*) Find expressions for internal energy and entropy in the form of ideal gas results plus corrections.

11.14 Calculate the internal energy of 1 mole of CO_2 at 20 C, 30 atm pressure, using the Lennard-Jones (6–12) potential (approximate the necessary derivatives of B and C). What per cent error would be introduced by assuming ideal gas behavior?

11.15 At moderate density, the PVT behavior of a gas may be represented by the virial equation of state neglecting terms higher than the second virial coefficient $B(T)$.

(*a*) In such a region, show that the fractional standard deviation in number density is given by

$$\frac{\sigma_N}{\bar{N}} = \left[\frac{\overline{(N-\bar{N})^2}}{\bar{N}^2}\right]^{1/2} = \left[\frac{1}{nN_0(2z-1)}\right]^{1/2}$$

where z is the compressibility factor Pv/RT.

(*b*) Calculate the second virial coefficient for methane, CH_4, at −40 C, using the Lennard-Jones (6–12) potential, and determine $\sigma_N/\bar{N}$ for 1 cm^3 of CH_4 at −40 C, 20 atm.

11.16 In the development of the equation of state for a mixture, show that inversions of the series, Eqs. 11.64 and 11.65, result in Eqs. 11.66 and 11.67 for $\mathfrak{z}_A$ and $\mathfrak{z}_B$.

11.17 Carry through the development of the mixture equation of state, Eq. 11.68, as indicated in the text.

11.18 Plot the second virial coefficient vs. composition for mixtures of CO_2 and N_2 at 200 K.

11.19 Consider a mixture of 50% argon, 50% nitrogen (mole basis) at 180 K, 20 atm pressure. What mass is contained in a 200 cm^3 vessel at this condition? What per cent error would be introduced by assuming an ideal gas mixture?

11.20 Consider the elementary model of the liquid phase, with Z_N given by Eq. 11.78, v_F given by Eq. 11.79, $\nu = e$, and

$$\tfrac{1}{2}N_0\varphi(0) = -\Delta u_{\text{vap}}$$

(*a*) Show that

$$u_{\text{liq}} = \tfrac{3}{2}RT - \Delta u_{\text{vap}}$$

(*b*) Equating molal Gibbs functions for the liquid and vapor (monatomic ideal gas), and neglecting v_{liq}, show that the vapor pressure is

$$P_{\text{vap}} = \frac{kT}{v_F}\, e^{-\Delta h_{\text{vap}}/RT}$$

11.21 It has been found that, for many liquids, the vaporization energy can be approximated by

$$\Delta u_{\text{vap}} = \frac{f(T)}{v}$$

Using this relation and the liquid model discussed in Problem 11.20, show that the equation of state for the liquid is

$$\left(P + \frac{f(T)}{v}\right)(v - 0.7816 b_0^{1/3} v^{2/3}) = RT$$

where b_0 is of the form of Eq. 11.46. This equation of state is termed the Eyring equation. Discuss the form of this expression in comparison with the form of the van der Waals equation (Problem 10.7).

12 Irreversible Processes

In the preceding chapters, we have developed the fundamentals of equilibrium statistical thermodynamics and have examined the properties and behavior of substances that exist in a state of thermodynamic equilibrium. In this, the final chapter, we shall extend our considerations to thermodynamic systems in non-equilibrium states and to irreversible processes. The discussion will be principally from the macroscopic point of view, but the microscopic viewpoint will be used in making certain developments. In connection with irreversible flows, we shall also consider briefly the transport processes (heat, mass, and momentum) and an elementary molecular theory of transport properties.

The subject of non-equilibrium thermodynamics (or thermodynamics of irreversible processes) finds particular application in the analysis of coupled irreversible flows. This point will be demonstrated by an elementary analysis of the thermomechanical effect and the related phenomenon, thermomolecular pressure difference, and also by an analysis of thermoelectric phenomena.

12.1. Forces, Flows, and Entropy Production

Classical thermodynamics presumes that a system exists in a state of thermodynamic equilibrium. For a homogeneous phase, the thermodynamic property relation can be written as

$$dU = T\,dS - P\,dV + \sum_i \mu_i\,dn_i + \tau\,dl + \mathscr{E}\,d\mathscr{Z} + \cdots \tag{12.1}$$

This relation can be viewed as expressing all the ways in which the internal energy U of a substance can be changed in a quasi-equilibrium or reversible process (for which the substance passes through a series of equilibrium states). The first term in Eq. 12.1, $T\,dS$, is the result of an internally reversible heat transfer to the substance, and $-P\,dV$ represents a reversible boundary movement work done on the system. Similarly, $\mu_i \, dn_i$ represents the contribution of a change in the amount of component i present, and $\tau \, dl$ is the energy change due to a change in length when tension τ is significant. The term $\mathscr{E} \, d\mathscr{Z}$ is the internal energy change associated with a change in electric charge $\mathscr{Z}$ under the influence of an electric field $\mathscr{E}$. There may also be other terms of importance in Eq. 12.1 in particular applications. The important point to be understood in examining Eq. 12.1 is that each term or contribution in the property relation is the product of a thermodynamic potential or driving force (an intensive property) and a change in that extensive property most closely associated with the potential.

In this chapter, our principal concern is with the entropy changes of a system. Let us therefore rewrite the property relation in the form

$$dS = \frac{1}{T} dU + \frac{P}{T} dV - \sum_i \frac{\mu_i}{T} dn_i - \frac{\mathscr{E}}{T} d\mathscr{Z} + \cdots \tag{12.2}$$

where additional terms are included as necessary for each particular application. We note that the entropy change for a reversible process can also be viewed as a summation of contributions, each of which consists of a potential times a change in an associated extensive property.

As discussed in Chapter 4, entropy is defined in classical thermodynamics by the expression

$$dS = \left(\frac{\delta Q}{T}\right)_{\text{rev}} \tag{12.3}$$

Our interest in this chapter, however, is not with equilibrium states and reversible processes, but with irreversible processes. The property relation, Eq. 12.1 or 12.2, was developed for reversible changes. We assume, however, that this relation is still valid for irreversible changes, at least if the system remains sufficiently close to an equilibrium state that entropy can be expressed explicitly as a function of $U, V, n, \ldots$. This assumption is found to be reasonable for most applications. Also, for an irreversible change, Eq. 12.3 is replaced (from the inequality of Clausius) by the expression

$$dS > \left(\frac{\delta Q}{T}\right)_{\text{irr}} \tag{12.4}$$

In order to eliminate the inequality in this relation, we write

$$dS = \frac{\delta Q}{T} + d_{\text{int}}S \tag{12.5}$$

in which $d_{\text{int}}S$† is the irreversible entropy production (or internal entropy

† In many books this quantity $d_{\text{int}}S$ is written as $\delta LW/T$, where δLW is called the lost work associated with internal irreversibilities.

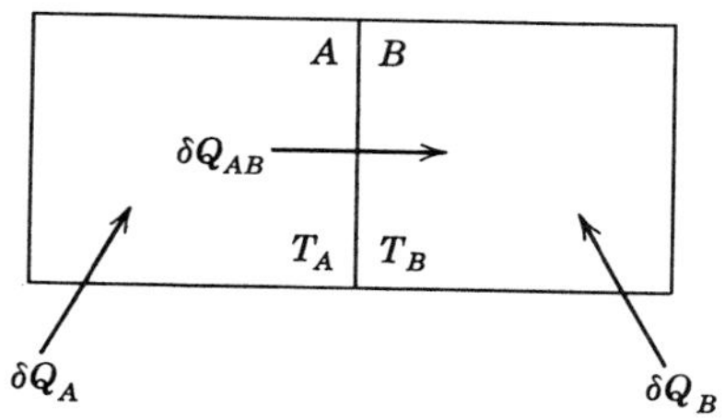

Figure 12.1 An irreversible heat flow between two systems.

generation). From Eqs. 12.3 and 12.4, we find that $d_{\text{int}}S$ is equal to zero for an internally reversible process and is greater than zero for an internally irreversible process.

To demonstrate the nature of the entropy production term $d_{\text{int}}S$ for an irreversible process, let us analyze two particular examples of such processes. We consider first the two systems A and B shown in Figure 12.1. System A is at a uniform temperature T_A, and system B at a uniform temperature T_B, presumed here to be less than T_A. Systems A and B are then brought into thermal communication as shown in Figure 12.1, and there is a flow of heat δQ_{AB} from A to B. At the same time, systems A and B receive δQ_A and δQ_B, respectively, from the surroundings. Therefore the entropy changes for A and B are

$$dS_A = \frac{\delta Q_A - \delta Q_{AB}}{T_A}$$

$$dS_B = \frac{\delta Q_B + \delta Q_{AB}}{T_B}$$

Now, considering a system comprised of A and B together, we have

$$dS_{A+B} = \frac{\delta Q_A}{T_A} + \frac{\delta Q_B}{T_B} + \delta Q_{AB}\left(\frac{1}{T_B} - \frac{1}{T_A}\right) = \frac{\delta Q_A}{T_A} + \frac{\delta Q_B}{T_B} + d_{\text{int}}S$$

where

$$\frac{\delta Q_A}{T_A} + \frac{\delta Q_B}{T_B}$$

represents the reversible entropy change for the combined system $A + B$, while $d_{\text{int}}S$ represents the irreversible entropy production inside the system boundary, in accordance with the definition, Eq. 12.5. Thus we find that

$$d_{\text{int}}S = \delta Q_{AB}\left(\frac{1}{T_B} - \frac{1}{T_A}\right) \tag{12.6}$$

For the combined system $A + B$, the transfer δQ_{AB} occurs inside the boundary, from one part of the system to another, and should more

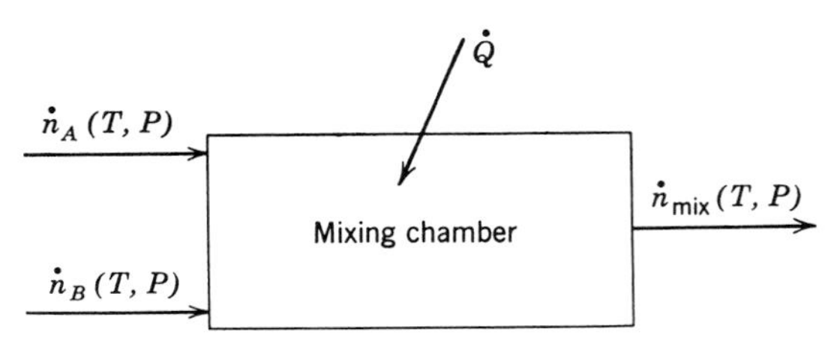

Figure 12.2 The irreversible mixing of two substances.

properly be termed a transfer of internal energy δU_{AB}. In these terms we note the similarity between this expression for entropy production in an irreversible process and the system entropy change in a reversible process as given by the first term of Eq. 12.2. However, the quantity $d_{\text{int}}S$ is expressed in terms of a difference or imbalance in thermodynamic potential $(1/T)$ times a change in the associated extensive property (dU). Thus we conclude that a finite difference in a driving force or potential between two systems results in an irreversible flow, which would tend to eliminate the imbalance in potential and bring the two systems to equilibrium.

It is also noted that $d_{\text{int}}S$ as given by Eq. 12.6 is greater than zero, since $T_A > T_B$ and $\delta Q_{AB} > 0$. Had T_A been less than T_B, then the heat transfer would have been in the opposite direction, $\delta Q_{AB} < 0$, and again $d_{\text{int}}S > 0$. As the temperature difference approaches zero, the irreversible entropy production approaches zero, and the only terms remaining in the entropy equation are the reversible contributions.

Since the applications of the equations of non-equilibrium thermodynamics are frequently those involving rate processes, it is desirable to express the irreversible entropy production for a process on a rate basis. Thus Eq. 12.6 can be written

$$\frac{d_{\text{int}}S}{dt} = \frac{\delta Q_{AB}}{\delta t}\left(\frac{1}{T_B} - \frac{1}{T_A}\right) = \dot{Q}_{AB}\left(\frac{1}{T_B} - \frac{1}{T_A}\right) \tag{12.7}$$

where $d_{\text{int}}S/dt$ is the rate of irreversible entropy production, and $\dot{Q}_{AB}$ is the rate of heat transfer from A to B, also called the heat flow, which is a result of the imbalance in the temperature.

The second example of an irreversible process that we will consider is the process of mixing two substances A and B at constant temperature and pressure, as shown in Figure 12.2. In this process, substance A at T, P enters the mixing chamber at the molal rate $\dot{n}_A$, substance B at T, P enters at the rate $\dot{n}_B$, and the mixture at T, P leaves the chamber at the rate $\dot{n}_{\text{mix}}$. The first law of thermodynamics for this process is

$$\dot{Q} + \dot{n}_A h_A + \dot{n}_B h_B = \dot{n}_{\text{mix}} h_{\text{mix}} = \dot{n}_A \bar{H}_A + \dot{n}_B \bar{H}_B$$

where $\tilde{H}_A$, $\tilde{H}_B$ are the partial molal enthalpies of components A and B in the mixture and are defined as

$$\tilde{H}_i = \left(\frac{\partial H}{\partial n_i}\right)_{T,P,n_{j\neq i}}$$

The entropy equation for this constant-temperature mixing process is

$$0 = \frac{\dot{Q}}{T} + \frac{d_{\text{int}}S}{dt} + \dot{n}_A s_A + \dot{n}_B s_B - \dot{n}_{\text{mix}} s_{\text{mix}}$$

However,

$$\dot{n}_{\text{mix}} s_{\text{mix}} = \dot{n}_A \tilde{S}_A + \dot{n}_B \tilde{S}_B$$

where $\tilde{S}_A$, $\tilde{S}_B$ are the partial molal entropies of the components. If we now combine these equations, the rate of irreversible entropy production is found to be

$$\begin{aligned}\frac{d_{\text{int}}S}{dt} &= \dot{n}_A\left[\left(\frac{h_A - \tilde{H}_A}{T}\right) - (s_A - \tilde{S}_A)\right] + \dot{n}_B\left[\left(\frac{h_B - \tilde{H}_B}{T}\right) - (s_B - \tilde{S}_B)\right] \\ &= \dot{n}_A\left(\frac{\mu_A - \tilde{\mu}_A}{T}\right) + \dot{n}_B\left(\frac{\mu_B - \tilde{\mu}_B}{T}\right) \\ &= \sum_i \left(\frac{\mu_i - \tilde{\mu}_i}{T}\right)\dot{n}_i \qquad (12.8)\end{aligned}$$

In this expression, μ_i is the chemical potential of pure i at T, P, while $\tilde{u}_i$ is that for component i in the mixture at T, P. Once again we find a relation for the irreversible entropy production that is analogous to the corresponding term of the thermodynamic property relation, Eq. 12.2. The difference between the two is that the irreversible entropy production is a consequence of a difference or imbalance in the thermodynamic potential.

In the first example, we found from Eq. 12.7 that the flow $\dot{Q}$ is associated with an imbalance in $1/T$, and, in our second example, we have seen from Eq. 12.8 that an imbalance in μ_i/T results in the flow $\dot{n}_i$. Similar results are found for each of the force-flow combinations of the thermodynamic property relation. We conclude, therefore, that the rate of irreversible entropy production can be expressed in the general form

$$\frac{d_{\text{int}}S}{dt} = \sum_i X_i J_i \qquad (12.9)$$

in which X_i represents any driving force of type i (not necessarily the component i of a mixture), an imbalance in thermodynamic potential, while J_i is the associated irreversible flow.

12.2. Transport Processes and Properties

In the preceding section, we discussed irreversible flows that result from imbalances in thermodynamic potentials, which we called the driving forces. The subject of transport processes, which includes the irreversible flows of heat, mass, and momentum, is a large and complex field, and it is not our intention to study this subject in great depth here. However, it is of interest to discuss the relation of each of these flows to its driving force, and also to examine at least an elementary molecular theory of the transport properties, which are the coefficients related to the irreversible flows. These coefficients are the following: the thermal conductivity $\mathscr{K}$, the heat flow per unit area due to a unit temperature gradient; the coefficient of diffusion for a component i, $\mathscr{D}_i$, the flow of molecules per unit area due to a unit gradient in number density of component i; and the coefficient of viscosity, η, the flow of y-directional momentum per unit area due to a unit gradient in y-directional velocity.

In our elementary molecular model, we consider a tank of volume V containing N molecules at temperature T. These molecules are assumed to behave like rigid spheres of diameter σ (the intermolecular potential shown in Figure 11.2). Let us assume that all N molecules travel at the average molecular speed $\overline{\mathsf{V}}$, which, from Eq. 3.72, is

$$\overline{\mathsf{V}} = \left(\frac{8kT}{\pi m}\right)^{1/2}$$

Finally, we assume in our model that all molecules move in a direction parallel to one of the coordinate axes, x, y, z. That is, at any instant of time, $N/6$ molecules are moving in the positive x-direction, $N/6$ in the negative x-direction, . . . , $N/6$ in the negative z-direction.

Let us now examine a particular molecule i that is moving at velocity $\overline{\mathsf{V}}$ in the positive x-direction. Molecule i has a relative velocity of zero to all other positive x-direction molecules and therefore will not encounter any other positive x-direction molecules during a unit time interval. However, molecule i has a relative velocity of $2\overline{\mathsf{V}}$ to all molecules moving in the negative x-direction. If we assume no deflection upon collision, molecule i would, during a unit interval of time, collide with all those negative x-direction molecules that initially lie inside a cylinder of cross-sectional area $\pi\sigma^2$ and length $2\overline{\mathsf{V}}$. This cylinder occupies a fraction

$$\frac{\pi\sigma^2 2\overline{\mathsf{V}}}{V}$$

of the total volume, and, since there are $N/6$ negative x-direction molecules, molecule i suffers

$$\frac{N}{6}\left(\frac{\pi\sigma^2 2\bar{V}}{V}\right) = \tfrac{1}{3}\rho_N \pi\sigma^2 \bar{V}$$

collisions with negative x-direction molecules during the unit time. We use the symbol ρ_N here to denote the number density N/V.

In a similar manner, molecule i has a relative velocity $\sqrt{2}\,\bar{V}$ to all $N/6$ molecules moving in the positive y-direction, and therefore suffers

$$\frac{\sqrt{2}}{6}\,\rho_N \pi\sigma^2 \bar{V}$$

collisions with this type during the unit time. This expression also represents the number of collisions of i with the negative y-, positive z-, and negative z-direction molecules per unit time. Therefore the total number of collisions, Γ, suffered by i per unit time is

$$\Gamma = C\rho_N \pi\sigma^2 \bar{V} \tag{12.10}$$

where

$$C = \tfrac{1}{3} + 4\left(\frac{\sqrt{2}}{6}\right) = 1.276 \tag{12.11}$$

A molecule having speed $\bar{V}$ travels a distance numerically equal to $\bar{V}$ in a unit time, during which it suffers Γ collisions. Therefore the average distance traveled between collisions, called the mean free path Λ, is

$$\Lambda = \frac{\bar{V}}{\Gamma} = \frac{1}{C\rho_N \pi\sigma^2} \tag{12.12}$$

The product $\pi\sigma^2$ in this expression is the collision cross section.

The molecular model described above is, of course, vastly oversimplified. A more reasonable model results if we consider that the rigid sphere molecules move in all directions, not only those parallel to the coordinate axes, and also have a Maxwell-Boltzmann velocity distribution (Eq. 3.64) instead of a uniform speed equal to the average value, $\bar{V}$. When this improved model is analyzed, it is found to result in an expression of the same form as Eq. 12.10, and the mean free path is again given by Eq. 12.12. However, the constant C in both these equations is not given by Eq. 12.11 but is instead found to be

$$C = \sqrt{2} = 1.414 \tag{12.13}$$

which should be a better value for C than that found by using the oversimplified model. Correlation with experimental data shows that it is, although the assumption of rigid sphere molecules still introduces a considerable error into the results.

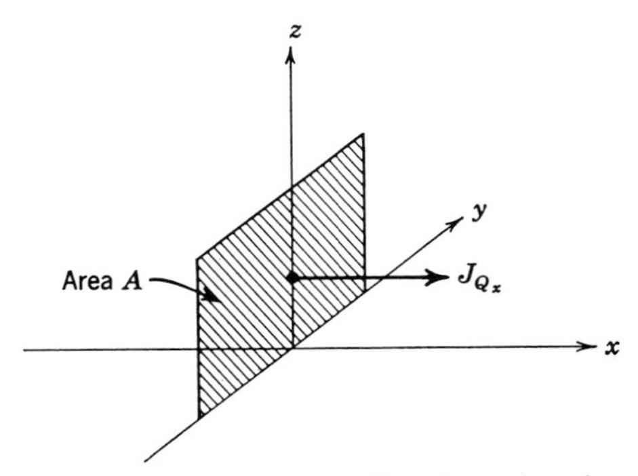

Figure 12.3 An x-directional heat flow through a plane area.

Having developed the basic concept of the mean free path, we are now in a position to examine the transport processes, at least from a simplified point of view. We shall analyze here the heat flow through a plane area resulting from a temperature gradient. The results for the other transport processes, mass, and momentum, will then be deduced from the similarities between these three processes.

Consider a tank of volume V containing N atoms of a monatomic ideal gas in which an x-directional temperature gradient has been established. We wish to determine the net energy transfer in the x-direction, J_{Q_x}, through a plane of area A normal to the x-direction, as shown in Figure 12.3. It is assumed that, at any time, there are $N/6$ particles moving in the positive x-direction, $N/6$ in the negative x-direction, . . . , and also that all have the average speed $\bar{\mathsf{V}}$. Therefore the net heat flow in the x-direction can be expressed in terms of the energy carried by particles that cross plane A moving in the positive x-direction minus that carried by those particles that move across A in the negative x-direction. That is,

$$J_{Q_x} = J_{Q_{+x}} - J_{Q_{-x}} \tag{12.14}$$

The first term in Eq. 12.14, $J_{Q_{+x}}$, can be expressed as

$$J_{Q_{+x}} = \left(\frac{\text{particles}}{\text{sec}}\right)_{+x}\left(\frac{\text{energy}}{\text{particle}}\right)_{+x} \tag{12.15}$$

But

$$\left(\frac{\text{particles}}{\text{sec}}\right)_{+x} = \left(\frac{\text{particles}}{\text{volume}}\right)_{+x}(\text{area})(\text{velocity})$$

$$= \frac{N}{6V}A\bar{\mathsf{V}} = \tfrac{1}{6}\rho_N A\bar{\mathsf{V}} \tag{12.16}$$

Also, the energy per particle of those particles moving in the positive x-direction is that associated with a plane at $x = -\Lambda$, the location at which the $+x$ particles underwent their last collisions. This can be written in terms of the equilibrium properties at $x = 0$ as

$$\left(\frac{\text{energy}}{\text{particle}}\right)_{+x} = \left(\frac{U}{N}\right)_{x=0} - \Lambda\left[\frac{d(U/N)}{dx}\right]_{x=0}$$

$$= C_{v_m}T - \Lambda C_{v_m}\frac{dT}{dx} \tag{12.17}$$

where C_{v_m} is the specific heat per particle of mass m. Therefore, substituting Eqs. 12.16 and 12.17 into Eq. 12.15, we have

$$J_{Q_{+x}} = \tfrac{1}{6}\rho_N A\overline{\mathrm{V}}\left[C_{v_m}T - \Lambda C_{v_m}\frac{dT}{dx}\right] \tag{12.18}$$

In a similar manner, we find the term $J_{Q_{-x}}$ to be

$$J_{Q_{-x}} = \tfrac{1}{6}\rho_N A\overline{\mathrm{V}}\left[C_{v_m}T + \Lambda C_{v_m}\frac{dT}{dx}\right] \tag{12.19}$$

Therefore the net x-direction heat flow J_{Q_x}, the observed macroscopic flow, is, from Eqs. 12.14, 12.18, and 12.19,

$$J_{Q_x} = \tfrac{1}{6}\rho_N A\overline{\mathrm{V}}\left[-2\Lambda C_{v_m}\frac{dT}{dx}\right]$$

$$= -\tfrac{1}{3}A\overline{\mathrm{V}}\Lambda\left(\rho_N C_{v_m}\frac{dT}{dx}\right) \tag{12.20}$$

The quantity in the brackets in Eq. 12.20 represents the x-directional gradient in energy per unit volume.

If we make a similar analysis of the mass transport process, the quantity of interest is not the energy density but is instead the number density of a component i, ρ_{N_i}. Thus it is reasonable to anticipate that the result for mass transport will be of the form of Eq. 12.20, but the quantity in the brackets will be replaced by the x-directional gradient in number density. It is found that the x-direction mass flow of component i is given as

$$J_{N_{ix}} = -\tfrac{1}{3}A\overline{\mathrm{V}}\Lambda\left(\frac{d\rho_{N_i}}{dx}\right) \tag{12.21}$$

Similarly, in analyzing the momentum transport process to determine the x-directional flow of y-direction momentum, $J_{P_{yx}}$, we find

$$J_{P_{yx}} = -\tfrac{1}{3}A\overline{\mathrm{V}}\Lambda\left(\rho_N m\frac{d\mathrm{V}_y}{dx}\right) \tag{12.22}$$

The transport process equations are commonly expressed in terms of the transport properties as

$$J_{Q_x} = -\mathscr{K} A \frac{dT}{dx} \tag{12.23}$$

$$J_{N_i x} = -\mathscr{D} A \frac{d\rho_{N_i}}{dx} \tag{12.24}$$

$$J_{P_{yx}} = -\eta A \frac{d\mathsf{V}_y}{dx} \tag{12.25}$$

where $\mathscr{K}$ is the thermal conductivity, which for our simplified model is, by comparison of Eqs. 12.20 and 12.23,

$$\mathscr{K} = \tfrac{1}{3}\overline{\mathsf{V}}\Lambda\rho_N C_{v_m} \tag{12.26}$$

Similarly, the coefficient of diffusion, $\mathscr{D}$, is found from Eqs. 12.21 and 12.24 to be

$$\mathscr{D} = \tfrac{1}{3}\overline{\mathsf{V}}\Lambda \tag{12.27}$$

while the viscosity η for this model is, from Eqs. 12.22 and 12.25,

$$\eta = \tfrac{1}{3}\overline{\mathsf{V}}\Lambda\rho_N m \tag{12.28}$$

These simplified theory expressions for the transport properties, Eqs. 12.26 to 12.28, can be evaluated by using Eq. 3.72 for the average molecular speed $\overline{\mathsf{V}}$ and Eq. 12.12 for the mean free path Λ. In the latter, the constant C is found preferably from Eq. 12.13. However, when these results are compared with experimental data for the transport properties, the agreement is quite poor. This is to be expected because of the very simplified model to analyze the transport processes.

The rigorous theory of transport processes† is beyond the scope of this book. In this theory, if we again assume a rigid spheres potential, we find that the transport properties are given by expressions of the same form as Eqs. 12.26 to 12.28, except that the coefficient of each of the three equations is different and is not equal to $\frac{1}{3}$. (The factor of $\frac{1}{3}$ is replaced in Eq. 12.26 for $\mathscr{K}$ by $25\pi/64$, in Eq. 12.27 for $\mathscr{D}$ by $3\pi/16$, and in Eq. 12.28 for η by $5\pi/32$.) This improves the correlation with experimental data, but the agreement is still only fair. In order to achieve a good correlation with experimental values, it is necessary to utilize a more realistic function, such as the Lennard-Jones (6–12) potential, Eq. 11.42. The incorporation of this potential function into the rigorous theory equations results in a very good correlation of the transport properties for simple molecules.

† See, for example, J. O. Hirschfelder, C. F. Curtiss, and R. B. Bird, *Molecular Theory of Gases and Liquids*, John Wiley and Sons, New York, second printing, corrected, with notes added, March, 1964.

12.3. Coupled Flows and the Phenomenological Relations

The developments of the subject of non-equilibrium thermodynamics are particularly suitable to the analysis of coupled irreversible flows. We have seen in Section 12.1 that an irreversible flow is the result of a corresponding driving force, a difference in some thermodynamic potential between two systems or between two parts of a single system. However, if two forces are maintained simultaneously, it is found that each of the two resulting flows is somewhat different than would occur as a result of its related force alone. That is, the interaction of the two effects results in cross phenomena, such that both flows are dependent upon both forces.

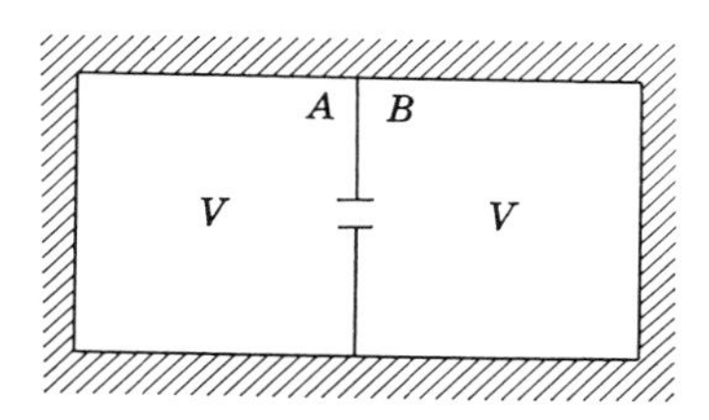

Figure 12.4 The communication of two systems through a small hole.

In a coupled flow situation, we assume that the flows are linearly related to the forces. Thus, for each flow J_i, we can write

$$J_i = \sum_k L_{ik} X_k \tag{12.29}$$

where the coefficients L_{ik} are called phenomenological coefficients. This assumption proves to be reasonable for relatively small forces X_k or, in other words, for relatively small departures from equilibrium. The set of equations relating the flows J_i to the various forces X_k according to Eq. 12.29 is called the set of phenomenological relations. The coefficients L_{ik} of each term remain to be determined from the particular physical application. The term for which $k = i$ is the force directly (and familiarly) related to the flow, for which the coefficient is related to a familiar physical quantity. Such quantities as thermal conductivity, diffusivity, viscosity, and electrical conductivity (or resistance) can be grouped in this category. Those terms for which $k \neq i$ represent the cross phenomena resulting from the coupled flows and are in general not so familiar to us. The problem of evaluating these coefficients is more complex, but fortunately it is simplified somewhat by the Onsager reciprocal relations, which are discussed in Section 12.4.

It is important to note here that, when Eq. 12.29 is used the forces and flows must be properly chosen, so that the rate of irreversible entropy production as given by Eq. 12.9 is satisfied. Such a choice is not always readily apparent, however, and is best illustrated by an example. Let us consider the simultaneous flow of heat and mass between two systems A and B, as indicated in Figure 12.4. The tank is well insulated, and we assume that there is no heat transfer with the surroundings. Each side of the tank has a volume V and initially contains n moles of a given gas, which has an internal energy U. Therefore the entropy of the gas in each side at this equilibrium state is specified, since

$$S = S(U, V, n)$$

Now let us consider the system to be in a constrained or non-equilibrium state, such that the internal energy and the number of moles in part A differ from the equilibrium values by ΔU and Δn, respectively. Consequently, the corresponding variations of the gas in part B are $-\Delta U$ and $-\Delta n$, in accordance with the first law of thermodynamics and the conservation of mass. We assume that this state is not too far removed from equilibrium, so that the departure of entropy, ΔS, from the equilibrium value can be expressed by a Taylor series in ΔU and Δn. Therefore, for part A,

$$\Delta S_A = \left(\frac{\partial S}{\partial U}\right)_n (\Delta U) + \left(\frac{\partial S}{\partial n}\right)_U (\Delta n) + \frac{1}{2}\left(\frac{\partial^2 S}{\partial U^2}\right)_n (\Delta U)^2 + \left(\frac{\partial^2 S}{\partial U\,\partial n}\right)(\Delta U)(\Delta n) + \frac{1}{2}\left(\frac{\partial^2 S}{\partial n^2}\right)_U (\Delta n)^2 + \cdots \quad (12.30)$$

In a similar manner for part B with variations $-\Delta U$ and $-\Delta n$, we have

$$\Delta S_B = \left(\frac{\partial S}{\partial U}\right)_n (-\Delta U) + \left(\frac{\partial S}{\partial n}\right)_U (-\Delta n) + \frac{1}{2}\left(\frac{\partial^2 S}{\partial U^2}\right)_n (-\Delta U)^2 + \left(\frac{\partial^2 S}{\partial U\,\partial n}\right)(-\Delta U)(-\Delta n) + \frac{1}{2}\left(\frac{\partial^2 S}{\partial n^2}\right)_U (-\Delta n)^2 + \cdots \quad (12.31)$$

Adding these expressions, we have, for a system of total mass $2n$,

$$2\,\Delta S = \Delta S_A + \Delta S_B$$

or, for a system of mass n, the variation ΔS in the entropy is

$$\Delta S = \frac{1}{2}\left(\frac{\partial^2 S}{\partial U^2}\right)_n (\Delta U)^2 + \left(\frac{\partial^2 S}{\partial U\,\partial n}\right)(\Delta U)(\Delta n) + \frac{1}{2}\left(\frac{\partial^2 S}{\partial n^2}\right)_U (\Delta n)^2 \quad (12.32)$$

If we now differentiate Eq. 12.32 with respect to time, the result is

$$\frac{d_{\text{int}}S}{dt} = \frac{d\,(\Delta S)}{dt} = \frac{d\,(\Delta U)}{dt}\left[\left(\frac{\partial^2 S}{\partial U^2}\right)_n (\Delta U) + \left(\frac{\partial^2 S}{\partial U\,\partial n}\right)(\Delta n)\right] + \frac{d\,(\Delta n)}{dt}\left[\left(\frac{\partial^2 S}{\partial U\,\partial n}\right)(\Delta U) + \left(\frac{\partial^2 S}{\partial n^2}\right)_U (\Delta n)\right] \quad (12.33)$$

This result can also be written in the form of Eq. 12.9 as

$$\frac{d_{\text{int}}S}{dt} = \left[\Delta\left(\frac{\partial S}{\partial U}\right)_n\right]\frac{d(\Delta U)}{dt} + \left[\Delta\left(\frac{\partial S}{\partial n}\right)_U\right]\frac{d(\Delta n)}{dt} = X_U J_U + X_n J_n \tag{12.34}$$

In this expression the forces and flows are

$$X_U = \Delta\left(\frac{\partial S}{\partial U}\right)_n \tag{12.35}$$

$$J_U = \frac{d(\Delta U)}{dt} \tag{12.36}$$

$$X_n = \Delta\left(\frac{\partial S}{\partial n}\right)_U \tag{12.37}$$

$$J_n = \frac{d(\Delta n)}{dt} \tag{12.38}$$

However, since V is a constant, we find, using Eq. 12.2,

$$X_U = \Delta\left(\frac{\partial S}{\partial U}\right)_n = \Delta\left(\frac{1}{T}\right) \tag{12.39}$$

$$X_n = \Delta\left(\frac{\partial S}{\partial n}\right)_U = -\Delta\left(\frac{\mu}{T}\right) \tag{12.40}$$

We have developed a relation, Eq. 12.33, for the irreversible entropy production for this physical problem, and we have found that, if this relation is to be of the form of Eq. 12.9, then the flows are given by Eqs. 12.36 and 12.38 and the forces by Eqs. 12.39 and 12.40. (There are other choices for forces and flows that will also satisfy Eq. 12.9.) The result found for the driving forces, expressed by Eqs. 12.39 and 12.40, is not an unexpected one (see Eqs. 12.8 and 12.9).

For the example problem considered here, the phenomenological relations are written as

$$J_U = L_{UU}X_U + L_{Un}X_n \tag{12.41}$$

$$J_n = L_{nU}X_U + L_{nn}X_n \tag{12.42}$$

in which the four phenomenological coefficients must be related to the physical problem, as mentioned previously. This matter is considered in Section 12.5, after the development of the Onsager reciprocal relations, which enables us to eliminate one of the coefficients.

At present, however, let us generalize the procedure followed in the development of the relations of our example. The basic variables used in this analysis were the departures of certain parameters (U, n) from their equilibrium values. Therefore, in general, we consider the properties A_i, which at equilibrium have the values $A_{i_{\rm eq}}$. Our basic variables, then, are the instantaneous departures of these properties from their equilibrium values, or

$$a_i = A_i - A_{i_{\rm eq}} \tag{12.43}$$

Thus a_1 and a_2 were in our example the quantities ΔU and Δn. The flows, therefore, are expressed as

$$J_i = \frac{da_i}{dt} = \dot{a}_i \tag{12.44}$$

In the general case, we assume that the entropy variation can be given as a second-order expression in the state variables, so that we can write

$$\Delta S = -\tfrac{1}{2} \sum_i \sum_k C_{ik} a_i a_k \tag{12.45}$$

In this expression, the various C_{ik} are the appropriate coefficients, which for our special example are found from a term-by-term comparison of Eqs. 12.32 and 12.45. The general forces X_i are defined as

$$X_i = \frac{\partial\, \Delta S}{\partial a_i} = -\sum_k C_{ik} a_k \tag{12.46}$$

and the rate of irreversible entropy production is found by differentiating Eq. 12.45. The result is

$$\frac{d\,\Delta S}{dt} = -\sum_i \sum_k C_{ik} \dot{a}_i a_k = \sum_i X_i J_i \tag{12.47}$$

which is in agreement with the expression given in Eq. 12.9.

12.4. The Onsager Reciprocal Relations

In the previous section, we considered coupled flows and assumed that each of the flows in a non-equilibrium system can be represented as resulting from all the forces present, according to Eq. 12.29,

$$J_i = \sum_k L_{ik} X_k$$

In this expression the L_{ik} are the phenomenological coefficients, which depend upon the physical process. In this section, we derive the Onsager reciprocal relations, a powerful result that enables us to equate cross coefficients in the phenomenological equations.

We consider first the fundamentals of fluctuation theory, which deals with the instantaneous fluctuations in a system about the most probable distribution. The difference in entropy for some instantaneous non-equilibrium state for which the thermodynamic probability is w (the most probable distribution has the larger value w_{mp}) is

$$\Delta S = S - S_{\mathrm{eq}} = k \ln \frac{w}{w_{\mathrm{mp}}} \tag{12.48}$$

The mathematical probability for the existence of such a state at any time can therefore be expressed as

$$P = \frac{w}{\sum w} = \frac{w/w_{\mathrm{mp}}}{\sum (w/w_{\mathrm{mp}})} = \frac{e^{\Delta S/k}}{\sum e^{\Delta S/k}} \tag{12.49}$$

Let us turn now to a continuous function representation of possible non-equilibrium states. In terms of the non-equilibrium variables discussed in Section 12.3 and defined by Eq. 12.43,

$$a_i = A_i - A_{i_{\mathrm{eq}}}$$

the mathematical probability of observing a system in a non-equilibrium state specified by a set of variables in the range a_1 to $a_1 + da_1$, a_2 to $a_2 + da_2, \ldots, a_n$ to $a_n + da_n$ is, in a manner analogous to Eq. 12.49,

$$P\, da_1\, da_2 \cdots da_n = \frac{e^{\Delta S/k}\, da_1\, da_2 \cdots da_n}{\iint \cdots \int e^{\Delta S/k}\, da_1\, da_2 \cdots da_n} \tag{12.50}$$

For the purposes of our development, we need to know the average value of a product of the type $a_i X_j$, where a_i is a non-equilibrium state variable, Eq. 12.43, and X_j a driving force that, from Eqs. 12.46 and 12.48, can be expressed as

$$X_j = \frac{\partial\, \Delta S}{\partial a_j} = k \frac{\partial \ln (w/w_{\mathrm{mp}})}{\partial a_j} \tag{12.51}$$

However, since both w_{mp} and Σw are constants, Eq. 12.51 can alternatively be written

$$X_j = k \frac{\partial \ln P}{\partial a_j} \tag{12.52}$$

Thus the average of a product $a_i X_j$ is found, by utilizing Eqs. 12.50 and 12.52, to be

$$\begin{aligned}
\overline{a_i X_j} &= \iint \cdots \int a_i X_j P\, da_1\, da_2 \cdots da_n \\
&= k \iint \cdots \int a_i \frac{\ln P}{\partial a_j} P\, da_1\, da_2 \cdots da_n \\
&= k \iint \cdots \int da_1 \cdots da_{j-1}\, da_{j+1} \cdots da_n \int a_i \frac{\partial P}{\partial a_j}\, da_j
\end{aligned} \tag{12.53}$$

The last integral of Eq. 12.53 can be integrated by parts; the result is

$$\int_{-\infty}^{+\infty} a_i \frac{\partial P}{\partial a_j}\, da_j = (a_i P)\Big|_{-\infty}^{+\infty} - \int_{-\infty}^{+\infty} P \frac{\partial a_i}{\partial a_j}\, da_j \tag{12.54}$$

The mathematical probability P approaches zero very rapidly as a_i becomes large (in either direction), so that the first term in Eq. 12.54 equals zero. Furthermore, since a_i and a_j are independent variables, it follows that

$$\frac{\partial a_i}{\partial a_j} = \delta_{ij} \tag{12.55}$$

where δ_{ij} is the Krönecker delta, which has the following characteristics:

$$\begin{aligned} \delta_{ij} &= 1, \quad i = j \\ &= 0, \quad i \neq j \end{aligned} \tag{12.56}$$

Consequently, Eq. 12.54 becomes

$$\int_{-\infty}^{+\infty} a_i \frac{\partial P}{\partial a_j}\, da_j = 0 - \int_{-\infty}^{+\infty} P\delta_{ij}\, da_j \tag{12.57}$$

Substituting Eq. 12.57 into Eq. 12.53, we have

$$\begin{aligned} \overline{a_i X_j} &= -k \int\int \cdots \int da_1 \cdots da_{j-1}\, da_{j+1} \cdots da_n \int P\delta_{ij}\, da_j \\ &= -k\delta_{ij} \int\int \cdots \int P\, da_1 \cdots da_n \\ &= -k\delta_{ij} \end{aligned} \tag{12.58}$$

Recalling the basic assumptions made in the ensemble developments in Chapter 10, we note that the average $\overline{a_i X_j}$ given by Eq. 12.58 can be regarded either as the time-average value for a single system of interest or as an ensemble average over all the systems making up the ensemble at one time.

We now introduce the concept of microscopic reversibility. Consider a certain fluctuation a_i at some time t, followed by a fluctuation a_j at an interval of time τ later. The condition of microscopic reversibility requires that

$$\overline{a_i(t)a_j(t + \tau)} = \overline{a_i(t)a_j(t - \tau)} \tag{12.59}$$

That is, fluctuations toward and away from equilibrium occur, on the average, with the same frequency. This requirement, Eq. 12.59, can alternatively be expressed by the statement that the order in which fluctuations a_i, a_j occur does not influence the average of their product, or

$$\overline{a_i(t)a_j(t + \tau)} = \overline{a_j(t)a_i(t + \tau)} \tag{12.60}$$

The average quantities of Eq. 12.60 can also be determined by a different averaging technique, in which we first fix all the parameters (including a_i, a_j) at time t and determine the averages for a_i and a_j at $t + \tau$ under these conditions. Then the average is taken over the previously fixed parameters. This averaging process is represented mathematically as

$$\overline{a_i(t)\overline{a_j(t + \tau)}} = \overline{a_j(t)\overline{a_i(t + \tau)}} \tag{12.61}$$

It is apparent that we can utilize this same double-averaging technique after first subtracting the product $a_i(t)a_j(t)$ from both sides of the equation. The result of this procedure is

$$\overline{a_i(t)\overline{[a_j(t + \tau) - a_j(t)]}} = \overline{a_j(t)\overline{[a_i(t + \tau) - a_i(t)]}} \tag{12.62}$$

Let us now divide both sides of Eq. 12.62 by τ, and let τ become very small, so that

$$\lim_{\tau \to \tau_0} \overline{\left[\frac{a_j(t + \tau) - a_j(t)}{\tau}\right]} = \overline{\dot{a}_j(t)} \tag{12.63}$$

The characteristic molecular time τ_0 in this expression is of the order of time between molecular collisions, and consequently $\overline{\dot{a}_j(t)}$ is to be interpreted as an average rate of regression or damping of a fluctuation. The quantity $\dot{a}_i(t)$ is defined by an equation analogous to Eq. 12.63 and must be similarly interpreted. Thus we find that

$$\overline{a_i(t)\overline{\dot{a}_j(t)}} = \overline{a_j(t)\overline{\dot{a}_i(t)}} \tag{12.64}$$

At this point, we make the important assumption that connects the theory of microscopic fluctuations discussed above and the theory of macroscopic irreversible processes. We assume that the flow in a macroscopic irreversible process occurring in a system not too far removed from equilibrium is equivalent to, and can be described by, the same expression as the average rate of regression of a fluctuation $\overline{\dot{a}_i(t)}$. That is, at any time t we can write

$$\overline{\dot{a}_i} = J_i = \sum_k L_{ik} X_k \tag{12.65}$$

for any variable i. Therefore, substituting Eq. 12.65 into Eq. 12.64, we have the relation

$$\overline{a_i \sum_k L_{jk} X_k} = \overline{a_j \sum_k L_{ik} X_k}$$

or

$$\sum_k L_{jk} \overline{a_i X_k} = \sum_k L_{ik} \overline{a_j X_k} \tag{12.66}$$

But, from Eq. 12.58, we find that this relation reduces to

$$-k \sum_k L_{jk}\delta_{ik} = -k \sum_k L_{ik}\delta_{jk}$$

which, from the definition of the Krönecker delta, Eq. 12.56, becomes

$$L_{ji} = L_{ij} \tag{12.67}$$

Equation 12.67 is the statement of the Onsager reciprocal relations for any i, j in a system. This powerful relation, which equates the cross coefficients in a set of phenomenological relations, thus provides an important reduction in the difficult problem of relating phenomenological coefficients to physical parameters for irreversible coupled flow applications.

12.5. The Thermomechanical Effect and Thermomolecular Pressure Difference

Let us now return to the example problem considered in Section 12.3, in which two parts A and B of a tank were permitted to communicate through a small hole (or a capillary tube, or possibly a permeable membrane). It was determined there that the coupled flows of energy and mass can be expressed by Eqs. 12.41 and 12.42,

$$J_U = L_{UU}X_U + L_{Un}X_n$$

$$J_n = L_{nU}X_U + L_{nn}X_n$$

where the forces X_U and X_n driving these flows are given by Eqs. 12.39 and 12.40. If we assume that ΔT is small, Eq. 12.39 becomes

$$X_U = \Delta\left(\frac{1}{T}\right) = -\frac{\Delta T}{T^2} \tag{12.68}$$

and, for small ΔT and ΔP, Eq. 12.40 becomes

$$\begin{aligned} X_n = -\Delta\left(\frac{\mu}{T}\right) &= -\frac{T\Delta\mu - \mu\,\Delta T}{T^2} \\ &= \frac{h\,\Delta T}{T^2} - \frac{v\,\Delta P}{T} \end{aligned} \tag{12.69}$$

Now, utilizing the Onsager reciprocal relation, Eq. 12.67, which for this problem is

$$L_{Un} = L_{nU} \tag{12.70}$$

and also Eqs. 12.68 and 12.69, we find that the phenomenological relations, Eqs. 12.41 and 12.42, can be rearranged to the form

$$J_U = \left(\frac{L_{Un}h - L_{UU}}{T^2}\right)\Delta T + \left(\frac{-L_{Un}v}{T}\right)\Delta P \qquad (12.71)$$

$$J_n = \left(\frac{L_{nn}h - L_{Un}}{T^2}\right)\Delta T + \left(\frac{-L_{nn}v}{T}\right)\Delta P \qquad (12.72)$$

To continue the analysis of this problem we consider two special cases, the first of which is a uniform temperature system. For this case, $\Delta T = 0$, so that, from Eqs. 12.71 and 12.72,

$$\left(\frac{J_U}{J_n}\right)_{\Delta T=0} = \frac{L_{Un}}{L_{nn}} = U^* \qquad (12.73)$$

where U^* is called the energy transport parameter and is the energy carried per unit mass (mole) as a result of the maintenance of a pressure difference at uniform temperature. This energy transfer due to the pressure difference is termed the thermomechanical effect.

The second special case to be considered here is the so-called stationary state of the system, the steady state condition at which there is an energy flow but no net flow of mass. That is, $J_n = 0$ but $J_U \neq 0$. For this special case, we find, from Eq. 12.72 with $J_n = 0$, that

$$\left(\frac{\Delta P}{\Delta T}\right)_{J_n=0} = \frac{h - L_{Un}/L_{nn}}{vT} \qquad (12.74)$$

But, from Eq. 12.73, this becomes

$$\frac{\Delta P}{\Delta T} = \frac{h - U^*}{vT} = \frac{-Q^*}{vT} \qquad (12.75)$$

where Q^* is called the heat transport parameter, defined as

$$Q^* = U^* - h \qquad (12.76)$$

Equation 12.75 expresses the thermomolecular pressure difference that is established as a result of the maintenance of a temperature gradient.

The thermomechanical effect and thermomolecular pressure difference have been observed experimentally in a variety of systems. One example is the effect called thermo-osmosis, which occurs when the two parts communicate through a membrane. We consider here one particular application of these results, the situation resulting when tanks A and B contain a monatomic ideal gas and communicate through a very small hole (or a capillary tube). We assume that the diameter of this hole is small compared to the mean free path of the atoms in the tanks. Thus every atom that arrives at the hole with a component of velocity normal to the opening passes freely through to the other side of the tank. Using this assumption and the Maxwell-Boltzmann velocity distribution, Eq. 3.64, we find that the average kinetic energy of the atoms passing through the opening is (see Problem 3.15)

$$\overline{\tfrac{1}{2}m\mathrm{V}^2} = 2kT \tag{12.77}$$

Therefore the energy transport parameter U^*, the energy that would be carried per mole of gas at uniform temperature, is

$$U^* = N_0(2kT) = 2RT \tag{12.78}$$

Also, for a monatomic ideal gas, we have, from Eq. 6.26,

$$h = \tfrac{5}{2}RT$$

and, from Eq. 6.24,

$$Pv = RT$$

Using these results in Eq. 12.75, we find the thermomolecular pressure difference in this case to be

$$\frac{\Delta P}{\Delta T} = \frac{\tfrac{5}{2}RT - 2RT}{(RT/P)T} = \frac{1}{2}\frac{P}{T}$$

or

$$\frac{P_A}{\sqrt{T_A}} = \frac{P_B}{\sqrt{T_B}} \tag{12.79}$$

This relation has been confirmed experimentally for the conditions assumed in this development; in this situation the gas is called a Knudsen gas.

It is of interest to note that, if the hole between tanks A and B is large compared to the mean free path, then there is a bulk flow of gas through the hole during which process atoms undergo numerous collisions. Thus Eq. 12.77 is no longer valid, and to evaluate U^* we must include a work term Pv in addition to the ordinary internal energy of the gas. The energy transport parameter is in this case

$$U^* = u + Pv = \tfrac{3}{2}RT + RT = \tfrac{5}{2}RT \tag{12.80}$$

and we find from Eq. 12.75 that

$$\Delta P = 0 \tag{12.81}$$

This is, of course, the commonly observed situation, for which there is no thermomolecular pressure difference, and in this case the gas is termed an ordinary Boyle gas. There are also situations between the two extremes discussed here, Knudsen gas and Boyle gas, for which neither model is appropriate, and the analysis of such a problem becomes very difficult.

12.6. Thermoelectric Effects

In this section we consider the simultaneous flows of heat and electricity, and as an application of the results of this analysis we examine the thermoelectric effects. A heat flow resulting from a temperature gradient has been discussed previously in this chapter. In a similar manner, let us reconsider the thermodynamic property relation in the form of Eq. 12.2, in which the term $-(\mathscr{E}/T)\,d\mathscr{Z}$ was included to account for the change in entropy associated with reversible electrical work. In accordance with our discussion of forces and flows in Section 12.1, we may conclude that an imbalance in electrical potential $\mathscr{E}$ results in an irreversible flow of electric charge $\mathscr{Z}$. The rate of flow $d\mathscr{Z}/dt$ is then denoted by the symbol J_I, the electrical current, or alternatively by its familiar symbol I.

The analysis of the coupled flows of heat and electricity as a consequence of gradients ΔT and $\Delta\mathscr{E}$ in a metal bar is similar to the analysis made in Section 12.3. It is reasonable to anticipate the result of such an analysis, which is, for small gradients,

$$\frac{d_{\text{int}}S}{dt} = -J_Q\frac{\Delta T}{T^2} - I\frac{\Delta\mathscr{E}}{T} \tag{12.82}$$

This equation can also be written in the form

$$\frac{d_{\text{int}}S}{dt} = -J_S\frac{\Delta T}{T} - I\frac{\Delta\mathscr{E}}{T} \tag{12.83}$$

where the entropy flow J_S is, by definition,

$$J_S = \frac{J_Q}{T} \tag{12.84}$$

Thus Eq. 12.83 is of the general form developed in Eq. 12.9,

$$\frac{d_{\text{int}}S}{dt} = J_S X_S + I X_I \tag{12.85}$$

where we have coupled flows of entropy and electricity. The forces X_S, X_I driving the flows are, from Eqs. 12.83 and 12.85,

$$X_S = -\frac{\Delta T}{T} \tag{12.86}$$

$$X_I = -\frac{\Delta\mathscr{E}}{T} \tag{12.87}$$

The phenomenological relations for the coupled flows J_S, I are written as

$$J_S = L_{SS}X_S + L_{SI}X_I \tag{12.88}$$

$$I = L_{IS}X_S + L_{II}X_I \tag{12.89}$$

If we utilize the Onsager reciprocal relation

$$L_{SI} = L_{IS} \tag{12.90}$$

and Eqs. 12.86 and 12.87, Eqs. 12.88 and 12.89 become

$$J_S = \left(\frac{-L_{SS}}{T}\right)\Delta T + \left(\frac{-L_{SI}}{T}\right)\Delta\mathscr{E} \tag{12.91}$$

$$I = \left(\frac{-L_{SI}}{T}\right)\Delta T + \left(\frac{-L_{II}}{T}\right)\Delta\mathscr{E} \tag{12.92}$$

Let us now consider two special cases, as we did in the analysis of the preceding section. The first case is that of uniform temperature in the system, for which we find

$$\left(\frac{J_S}{I}\right)_{\Delta T=0} = \frac{L_{SI}}{L_{II}} = S^* \tag{12.93}$$

where S^*, the entropy transport parameter, is defined by Eq. 12.93 and can be considered the entropy flow per unit electrical current for a uniform temperature system.

The second special case to be analyzed is the stationary state of the system, the steady state condition for which there is no net electrical current. That is, $I = 0$ while $J_S \neq 0$. For this case, we find from Eqs. 12.92 and 12.93 that

$$\left(\frac{\Delta\mathscr{E}}{\Delta T}\right)_{I=0} = -\frac{L_{SI}}{L_{II}} = -S^* \tag{12.94}$$

As application of the developments concerning the simultaneous flows of heat (or entropy) and electricity, we consider the thermoelectric phenomena, the Seebeck, Peltier, and Thomson effects.

Seebeck Effect. The Seebeck effect is the phenomenon utilized when we perform temperature measurements by means of a thermocouple. Consider the circuit shown in Figure 12.5, which is comprised of two dissimilar metals A and B. The junctions between these metals are located at different temperatures T_{H} and T_{L}. As a result of this temperature difference, an electrical potential will be developed across the open circuit 4-1, and the magnitude of this potential can be correlated with the temperature difference $T_{\mathrm{H}} - T_{\mathrm{L}}$ for any pair of materials making up the thermocouple.

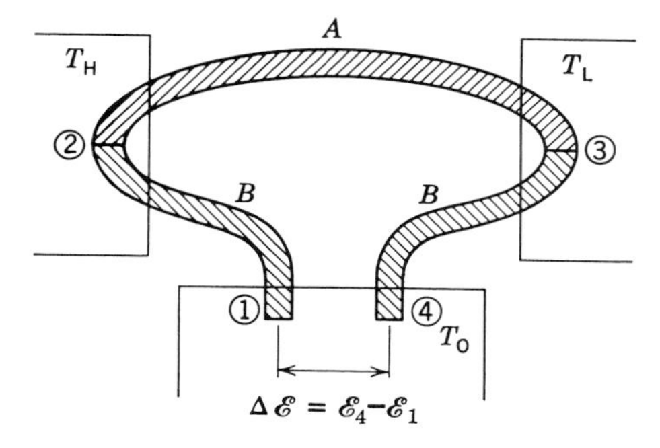

Figure 12.5 Illustration of the Seebeck effect.

Equation 12.94 is directly applicable to the analysis of this problem, since there is no flow of electrical current I. Therefore, integrating Eq. 12.94 around the circuit 1-2-3-4 (where points 1 and 4 are at temperature T_0, point 2 is at T_H, and point 3 at T_L), we have

$$
\begin{aligned}
\Delta\mathscr{E} &= \mathscr{E}_4 - \mathscr{E}_1 = (\mathscr{E}_4 - \mathscr{E}_3) + (\mathscr{E}_3 - \mathscr{E}_2) + (\mathscr{E}_2 - \mathscr{E}_1) \\
&= -\int_{T_L}^{T_0} S_B{}^* \, dT - \int_{T_H}^{T_L} S_A{}^* \, dT - \int_{T_0}^{T_H} S_B{}^* \, dT \\
&= \int_{T_L}^{T_H} S_A{}^* \, dT - \int_{T_L}^{T_H} S_B{}^* \, dT \\
&= \int_{T_L}^{T_H} (S_A{}^* - S_B{}^*) \, dT
\end{aligned}
\tag{12.95}
$$

We find that the measured emf, $\Delta\mathscr{E}$, is a function of the junction temperatures T_H and T_L, but not of the temperature T_0. This value is also seen to be fundamentally dependent upon the characteristics of the materials A and B chosen for the thermocouple, as evidenced by the entropy transport parameters. This basic quantity of interest, the difference in S^* for the materials, is called the Seebeck coefficient α and is defined as

$$
\alpha_{AB} = (S_A{}^* - S_B{}^*) = \frac{d\mathscr{E}_{AB}}{dT} \tag{12.96}
$$

We note that the Seebeck coefficient is equal to the rate of change of electrical potential with temperature, which is an indication of the sensitivity of the thermocouple.

Peltier Effect. The Peltier effect is the basic phenomenon utilized in the construction of thermoelectric refrigeration and power generation devices.

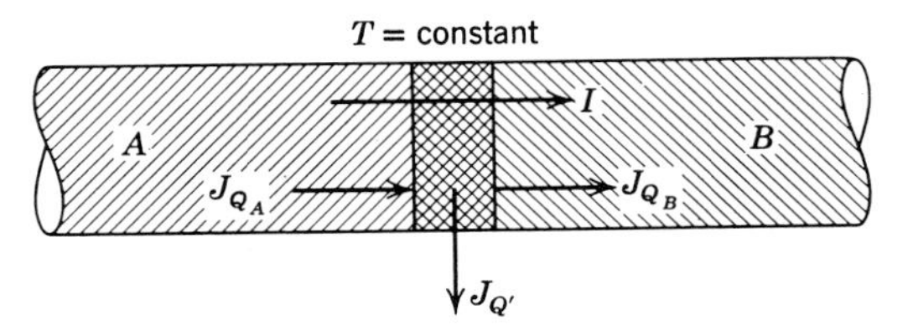

Figure 12.6 Illustration of the Peltier effect.

Consider Figure 12.6, in which an electric current flows in a bar of uniform temperature consisting of two metals A and B. The junction at which these materials are connected has an electrical resistance $\mathscr{R}$, and it is therefore necessary to remove heat in order to keep the junction at temperature T. If A and B are identical materials, then the quantity of heat required to be removed in the familiar Joule heat, $I^2\mathscr{R}$. When A and B are dissimilar, however, it has been observed that this quantity of heat, $J_{Q'}$ removed is somewhat different (either greater or less) than $I^2\mathscr{R}$ because of the Peltier effect.

Let us write the first law of thermodynamics for the junction, which, for the directions indicated in Figure 12.6, is

$$J_{Q'} = (I^2\mathscr{R} + J_{Q_A})_{\Delta T=0} - (J_{Q_B})_{\Delta T=0} \tag{12.97}$$

The Peltier heat J_{Q_π} is by definition the quantity of heat, in addition to the Joule heat $I^2\mathscr{R}$, that must be removed from the junction in order to maintain the junction at constant temperature T. That is,

$$J_{Q_\pi} = \pi_{AB}I = J_{Q'} - I^2\mathscr{R} = (J_{Q_A})_{\Delta T=0} - (J_{Q_B})_{\Delta T=0} \tag{12.98}$$

In this expression, π_{AB} is called the Peltier coefficient for the material combination, AB. We note that Eq. 12.93 is directly applicable to the J_Q for each material, since the temperature remains constant. For material A,

$$\left(\frac{J_{S_A}}{I}\right)_{\Delta T=0} = S_A{}^* = \left(\frac{J_{Q_A}}{IT}\right)_{\Delta T=0}$$

or

$$(J_{Q_A})_{\Delta T=0} = S_A{}^*IT \tag{12.99}$$

Similarly, for material B, we have

$$(J_{Q_B})_{\Delta T=0} = S_B{}^*IT \tag{12.100}$$

Substituting Eqs. 12.99 and 12.100 into Eq. 12.98, we find that the Peltier coefficient can be represented in terms of the entropy transport parameters as

$$\pi_{AB} = T(S_A{}^* - S_B{}^*) \tag{12.101}$$

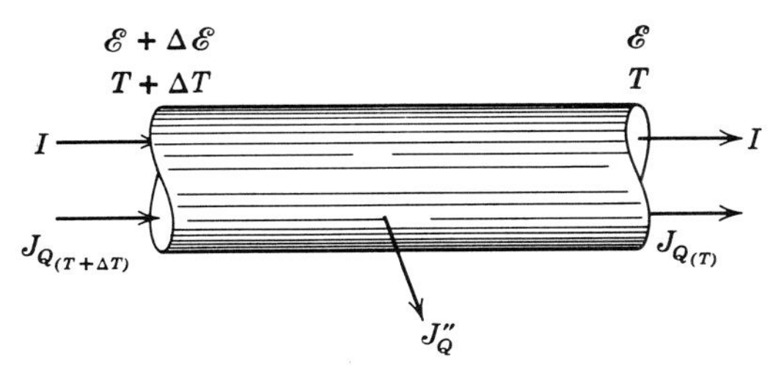

Figure 12.7 Illustration of the Thomson effect.

Thomson Effect. The third thermoelectric effect that we consider is the Thomson effect, which can be described in terms of the situation shown in Figure 12.7. Suppose that a temperature gradient and an electrical potential gradient are maintained simultaneously in a bar of electrical resistance $\mathscr{R}$. Thus there will be irreversible flows of heat, J_Q, and electrical current I as shown in Figure 12.7. The heat transfer quantity $J_{Q''}$ shown includes not only the irreversible Joule heat $I^2\mathscr{R}$, but also an additional amount, the Thomson heat, which is the consequence of the coupled flows. Writing the first law of thermodynamics for the bar, we have, for directions as indicated in Figure 12.7,

$$J_{Q_{(T+\Delta T)}} - J_{Q_{(T)}} - J_{Q''} = -I(\mathscr{E} + \Delta\mathscr{E}) + I\mathscr{E}$$

or

$$J_{Q''} = I\,\Delta\mathscr{E} + J_{Q_{(T+\Delta T)}} - J_{Q_{(T)}} \tag{12.102}$$

From Eq. 12.93, as in the development of the Peltier heat relations, we find

$$J_{Q_{(T)}} = S^*IT \tag{12.103}$$

$$J_{Q_{(T+\Delta T)}} = \left(S^* + \frac{dS^*}{dT}\Delta T\right)I(T + \Delta T) \tag{12.104}$$

Substituting these quantities into Eq. 12.102, we find

$$\begin{aligned} J_{Q''} &= I\left[\Delta\mathscr{E} + \left(S^* + \frac{dS^*}{dT}\Delta T\right)(T + \Delta T) - S^*T\right] \\ &= I\left[\Delta\mathscr{E} + S^*\,\Delta T + T\frac{dS^*}{dT}\Delta T\right] \end{aligned} \tag{12.105}$$

where the term containing the higher-order product $(\Delta T)^2$ has been neglected.

In order to simplify this expression, we rearrange Eq. 12.92 to the form

$$\Delta\mathscr{E} = \frac{IT}{L_{II}} - \frac{L_{SI}}{L_{II}}\Delta T \tag{12.106}$$

To evaluate the phenomenological coefficients, we consider the case of constant temperature, for which Eq. 12.106 must reduce to Ohm's law. Thus

$$\Delta\mathscr{E} = -\frac{IT}{L_{II}} = I\mathscr{R}$$

or

$$\mathscr{R} = \frac{-T}{L_{II}} \tag{12.107}$$

Using this expression and also Eq. 12.93 in Eq. 12.106, we obtain

$$\Delta\mathscr{E} = I\mathscr{R} - S^* \,\Delta T \tag{12.108}$$

Therefore Eq. 12.105 can be written in the form

$$J_{Q''} = I^2\mathscr{R} + IT\frac{dS^*}{dT}\Delta T \tag{12.109}$$

From this relation we see that $J_{Q''}$ is indeed a result of the irreversible Joule heat $I^2\mathscr{R}$ and also includes another contribution that depends upon the characteristics (S^*) of the material. This second term is called the Thomson heat J_{Q_σ}, and by convention we consider Thomson heat positive when it must be transferred to the bar to keep the temperature constant. Thus

$$J_{Q_\sigma} = \sigma I \,\Delta T = -(J_{Q''} - I^2\mathscr{R}) = -IT\frac{dS^*}{dT}\Delta T \tag{12.110}$$

where σ is the Thomson coefficient of the material. It follows from this equation that σ is related to the entropy transport parameter by

$$\sigma = -T\left(\frac{dS^*}{dT}\right) \tag{12.111}$$

The difference in Thomson coefficient for two materials A and B is therefore found to be

$$(\sigma_A - \sigma_B) = -T\frac{d}{dT}(S_A{}^* - S_B{}^*) \tag{12.112}$$

In considering the three thermoelectric effects, we have found that each of the coefficients α, π, σ can be expressed in terms of the entropy transport parameters. Therefore we can relate the coefficients to one another. From Eqs. 12.96 and 12.101, we have

$$\frac{\pi_{AB}}{T} = \alpha_{AB} = \frac{d\mathscr{E}_{AB}}{dT} \tag{12.113}$$

and, from Eqs. 12.96 and 12.112,

$$(\sigma_A - \sigma_B) = -T\frac{d\alpha_{AB}}{dT} = \frac{d^2\mathscr{E}_{AB}}{dT^2} \tag{12.114}$$

These two relations among the three important thermoelectric parameters are called the Kelvin relations, and they enable us to calculate any two of the coefficients in terms of the third.

The three reversible thermoelectric phenomena, the Seebeck, Peltier, and Thomson effects, are normally overshadowed in any system by the irreversible effects, Joule heat (whenever an electric current flows), and ordinary heat conduction (whenever a temperature gradient exists). Until recently, the principal application of thermoelectric effects has been in the use of thermocouples for temperature measurement. However, the development of semiconductor materials, with desirable characteristics for thermoelectric applications, in recent years has also made thermoelectric refrigeration and power generation appear quite promising for certain specialized applications.

PROBLEMS

12.1 Calculate the frequency of collision and the mean free path for a nitrogen molecule at 0 C and a pressure of
(*a*) 10 atm.
(*b*) 10^{-4} atm.
Use $\sigma = 3.681$ Å for N_2.

12.2 The rate of collisions of neon atoms with the walls of a 1-liter tank at 25 C, 1 atm, was found in Problem 3.14. Compare that number with the number of collisions per second of one of the neon atoms with other atoms in the tank. Assume a collision diameter of 2.58 Å.

12.3 A number of problems in Chapter 3 dealt with the subject of molecular effusion, where it was assumed that every particle striking a "small" hole in the wall passed through. In Problem 3.18, air was treated as a pure substance with a molecular weight of 28.97. Using a collision diameter of 3.7 Å, show that in this case the hole is sufficiently "small" that such an assumption is reasonable.

12.4 From the developments of the simple theory and subsequent discussion of Section 12.3, show that the rigorous theory transport properties for the rigid-spheres potential can be expressed in the form

$$\mathscr{K} = \frac{25}{32}\left(\frac{\sqrt{\pi mkT}}{\pi\sigma^2}\right)\left(\frac{C_v}{M}\right)$$

$$\mathscr{D} = \frac{3}{8}\left(\frac{\sqrt{\pi mkT}}{\pi\sigma^2}\right)\left(\frac{1}{\rho}\right)$$

$$\eta = \frac{5}{16}\left(\frac{\sqrt{\pi mkT}}{\pi\sigma^2}\right)$$

12.5 Using the relations found in Problem 12.4, calculate $\mathscr{K}$, $\mathscr{D}$, η for argon at 300 K, 1 atm, for the rigid-spheres model. For argon, $\sigma = 3.64$ Å. Compare the results with the experimental values:

$$\mathscr{K} = 4.22 \times 10^{-5} \text{ cal/sec-cm-K}$$
$$\mathscr{D} = 0.186 \text{ cm}^2\text{/sec}$$
$$\eta = 2.271 \times 10^{-4} \text{ gm/cm-sec}$$

12.6 From the rigorous transport theory, it is found that the transport properties for the Lennard-Jones (6–12) potential are given, to a first approximation, by the corresponding rigid-spheres results divided by a reduced collision integral $\Omega^{(l,s)*}$. In particular,

$$\mathscr{D} = \frac{\mathscr{D}_{\text{rig-sph}}}{\Omega^{(1,1)*}}, \quad \mathscr{K} = \frac{\mathscr{K}_{\text{rig-sph}}}{\Omega^{(2,2)*}}, \quad \eta = \frac{\eta_{\text{rig-sph}}}{\Omega^{(2,2)*}}$$

The functions $\Omega^{(1,1)*}$, $\Omega^{(2,2)*}$ are tabulated below.

T^*	$\Omega^{(1,1)*}$	$\Omega^{(2,2)*}$
0.5	2.066	2.257
1.0	1.439	1.587
1.5	1.198	1.314
2.0	1.075	1.175
2.5	0.9996	1.093
3.0	0.9490	1.039
3.5	0.9120	0.9999
4.0	0.8836	0.9700
5.0	0.8422	0.9269

Using $\epsilon/k = 124$ K and $\sigma = 3.418$ Å for argon, compare $\mathscr{K}$, $\mathscr{D}$, η with the results of Problem 12.5. Note that σ is not the same value as in Problem 12.5. Thus, the rigid-spheres $\mathscr{K}$, $\mathscr{D}$, η to be used in this calculation are somewhat different from the results of that problem. Why are these σ's different? And why are ϵ/k and σ different from the values in Table A.8?

12.7 Calculate the viscosity of nitrogen at the conditions of Problem 12.1. Use $\epsilon/k = 91.5$ K, $\sigma = 3.681$ Å for the Lennard-Jones potential.

12.8 The coefficient of diffusion in a binary mixture, $\mathscr{D}_{AB}$, assuming the Lennard-Jones potential, is calculated using the expressions and procedures discussed in Problems 12.4 and 12.6. The molecular mass m is replaced by the reduced mass m_r, σ by σ_{AB}, and $\Omega_{AB}^{(1,1)*}$ is found using ϵ_{AB}. The interaction parameters σ_{AB} and ϵ_{AB} are given by Eq. 11.73. Calculate $\mathscr{D}_{AB}$ for a mixture of Ar and N_2 at 20 C, 1 atm, and compare the result with the experimental value 0.20 cm^2/sec. Use the pure substance Lennard-Jones parameters given in Problems 12.6 and 12.7.

12.9 The dimensionless Prandtl number,

$$\text{Pr} = \frac{C_p \eta}{\mathscr{K}}$$

is important in the analysis of simultaneous transfer of heat and momentum. Similarly, the dimensionless Schmidt number,

$$\text{Sc} = \frac{\eta}{\rho\mathscr{D}}$$

finds considerable application in the analysis of simultaneous transfer of momentum and mass. Using $\epsilon/k = 113$ K, $\sigma = 3.433$ Å, calculate Pr and Sc for diatomic oxygen at 25 C, 1 atm.

12.10 The viscosity of O_2 at 0 C, 1 atm, is 1.918×10^{-4} gm/cm-sec.

(*a*) Calculate the rigid-spheres collision diameter of the molecule from these data.

(*b*) The constant b in the van der Waals equation of state (Problem 10.7) can be interpreted as the rigid-spheres molecular volume $\frac{2}{3}\pi N_0\sigma^3$. Calculate σ from the value of b for O_2, and compare with that found in part (*a*).

12.11 An experimental setup consists of two bulbs connected by a fine capillary tube. Each has a volume of 100 cm³ and contains helium. Bulb A is maintained at 20 C, and bulb B at -10 C. If the pressure in A is found to be 120 mm Hg, what is the pressure in B? If $\sigma = 2.18$ Å is assumed for helium, approximately what diameter must the capillary be for the helium to behave like a Knudsen gas?

12.12 A particular pair of *n*- and *p*-type semiconductors has a Seebeck coefficient $\alpha = 200\ \mu v/K$. For a junction resistance of 0.015 ohm and $T = 300$ K, what current flow (in the appropriate direction) will result in a net zero heat transfer to the junction? What happens if the direction of current flow is reversed?

12.13 The measured EMF for a particular copper-constantan thermocouple having its reference junction at 0 C is given below.

T, °C	$\mathscr{E}$, millivolts
0	0
40	1.61
80	3.36
120	5.23

It has been found that $\mathscr{E}$ can be represented by the formula

$$\mathscr{E} = A + BT + CT^2 + DT^3, \quad T = K$$

(*a*) Determine the constants A, B, C, D.

(*b*) Find expressions for α_{AB}, π_{AB}, $\sigma_A - \sigma_B$ as functions of temperature, and the values at 0 C and 100 C.

12.14 (*a*) Using entropy transport parameters, show that the thermoelectric EMF of each of the arrangements shown in Figure 12.8 is independent of the presence of material C. This result is known as the law of intermediate metals.

(*b*) Show that the result of part (*a*) permits the establishment of absolute values (relative to a neutral metal) of Seebeck coefficients for individual materials.

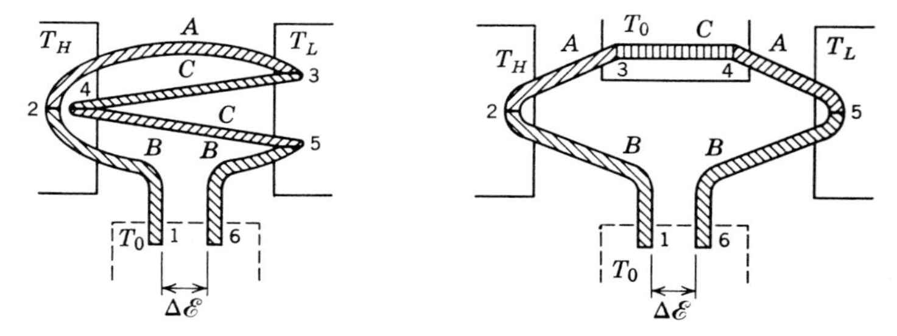

Figure 12.8 Sketch for Problem 12.14.

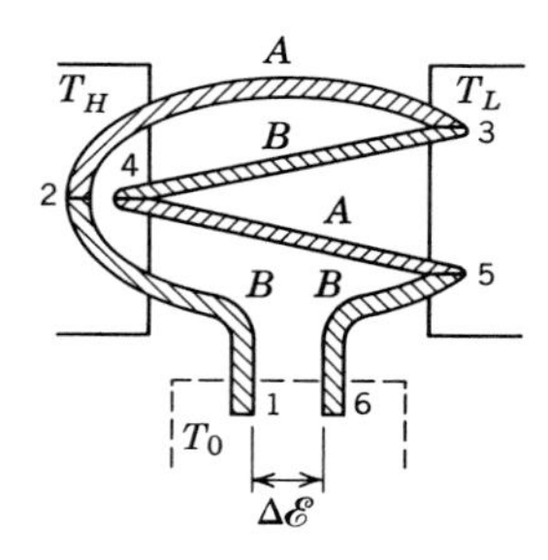

Figure 12.9 Sketch for Problem 12.15.

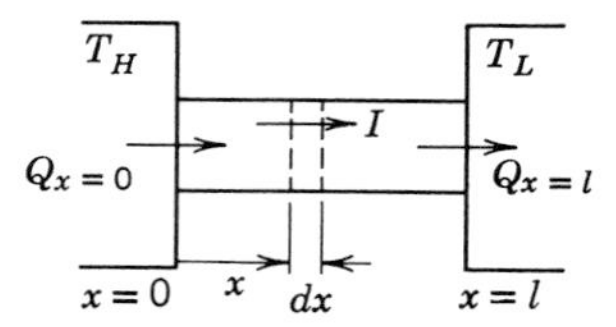

Figure 12.10 Sketch for Problem 12.16.

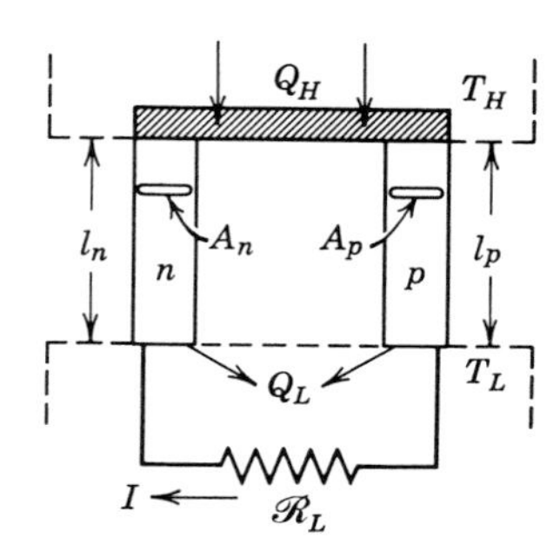

Figure 12.11 Sketch for Problem 12.17.

12.15 A thermopile is constructed by connecting thermocouples in series, as shown in Figure 12.9. Using entropy transport parameters, show that the EMF developed by the thermopile shown is twice that of a single thermocouple.

12.16 Consider the simultaneous flow of heat and electricity in a rod, as shown in Figure 12.10. The electrical resistance of a differential volume $A\,dx$ is $(\rho_e/A)\,dx$, where ρ_e is the resistivity. Consider only heat conduction and Joule heating, and assume constant $\mathscr{K}$, ρ_e:

(*a*) Show that for the element $A\,dx$, the differential equation for the steady state condition is

$$\frac{d^2T}{dx^2} + \left(\frac{I}{A}\right)^2 \frac{\rho_e}{\mathscr{K}} = 0$$

(*b*) Show that the temperature distribution is

$$T = T_H - (T_H - T_L)\left(\frac{x}{l}\right) - \frac{1}{2}\left(\frac{I}{A}\right)^2 \frac{\rho_e}{\mathscr{K}}$$

(*c*) Show that half of the total Joule heat $I^2\mathscr{R}$ flows to each junction.

12.17 Consider the thermoelectric power generator shown in Figure 12.11. The current flow is I, and the load resistance $\mathscr{R}_L$. Using the result of Problem 12.16*c* and assuming constant α, ρ_e, $\mathscr{K}$, show that the thermal efficiency can be expressed as

$$\eta_{\text{th}} = \frac{W_{\text{out}}}{Q_H} = \frac{I^2\mathscr{R}_L}{K\,\Delta T + \alpha I T_H - \frac{1}{2}I^2\mathscr{R}}$$

where

$$K = \frac{\mathscr{K}_n A_n}{l_n} + \frac{\mathscr{K}_p A_p}{l_p}$$

$$\mathscr{R} = \frac{\rho_{e_n} l_n}{A_n} + \frac{\rho_{e_p} l_p}{A_p}$$

12.18 Consider the thermoelectric refrigerator shown in Figure 12.12. Using the assumptions made in Problem 12.17, show that the coefficient of performance of the refrigerator can be expressed as

$$\beta = \frac{Q_L}{W_{\text{in}}} = \frac{\alpha I T_L - \frac{1}{2}I^2\mathscr{R} - K\,\Delta T}{\alpha I\,\Delta T + I^2\mathscr{R}}$$

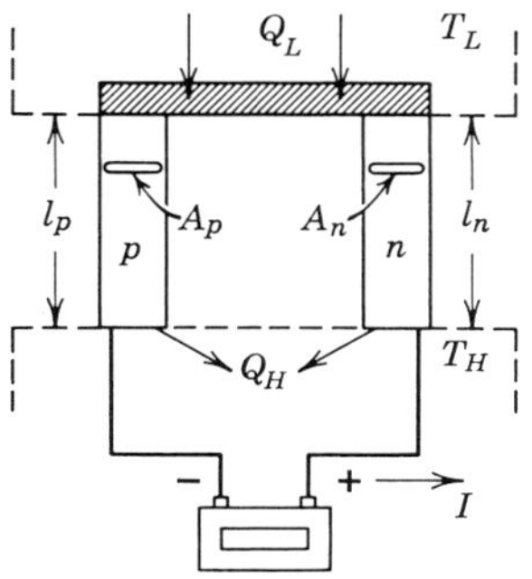

Figure 12.12 Sketch for Problem 12.18.

Appendix

TABLE A.1
Constants and Conversion Factors

Universal gas constant $R = 1.98726$ cal/gm mole-K
$= 0.082055$ lit-atm/gm mole-K
Avogadro's number $N_0 = 6.0232 \times 10^{23}$ 1/gm mole
Boltzmann's constant $k = R/N_0 = 1.3804 \times 10^{-16}$ erg/K
Planck's constant $h = 6.62517 \times 10^{-27}$ erg-sec
Speed of light $c = 2.998 \times 10^{10}$ cm/sec
$hc/k = 1.4388$ cm-K
Mass per atomic weight $m = 1.66 \times 10^{-24}$ gm
Electron mass $m_e = 9.1086 \times 10^{-28}$ gm
1 Angstrom (Å) $= 10^{-8}$ cm
1 dyne = 1 gm-cm/sec^2
1 liter = 1000.028 cm^3
1 erg = 1 dyne-cm
1 electron volt (ev) $= 1.6021 \times 10^{-12}$ erg
1 joule $= 10^7$ ergs
1 calorie (cal) = 4.184 joules
1 atmosphere (atm) $= 1.01325 \times 10^6$ dynes/cm^2 = 760 mm Hg
T (°K) = T (°C) + 273.15

2.54 cm = 1 inch
453.59 gm = 1 lbm
252 cal = 1 Btu
1 atm = 14.696 lbf/in.^{2}
1 °K = 1.8 °R
1 cal/gm mole-K = 1 Btu/lb mole-R

TABLE A.2
Atomic Weights

Name	Symbol	At. No.	International Atomic Weight 1959
Actinium	Ac	89	227
Aluminum	Al	13	26.98
Americium	Am	95	(243)
Antimony, stibium	Sb	51	121.76
Argon	Ar	18	39.944
Arsenic	As	33	74.92
Astatine	At	85	(210)
Barium	Ba	56	137.36
Berkelium	Bk	97	(249)
Beryllium	Be	4	9.013
Bismuth	Bi	83	208.99
Boron	B	5	10.82
Bromine	Br	35	79.916
Cadmium	Cd	48	112.41
Calcium	Ca	20	40.08
Californium	Cf	98	(251)
Carbon	C	6	12.011
Cerium	Ce	58	140.13
Cesium	Cs	55	132.91
Chlorine	Cl	17	35.457
Chromium	Cr	24	52.01
Cobalt	Co	27	58.94
Columbium, see *Niobium*			
Copper	Cu	29	63.54
Curium	Cm	96	(247)
Dysprosium	Dy	66	162.51
Einsteinium	E	99	(254)
Erbium	Er	68	167.27
Europium	Eu	63	152.0
Fermium	Fm	100	(253)
Fluorine	F	9	19.00
Francium	Fr	87	(223)
Gadolinium	Gd	64	157.26
Gallium	Ga	31	69.72
Germanium	Ge	32	72.60
Gold, aurum	Au	79	197.0
Hafnium	Hf	72	178.50
Helium	He	2	4.003
Holmium	Ho	67	164.94
Hydrogen	H	1	1.0080
Indium	In	49	114.82
Iodine	I	53	126.91
Iridium	Ir	77	192.2
Iron, ferrum	Fe	26	55.85
Krypton	Kr	36	83.80
Lanthanum	La	57	138.92
Lead, plumbum	Pb	82	207.21
Lithium	Li	3	6.940
Lutetium	Lu	71	174.99
Magnesium	Mg	12	24.32
Manganese	Mn	25	54.94
Mendelevium	Mv	101	(256)
Mercury, hydrargyrum	Hg	80	200.61
Molybdenum	Mo	42	95.95
Neodymium	Nd	60	144.27
Neon	Ne	10	20.183
Neptunium	Np	93	(237)
Nickel	Ni	28	58.71
Niobium (columbium)	Nb	41	92.91
Nitrogen	N	7	14.008
Nobelium	No	102	(254)
Osmium	Os	76	190.2
Oxygen	O	8	16.000
Palladium	Pd	46	106.4
Phosphorus	P	15	30.975
Platinum	Pt	78	195.09
Plutonium	Pu	94	(242)
Polonium	Po	84	210
Potassium, kalium	K	19	39.100
Praseodymium	Pr	59	140.92
Promethium	Pm	61	(147)
Protactinium	Pa	91	231
Radium	Ra	88	226
Radon	Rn	86	222
Rhenium	Re	75	186.22
Rhodium	Rh	45	102.91
Rubidium	Rb	37	85.48
Ruthenium	Ru	44	101.1
Samarium	Sm	62	150.35
Scandium	Sc	21	44.96
Selenium	Se	34	78.96
Silicon	Si	14	28.09
Silver, argentum	Ag	47	107.873
Sodium, natrium	Na	11	22.991
Strontium	Sr	38	87.63
Sulfur	S	16	32.066*
Tantalum	Ta	73	180.95
Technetium	Tc	43	(99)
Tellurium	Te	52	127.61
Terbium	Tb	65	158.93
Thallium	Tl	81	204.39
Thorium	Th	90	232.
Thulium	Tm	69	168.94
Tin, stannum	Sn	50	118.70
Titanium	Ti	22	47.90
Tungsten (wolfram)	W	74	183.86
Uranium	U	92	238.07
Vanadium	V	23	50.95
Xenon	Xe	54	131.30
Ytterbium	Yb	70	173.04
Yttrium	Y	39	88.91
Zinc	Zn	30	65.38
Zirconium	Zr	40	91.22

* Because of natural variations in the relative abundances of the isotopes of sulfur the atomic weight of this element has a range of ± 0.003.

From *Handbook of Chemistry and Physics*, 44th ed., C. D. Hodginan, editor, The Chemical Rubber Publishing Co., Cleveland, Ohio, 1962.

TABLE A.3
Series and Integrals

$$e^x = 1 + \frac{x}{1!} + \frac{x^2}{2!} + \frac{x^3}{3!} + \cdots$$

$$\ln(1 + x) = x - \frac{x^2}{2} + \frac{x^3}{3} - \frac{x^4}{4} + \cdots$$

$$(x + y)^n = x^n + nx^{n-1}y + \frac{n(n-1)}{2!}x^{n-2}y^2 + \cdots \qquad (y^2 < x^2)$$

n	$\int_0^\infty x^n e^{-ax^2}\,dx$	n	$\int_0^\infty x^n e^{-ax^2}\,dx$
0	$\frac{1}{2}\left(\frac{\pi}{a}\right)^{1/2}$	3	$\frac{1}{2a^2}$
1	$\frac{1}{2a}$	4	$\frac{3}{8}\left(\frac{\pi}{a^5}\right)^{1/2}$
2	$\frac{1}{4}\left(\frac{\pi}{a^3}\right)^{1/2}$	5	$\frac{1}{a^3}$

$$\int_{-\infty}^{+\infty} x^n e^{-ax^2}\,dx = 0 \qquad (n = 1, 3, 5, \ldots)$$

$$= 2\int_0^\infty x^n e^{-ax^2}\,dx \qquad (n = 0, 2, 4, \ldots)$$

$$\int_0^\infty \frac{x^3}{e^x - 1}\,dx = \frac{\pi^4}{15}$$

$$\int_0^\infty x^2 \ln(1 - e^{-x})\,dx = -\frac{\pi^4}{45}$$

Error function: $\operatorname{erf}(x) = \frac{2}{\sqrt{\pi}}\int_0^x e^{-x^2}\,dx$

x	erf (x)	x	erf (x)	x	erf (x)
0	0	0.7	0.6778	1.4	0.9523
0.1	0.1125	0.8	0.7421	1.5	0.9661
0.2	0.2227	0.9	0.7969	1.6	0.9764
0.3	0.3286	1.0	0.8427	1.7	0.9838
0.4	0.4284	1.1	0.8802	1.8	0.9891
0.5	0.5205	1.2	0.9103	1.9	0.9928
0.6	0.6039	1.3	0.9340	2.0	0.9953

TABLE A.4

Harmonic Oscillator Functions

x	$-\ln(1 - e^{-x})$	$\dfrac{x}{e^x - 1}$	$\dfrac{x^2e^x}{(e^x - 1)^2}$
0	∞	1.00000	1.00000
0.01	4.61017	0.99501	0.99999
0.02	3.92201	0.99003	0.99997
0.03	3.52152	0.98508	0.99993
0.04	3.23881	0.98013	0.99987
0.05	3.02063	0.97521	0.99979
0.06	2.84326	0.97030	0.99970
0.07	2.69406	0.96541	0.99959
0.08	2.56546	0.96053	0.99947
0.09	2.45261	0.95568	0.99933
0.10	2.35217	0.95083	0.99917
0.12	2.17966	0.94120	0.99880
0.14	2.03530	0.93163	0.99837
0.16	1.91152	0.92213	0.99787
0.18	1.80345	0.91270	0.99730
0.20	1.70777	0.90333	0.99667
0.22	1.62211	0.89403	0.99598
0.24	1.54472	0.88480	0.99521
0.26	1.47426	0.87563	0.99439
0.28	1.40970	0.86653	0.99349
0.30	1.35023	0.85749	0.99253
0.32	1.29517	0.84852	0.99151
0.34	1.24400	0.83962	0.99042
0.36	1.19626	0.83078	0.98927
0.38	1.15158	0.82200	0.98805
0.40	1.10963	0.81330	0.98677
0.42	1.07016	0.80466	0.98543
0.44	1.03293	0.79608	0.98402
0.46	0.99773	0.78757	0.98255
0.48	0.96439	0.77913	0.98102
0.50	0.93275	0.77075	0.97943
0.52	0.90269	0.76243	0.97777
0.54	0.87407	0.75418	0.97605
0.56	0.84679	0.74600	0.97427
0.58	0.82075	0.73788	0.97243
0.60	0.79587	0.72982	0.97053

TABLE A.4 (*Continued*)
Harmonic Oscillator Functions

x	$-\ln(1 - e^{-x})$	$\frac{x}{e^x - 1}$	$\frac{x^2 e^x}{(e^x - 1)^2}$
0.62	0.77207	0.72183	0.96857
0.64	0.74928	0.71390	0.96656
0.66	0.72743	0.70604	0.96448
0.68	0.70647	0.69824	0.96234
0.70	0.68634	0.69050	0.96015
0.72	0.66700	0.68283	0.95790
0.74	0.64839	0.67522	0.95559
0.76	0.63049	0.66768	0.95323
0.78	0.61324	0.66019	0.95081
0.80	0.59662	0.65277	0.94833
0.82	0.58059	0.64542	0.94580
0.84	0.56512	0.63812	0.94322
0.86	0.55019	0.63089	0.94058
0.88	0.53577	0.62372	0.93789
0.90	0.52184	0.61661	0.93515
0.92	0.50836	0.60956	0.93235
0.94	0.49533	0.60257	0.92951
0.96	0.48271	0.59565	0.92661
0.98	0.47050	0.58878	0.92367
1.00	0.45868	0.58198	0.92067
1.02	0.44722	0.57523	0.91763
1.04	0.43611	0.56855	0.91454
1.06	0.42535	0.56193	0.91140
1.08	0.41490	0.55536	0.90822
1.10	0.40477	0.54886	0.90499
1.12	0.39494	0.54241	0.90171
1.14	0.38540	0.53603	0.89839
1.16	0.37613	0.52970	0.89503
1.18	0.36713	0.52343	0.89162
1.20	0.35838	0.51722	0.88817
1.22	0.34988	0.51106	0.88468
1.24	0.34162	0.50497	0.88115
1.26	0.33359	0.49893	0.87758
1.28	0.32578	0.49295	0.87396
1.30	0.31819	0.48702	0.87031

TABLE A.4 (*Continued*)
Harmonic Oscillator Functions

x	$-\ln(1 - e^{-x})$	$\frac{x}{e^x - 1}$	$\frac{x^2 e^x}{(e^x - 1)^2}$
1.32	0.31079	0.48115	0.86663
1.34	0.30360	0.47534	0.86290
1.36	0.29660	0.46958	0.85914
1.38	0.28979	0.46388	0.85534
1.40	0.28316	0.45824	0.85151
1.42	0.27670	0.45264	0.84764
1.44	0.27040	0.44711	0.84374
1.46	0.26427	0.44163	0.83981
1.48	0.25830	0.43620	0.83584
1.50	0.25248	0.43083	0.83185
1.52	0.24681	0.42551	0.82782
1.54	0.24128	0.42024	0.82377
1.56	0.23590	0.41502	0.81968
1.58	0.23064	0.40986	0.81557
1.60	0.22552	0.40475	0.81143
1.62	0.22052	0.39970	0.80726
1.64	0.21565	0.39469	0.80307
1.66	0.21089	0.38973	0.79885
1.68	0.20625	0.38483	0.79461
1.70	0.20173	0.37998	0.79035
1.72	0.19731	0.37518	0.78606
1.74	0.19300	0.37042	0.78175
1.76	0.18880	0.36572	0.77742
1.78	0.18469	0.36107	0.77306
1.80	0.18068	0.35646	0.76869
1.82	0.17677	0.35191	0.76430
1.84	0.17295	0.34740	0.75989
1.86	0.16922	0.34294	0.75547
1.88	0.16557	0.33853	0.75103
1.90	0.16201	0.33416	0.74657
1.92	0.15854	0.32984	0.74209
1.94	0.15514	0.32557	0.73761
1.96	0.15182	0.32135	0.73310
1.98	0.14858	0.31717	0.72859
2.00	0.14541	0.31304	0.72406

TABLE A.4 (*Continued*)
Harmonic Oscillator Functions

x	$-\ln(1 - e^{-x})$	$\frac{x}{e^x - 1}$	$\frac{x^2 e^x}{(e^x - 1)^2}$
2.05	0.13781	0.30290	0.71270
2.10	0.13063	0.29304	0.70127
2.15	0.12385	0.28346	0.68979
2.20	0.11744	0.27414	0.67827
2.25	0.11138	0.26509	0.66672
2.30	0.10565	0.25629	0.65515
2.35	0.10023	0.24775	0.64358
2.40	0.09510	0.23945	0.63200
2.45	0.09025	0.23139	0.62044
2.50	0.08565	0.22356	0.60889
2.55	0.08130	0.21597	0.59737
2.60	0.07718	0.20861	0.58589
2.65	0.07327	0.20146	0.57445
2.70	0.06957	0.19453	0.56307
2.75	0.06606	0.18781	0.55174
2.80	0.06274	0.18129	0.54049
2.85	0.05959	0.17498	0.52930
2.90	0.05660	0.16886	0.51820
2.95	0.05376	0.16293	0.50719
3.00	0.05107	0.15719	0.49627
3.05	0.04852	0.15163	0.48545
3.10	0.04610	0.14624	0.47473
3.15	0.04380	0.14103	0.46413
3.20	0.04162	0.13598	0.45363
3.25	0.03955	0.13110	0.44326
3.30	0.03758	0.12638	0.43301
3.35	0.03572	0.12181	0.42289
3.40	0.03394	0.11739	0.41289
3.45	0.03226	0.11311	0.40304
3.50	0.03066	0.10898	0.39331
3.55	0.02915	0.10499	0.38373
3.60	0.02770	0.10113	0.37429
3.65	0.02634	0.09740	0.36499
3.70	0.02503	0.09380	0.35584
3.75	0.02380	0.09032	0.34684

TABLE A.4 (*Continued*)
Harmonic Oscillator Functions

x	$-\ln(1 - e^{-x})$	$\dfrac{x}{e^x - 1}$	$\dfrac{x^2 e^x}{(e^x - 1)^2}$
3.80	0.02263	0.08695	0.33799
3.85	0.02151	0.08371	0.32928
3.90	0.02045	0.08057	0.32073
3.95	0.01944	0.07755	0.31233
4.00	0.01849	0.07463	0.30409
4.05	0.01758	0.07181	0.29600
4.10	0.01671	0.06909	0.28806
4.15	0.01589	0.06647	0.28027
4.20	0.01511	0.06394	0.27264
4.25	0.01437	0.06150	0.26516
4.30	0.01366	0.05915	0.25783
4.35	0.01299	0.05688	0.25066
4.40	0.01235	0.05469	0.24364
4.45	0.01175	0.05258	0.23676
4.50	0.01117	0.05055	0.23004
4.55	0.01062	0.04859	0.22347
4.60	0.01010	0.04671	0.21704
4.65	0.00961	0.04489	0.21076
4.70	0.00914	0.04314	0.20462
4.75	0.00869	0.04145	0.19863
4.80	0.00826	0.03983	0.19277
4.85	0.00786	0.03827	0.18706
4.90	0.00748	0.03676	0.18149
4.95	0.00711	0.03531	0.17605
5.00	0.00676	0.03392	0.17074
5.1	0.00612	0.03128	0.16053
5.2	0.00553	0.02885	0.15083
5.3	0.00500	0.02659	0.14162
5.4	0.00453	0.02450	0.13290
5.5	0.00410	0.02257	0.12464
5.6	0.00371	0.02079	0.11683
5.7	0.00335	0.01914	0.10944
5.8	0.00303	0.01761	0.10247
5.9	0.00274	0.01621	0.09589
6.0	0.00248	0.01491	0.08968

Table A.4 (*Continued*)
Harmonic Oscillator Functions

x	$-\ln(1 - e^{-x})$	$\dfrac{x}{e^x - 1}$	$\dfrac{x^2 e^x}{(e^x - 1)^2}$
6.1	0.00225	0.01371	0.08383
6.2	0.00203	0.01261	0.07833
6.3	0.00184	0.01159	0.07315
6.4	0.00166	0.01065	0.06828
6.5	0.00150	0.00979	0.06371
6.6	0.00136	0.00899	0.05942
6.7	0.00123	0.00826	0.05539
6.8	0.00111	0.00758	0.05162
6.9	0.00101	0.00696	0.04808
7.0	0.00091	0.00639	0.04476
7.1	0.00083	0.00586	0.04166
7.2	0.00075	0.00538	0.03876
7.3	0.00068	0.00494	0.03605
7.4	0.00061	0.00453	0.03351
7.5	0.00055	0.00415	0.03115
7.6	0.00050	0.00381	0.02894
7.7	0.00045	0.00349	0.02687
7.8	0.00041	0.00320	0.02495
7.9	0.00037	0.00293	0.02316
8.0	0.00033	0.00268	0.02148
8.1	0.00030	0.00246	0.01993
8.2	0.00027	0.00225	0.01848
8.3	0.00025	0.00206	0.01713
8.4	0.00022	0.00189	0.01587
8.5	0.00020	0.00173	0.01471
8.6	0.00018	0.00158	0.01362
8.7	0.00016	0.00145	0.01261
8.8	0.00015	0.00133	0.01168
8.9	0.00014	0.00121	0.01081
9.0	0.00012	0.00111	0.01000
9.1	0.00011	0.00102	0.00925
9.2	0.00010	0.00093	0.00855
9.3	0.000091	0.00085	0.00791
9.4	0.000083	0.00078	0.00731
9.5	0.000075	0.00071	0.00676

Table A.4 (*Continued*)
Harmonic Oscillator Functions

x	$-\ln(1 - e^{-x})$	$\frac{x}{e^x - 1}$	$\frac{x^2 e^x}{(e^x - 1)^2}$
9.6	0.000068	0.00065	0.00624
9.7	0.000061	0.00059	0.00577
9.8	0.000055	0.00054	0.00533
9.9	0.000050	0.00050	0.00492
10.0	0.000045	0.00045	0.00454

TABLE A.5
Constants for Diatomic Molecules

Date	Substance	Ground State	r_ε, Å	ω_ε, cm^{-1}	D_0, ev
12/31/61	Br_2	$^1\Sigma_g^+$	2.284	323.2	1.971
3/31/61	Cl_2	$^1\Sigma_g^+$	1.988	561.1	2.473
12/31/60	F_2	$^1\Sigma$	1.409	923.1	1.592
3/31/61	H_2	$^1\Sigma_g^+$	0.7417	4405.3	4.477
9/30/61	I_2	$^1\Sigma_g^+$	2.667	214.5	1.544
3/31/61	N_2	$^1\Sigma_g^+$	1.0976	2357.6	9.757
3/31/61	O_2	$^3\Sigma_g^-$	1.2074	1580.2	5.115
3/31/61	CO	$^1\Sigma^+$	1.1281	2169.5	11.089
9/30/61	HBr	$^1\Sigma^+$	1.414	2649.6	3.754
3/31/61	HCl	$^1\Sigma^+$	1.2746	2989.6	4.426
12/31/63	HF	$^1\Sigma^+$	0.9168	4138.3	5.844
9/30/61	HI	$^1\Sigma^+$	1.604	2309.1	3.054
6/30/63	NO	$^2\Pi_{1/2}$, $^2\Pi_{3/2}$*	1.1508	1903.6	6.487

* The $^2\Pi_{3/2}$ state at $(E_e/hc) = 121.1$ cm^{-1} above the $^2\Pi_{1/2}$ state.

From JANAF THERMOCHEMICAL TABLES, The Dow Chemical Co., Midland, Mich. (Constants represent the naturally occurring isotopic mixture.)

TABLE A.6
Constants for Polyatomic Molecules

Date	Substance	Bond Distance, Å	Bond Angle, deg	ω_ε, cm^{-1}
3/31/61	CO_2	C—O: 1.926	O—C—O: 180	1342.9 667.3(2) 2349.3
6/30/61	CS_2	C—S: 1.553	S—C—S: 180	658 396.8(2) 1532.5
12/31/64	N_2O	N—N: 1.1282 N—O: 1.1842	N—N—O: 180	1276.5 589.2(2) 2223.7
3/31/61	H_2O	O—H: 0.9584	H—O—H: 104.45	3657.1 1594.6 3755.8
3/31/61	H_2S	S—H: 1.3455	H—S—H: 93.3	2614.6 1182.7 2627.5
12/31/60	NH_2	N—H: 1.025	H—N—H: 103	3400 1550 3450
9/30/64	NO_2	N—O: 1.197	O—N—O: 134.25	1357.8 756.8 1665.5
3/31/61	CH_4	C—H: 1.091	H—C—H: 109.47	2916.5 1534 (2) 3018.7(3) 1306 (3)

From JANAF THERMOCHEMICAL TABLES, The Dow Chemical Co., Midland, Mich. (Constants represent the naturally occurring isotopic mixture.)

TABLE A.7

The Reduced Second and Third Virial Coefficients $B^*(T^*)$ and $C^*(T^*)$ for the Lennard-Jones (6–12) Potential

$$B(T) = b_0 B^*(T^*)$$
$$C(T) = b_0^2 C^*(T^*)$$
$$T^* = kT/\epsilon,\ b_0 = \tfrac{2}{3}\pi N_0 \sigma^3$$

T^*	$B^*(T^*)$	$C^*(T^*)$	T^*	$B^*(T^*)$	(C^*T^*)
0.3	−27.881		2.6	−0.26613	0.3738
0.4	−13.799		2.8	−0.18451	0.3617
0.5	−8.7202		3.0	−0.11523	0.3523
0.6	−6.1980		3.2	−0.05579	0.3449
0.7	−4.7100	−3.3766	3.4	−0.00428	0.3389
0.8	−3.7342	−0.8495	3.6	+0.04072	0.3341
0.9	−3.0471	+0.0765	3.8	0.08033	0.3300
1.0	−2.5381	0.4297	4.0	0.11542	0.3266
1.1	−2.1464	0.5576	4.2	0.14668	0.3237
1.2	−1.8359	0.5924	4.4	0.17469	0.3212
1.3	−1.5841	0.5882	4.6	0.19990	0.3189
1.4	−1.3758	0.5683	4.8	0.22268	0.3169
1.5	−1.2009	0.5434	5.0	0.24334	0.3151
1.6	−1.0519	0.5180	6.0	0.32290	0.3077
1.7	−0.92362	0.4943	7.0	0.37609	0.3017
1.8	−0.81203	0.4728	8.0	0.41343	0.2962
1.9	−0.71415	0.4538	9.0	0.44060	0.2910
2.0	−0.62763	0.4371	10.0	0.46088	0.2861
2.2	−0.48171	0.4100	20.0	0.52537	0.2464
2.4	−0.36358	0.3894	30.0	0.52693	0.2195

Abridged from J. O. Hirschfelder, C. F. Curtiss, and R. B. Bird, *Molecular Theory of Gases and Liquids*, John Wiley and Sons, New York, Second Printing, Corrected, with Notes Added, March, 1964.

TABLE A.8
Force Constants for the Lennard-Jones (6–12) Potential from Experimental Virial Coefficient Data

Substance	ϵ/k, °K	σ, Å
Ne	35.8	2.75
Ar	119.0	3.41
Kr	173.0	3.59
Xe	225.3	4.07
N_2	95.05	3.698
O_2	117.5	3.58
CO	100.2	3.763
NO	131.0	3.17
CO_2	186.0	4.55
N_2O	193.0	4.54
CH_4	148.1	3.809
CF_4	152	4.70

From J. O. Hirschfelder, C. F. Curtiss, and R. B. Bird, *Molecular Theory of Gases and Liquids*, John Wiley and Sons, New York, Second Printing, Corrected, with Notes Added, March, 1964.

Some Selected References

F. C. Andrews: *Equilibrium Statistical Mechanics*, John Wiley and Sons, New York, 1963.

N. Davidson: *Statistical Mechanics*, McGraw-Hill Book Co., New York, 1962.

S. R. De Groot: *Thermodynamics of Irreversible Processes*, North Holland Publishing Co., Amsterdam, 1959.

M. Dole: *Introduction to Statistical Thermodynamics*, Prentice-Hall, Englewood Cliffs, N.J., 1954.

W. Feller: *An Introduction to Probability Theory and Its Applications, Vol. I*, Second Edition, John Wiley and Sons, New York, 1957.

G. Herzberg: *Atomic Spectra and Atomic Structure*, Dover Publications, New York, 1944.

G. Herzberg: *Molecular Spectra and Molecular Structure, I, Spectra of Diatomic Molecules*, Second Edition, D. Van Nostrand Co., Princeton, N.J., 1950.

T. L. Hill: *Statistical Thermodynamics*, Addison-Wesley Publishing Co., Reading, Mass., 1960.

J. O. Hirschfelder, C. F. Curtiss, and R. B. Bird: *Molecular Theory of Gases and Liquids*, 2nd printing with notes added, John Wiley and Sons, New York, 1964.

C. Kittel: *Elementary Statistical Physics*, John Wiley and Sons, New York, 1958.

J. F. Lee, F. W. Sears, and D. L. Turcotte: *Statistical Thermodynamics*, Addison-Wesley Publishing Co., Reading, Mass., 1963.

J. E. Mayer and M. G. Mayer: *Statistical Mechanics*, John Wiley and Sons, New York, 1940.

K. S. Pitzer: *Quantum Chemistry*, Prentice-Hall, Englewood Cliffs, N.J., 1953.

I. Prigogine: *Introduction to Thermodynamics of Irreversible Processes*, Second Edition, John Wiley and Sons, 1961.

E. Schrödinger: *Statistical Thermodynamics*, Cambridge University Press, 1948.

Answers to Selected Problems

2.3 (*a*) 3/10
(*b*) 5/18
2.6 (*a*) 3/4
(*b*) 7/26
2.9 15,600; 2600
2.12 0.00098; 0.00977; 0.04395; 0.11719; 0.20508; 0.24609
2.15 (*b*) 0.6827; 0.9545; 0.9973
2.18 10/243
2.21 4; 4

3.3 (*a*) 4.0; 2.46
(*b*) 2115; 671; 214
3.6 11.42; 3.61×10^3; 1.14×10^6
3.18 (*a*) 2.27×10^{13}/sec; 2.98×10^7/sec
(*b*) 2.26 micron Hg

4.9 1.63×10^{-21} cal/mole

5.6 7300 cm/sec; 1 cm/sec
5.12 (*a*) 2.42; 7.26; 14.52; 24.2; 36.3×10^{-14} erg
(*b*) 0.96; 2.88; 5.76; 9.6; 14.4×10^{-16} erg
5.15 (*a*) 0.875×10^{-12} erg
(*b*) 0.1115×10^{-12} erg
5.18 0.0288; 0.0463; 2.24

6.3 2.5; 1.5; 26.47
6.9 27.43 cal/mole-K
6.15 $(-4.95 \times 10^5 - 0.1)$ cal/mole; $(-1.06 \times 10^6 - 707)$ cal/mole
6.18 1.24%
6.21 (*b*) 5.67×10^{-5} erg/cm^2-sec-K^4

7.3 2.87 cal/mole-K; −8.6 cal/mole
7.6 −56,270 cal
7.12 (*a*) 1.265×10^{23}/cm^3; 1.475×10^{-11} erg; 1.07×10^5 K
(*b*) 1.26×10^{-19} erg; 9.13×10^{-4} K
7.18 (*c*) 28.6 amp/cm^2

8.3 (*b*) 1.87%; 0.9%; 0.19%
8.6 0.0795; 2.9; 6.1 cal/mole-K
8.12 −6043 cal/mole; −10.37 cal/mole-K
8.21 338 cal/mole
8.24 0; 0.132; 0; 0.245
8.27 9.81 cal/mole-K; 55.5 cal/mole-K

9.6 (*a*) $y_{N_2} = 0.727$
9.12 (*a*) −324,850 cal/mole
(*b*) 52.5
9.15 $y_{Ar} = 0.0089$; $y_{N_2} = 0.666$; $y_{O_2} = 0.048$; $y_{NO} = 0.0516$; $y_O = 0.2255$
9.21 (*a*) $(-1.06 \times 10^6 - 707)$ cal/mole; $(-1.06 \times 10^6 + 293{,}700)$ cal/mole; $(-1.06 \times 10^6 + 907{,}400)$ cal/mole; −418,000 cal/mole
(*b*) 22.4; 0.0072
(*c*) $y_{Ar} = 0.011$; $y_{Ar^+} = 0.484$; $y_{Ar^{++}} = 0.007$; $y_{e^-} = 0.498$

10.3 (*b*) $e^{-45.2 \times 10^{10}}$

10.6 (*a*) 6.26×10^{-9}
(*b*) 5.46×10^{-4}
10.9 1.14×10^{-12}
10.12 $1.656 \times 10^{-14}\ v^2$; $2.318 \times 10^{-16}\ v^2$

11.12 (*a*) 117.24 K; 3.467 Å
(*b*) 112.97 K; 3.561 Å
11.15 (*b*) 3.9×10^{-9}

12.6 4.23×10^{-5} cal/sec-cm-K; 0.184 cm^2/sec; 2.27×10^{-4} gm/cm-sec
12.9 0.558; 0.76
12.12 4.0 amp

Index

Activity, absolute, 283, 290
Angular momentum, 137, 147
Anharmonicity, 143
Antisymmetric wave function, 158
Atomic weights, table of, 350
Avogadro's number, 4

Bohr theory, 119
Boltzmann constant, 4
Boltzmann statistics, 56
  corrected, 65
Bose-Einstein statistics, 57

Canonical ensemble, 268
Cell model, 310
Chemical energy, 212
Chemical equilibrium, 239
Chemical potential, 107
Classical mechanics, 4
Cluster integral, 292
Collision diameter, 299
Combinations, 29
Combinatorial problems, 41
Communal entropy, 311
Configurational integral, 291
Corrected Boltzmann statistics, 65
Correction constants, 143
Coupled flows, 327
Cross section, 323

de Broglie, 121
Debye function, table of values, 193
Debye temperature, 193
Debye theory, 190
Degeneracy, 55
Degree of freedom, 7
Density, fluctuations in, 283
  number, 323
Dependent particles, 268
Diatomic molecule, 8, 202
  constants for, 358
  low temperature behavior, 219
Diffusion coefficient, 325, 343
Disorder, entropy and, 90
Dissociation energy, 143, 203
Distribution, 35
  equilibrium, 60
  thermodynamic, 53
Distribution function, 53
  Boltzmann, 62
  Bose-Einstein, 63
  canonical ensemble, 272
  Fermi-Dirac, 64
  grand canonical ensemble, 282
  Maxwell-Boltzmann, 73, 74
Driving force, 318
Dulong-Petit law, 189

Einstein, 121
Einstein solid, theory of, 188
Einstein temperature, 189
Electron gas, 196, 259
Electronic state, 119, 146
  atomic, 146
  molecular, 156
  properties, 175, 212
  table of, 154, 156
Energy, chemical, 203
  electronic, 154, 156
  rotational, 136
  translational, 129
  vibrational, 140

Energy, zero-point, 140, 144
Energy transport parameter, 335
Ensemble, 268, 278
Enthalpy, 100
  of formation, 212
Entropy, 89
  communal, 311
  disorder and, 90
  flow, 321
  information and, 113
  of mixing, 96
  production, 318, 321
Entropy transport parameter, 338
Equation of state, 2
  ideal gas, 69
  virial, 294
Equilibrium, chemical, 239
  constant, 242
  distribution, 60
  thermodynamic, 60, 89
Error function, 351
Euler-Maclaurin formula, 48

Fermi-Dirac statistics, 57
Fermi energy, 108, 197
First law of thermodynamics, 87
Flow, conjugate, 327
  entropy, 321
Fluctuations, 276
  density, 283
  energy, 277
Force constant, 299, 361
Free volume, 311
Frequency, 125
  of collision, 323
  vibrational, 139

Gas, diatomic, 202
  electron, 196
  ideal, 7, 69
  monatomic, 166
  photon, 177
  polyatomic, 227
  real, 289
Gas constant, 3
Gibbs function, 100
Grand canonical ensemble, 278
Grand partition function, 280, 282

Harmonic oscillator, 137
  table of functions, 352

Heat, 88
Heat transport parameter, 335
Heisenberg uncertainty principle, 122
Helmholtz function, 99
Hermite polynomial, 140
Homonuclear molecules, 205, 219
Hydrogen, ortho-para, 224

Ideal gas, 7, 69
  mixture, 78, 114
Independent events, 18
Independent particles, 7
Indistinguishable particles, 6, 42
Information, 110
Integrals, table of, 351
Intermolecular potential, 297
Internal energy, 69, 85
Internal modes, 7, 130
  partition function, 167
  properties, 167
Ionization, 259
Ionization potential, 149
Irreversible entropy production, 321
Irreversible process, 93
Isotropic substance, 69

Kinetic energy, 7, 67
Knudsen gas, 336

Lagrange multipliers, 45
Legendre polynomial, 136
Lennard-Jones potential, 300
  force constants, table of, 361
Linear molecule, 9, 227
Liquids, behavior of, 309

Mass, atomic, 4
  reduced, 133
Maxwell-Boltzmann distribution, 73, 74
Mean free path, 323
Mean value, 35
Metals, properties of, 160, 190
Mixtures, ideal gas, 78, 114
  real gas, 304
Mole, 3
Molecular constants, table of, 358
Molecular speed, 74
Moment of inertia, 133
  diatomic molecule, 133
  linear polyatomic molecule, 228

Moment of inertia, non-linear polyatomic molecule, 231
Monatomic gas, 67, 166
Morse potential, 145
Most probable distribution, 60
Mutually exclusive events, 23

Non-equilibrium thermodynamics, 10, 317
Non-linear molecule, 9, 230
Nuclear charge, 119
Nuclear spin, 99, 219

Onsager relations, 330

Partition function, 70
  chemical, 212
  electronic, 174, 212
  grand, 280, 282
  internal, 167, 204
  rotational, 205, 230
  system, 91, 272
  translational, 72, 171
  vibrational, 187
Pauli exclusion principle, 153, 158
Peltier effect, 339
Permutations, 25
Phase space, 5
Phenomenological relations, 327
Photon gas, 177
Planck constant, 4
Planck distribution, 181
Plasma, 259
Point centers of repulsion, 299
Polyatomic molecule, 9, 227
  constants for, 359
Potential energy, 7, 291
Pressure, 3, 69
Probability, 12
  compound, 17
  conditional, 15
  thermodynamic, 57
  total, 22

Quantum mechanics, 5, 119
Quantum number, 7
  azimuthal, 120, 147
  magnetic, 120, 147
  principal, 119, 147
  rotational, 136
Quantum number, spin, 120, 152
  translational, 128
  vibrational, 140

Radiation, 177
Rayleigh-Jeans formula, 181
Reduced mass, 133
Resistance, 340
Reversible process, 92
Rigid spheres, 299, 322
Rotation, 131
  characteristic temperature, 205
  degeneracy, 137
  energy, 136
  internal, 235
  partition function, 205, 230
  properties, 205
  wave function, 136
Rotational stretching, 143
Rotation-vibration interaction, 142, 213

Schrödinger wave equation, 123
Second law of thermodynamics, 92
Seebeck effect, 338
Shannon's formula, 113
Solids, band theory, 160
  behavior of, 186
  Debye, 190
  Einstein, 188
Specific heat, 87, 102
Specific volume, 101
Speed, molecular, 74
Spin, electron, 120, 152
  nuclear, 99, 219
Standard deviation, 36
Statistical mechanics, 9
Statistical thermodynamics, 9
Statistical weight, 55
Statistics, 34, 56
Stirling's formula, 47
Symmetric wave function, 158
Symmetry number, 205, 223
System, 53, 87
System partition function, 91, 272

Temperature, 3
  Debye, 193
  Einstein, 189
  rotational, 205
  vibrational, 187

Term symbols, 154, 156
Thermal conductivity, 326, 343
Thermocouple, 338
Thermodynamic probability, 57
Thermodynamics, 85
  first law of, 87
  second law of, 92
  third law of, 97
Thermoelectric phenomena, 337
Thermomechanical effect, 335
Thermomolecular pressure difference, 335
Third law of thermodynamics, 97
Thomson effect, 341
Translation, 126
  energy, 129, 169
  partition function, 72, 171
  properties, 171
  wave function, 129
Transport processes, 322

Uncertainty, 110
Undetermined multipliers, 45
Units, 3, 349

Variance, 36
Velocity, average, 75
Velocity, components, 67, 72
  most probable, 74
Vibration, 137
  characteristic temperature, 187
  corrections, 144
  energy, 140
  partition function, 187
  properties, 188, 210
  wave function, 141
Vibrational stretching, 144
Virial coefficients, 294
  for mixtures, 308
  Lennard-Jones, 302, 360
  point centers, 300
  rigid spheres, 299
Viscosity, 326, 343
Volume, 3

Wave function, 126
  antisymmetric, 158
  rotational, 136
  symmetric, 158
  translational, 129
  vibrational, 141
Wavelength, 121
Wave number, 140
Wien's formula, 182
Work, 88